Applied Mathematical Sciences
Volume 160

Editors
S.S. Antman J.E. Marsden L. Sirovich

Advisors
J.K. Hale P. Holmes J. Keener
J. Keller B.J. Matkowsky A. Mielke
C.S. Peskin K.R. Sreenivasan

Applied Mathematical Sciences

1. *John:* Partial Differential Equations, 4th ed.
2. *Sirovich:* Techniques of Asymptotic Analysis.
3. *Hale:* Theory of Functional Differential Equations, 2nd ed.
4. *Percus:* Combinatorial Methods.
5. *von Mises/Friedrichs:* Fluid Dynamics.
6. *Freiberger/Grenander:* A Short Course in Computational Probability and Statistics.
7. *Pipkin:* Lectures on Viscoelasticity Theory.
8. *Giacaglia:* Perturbation Methods in Non-linear Systems.
9. *Friedrichs:* Spectral Theory of Operators in Hilbert Space.
10. *Stroud:* Numerical Quadrature and Solution of Ordinary Differential Equations.
11. *Wolovich:* Linear Multivariable Systems.
12. *Berkovitz:* Optimal Control Theory.
13. *Bluman/Cole:* Similarity Methods for Differential Equations.
14. *Yoshizawa:* Stability Theory and the Existence of Periodic Solution and Almost Periodic Solutions.
15. *Braun:* Differential Equations and Their Applications, 3rd ed.
16. *Lefschetz:* Applications of Algebraic Topology.
17. *Collatz/Wetterling:* Optimization Problems.
18. *Grenander:* Pattern Synthesis: Lectures in Pattern Theory, Vol. I.
19. *Marsden/McCracken:* Hopf Bifurcation and Its Applications.
20. *Driver:* Ordinary and Delay Differential Equations.
21. *Courant/Friedrichs:* Supersonic Flow and Shock Waves.
22. *Rouche/Habets/Laloy:* Stability Theory by Liapunov's Direct Method.
23. *Lamperti:* Stochastic Processes: A Survey of the Mathematical Theory.
24. *Grenander:* Pattern Analysis: Lectures in Pattern Theory, Vol. II.
25. *Davies:* Integral Transforms and Their Applications, 2nd ed.
26. *Kushner/Clark:* Stochastic Approximation Methods for Constrained and Unconstrained Systems.
27. *de Boor:* A Practical Guide to Splines: Revised Edition.
28. *Keilson:* Markov Chain Models—Rarity and Exponentiality.
29. *de Veubeke:* A Course in Elasticity.
30. *Sniatycki:* Geometric Quantization and Quantum Mechanics.
31. *Reid:* Sturmian Theory for Ordinary Differential Equations.
32. *Meis/Markowitz:* Numerical Solution of Partial Differential Equations.
33. *Grenander:* Regular Structures: Lectures in Pattern Theory, Vol. III.
34. *Kevorkian/Cole:* Perturbation Methods in Applied Mathematics.
35. *Carr:* Applications of Centre Manifold Theory.
36. *Bengtsson/Ghil/Källén:* Dynamic Meteorology: Data Assimilation Methods.
37. *Saperstone:* Semidynamical Systems in Infinite Dimensional Spaces.
38. *Lichtenberg/Lieberman:* Regular and Chaotic Dynamics, 2nd ed.
39. *Piccini/Stampacchia/Vidossich:* Ordinary Differential Equations in $\mathbf{R}^n$.
40. *Naylor/Sell:* Linear Operator Theory in Engineering and Science.
41. *Sparrow:* The Lorenz Equations: Bifurcations, Chaos, and Strange Attractors.
42. *Guckenheimer/Holmes:* Nonlinear Oscillations, Dynamical Systems, and Bifurcations of Vector Fields.
43. *Ockendon/Taylor:* Inviscid Fluid Flows.
44. *Pazy:* Semigroups of Linear Operators and Applications to Partial Differential Equations.
45. *Glashoff/Gustafson:* Linear Operations and Approximation: An Introduction to the Theoretical Analysis and Numerical Treatment of Semi-Infinite Programs.
46. *Wilcox:* Scattering Theory for Diffraction Gratings.
47. *Hale/Magalhães/Oliva:* Dynamics in Infinite Dimensions, 2nd ed.
48. *Murray:* Asymptotic Analysis.
49. *Ladyzhenskaya:* The Boundary-Value Problems of Mathematical Physics.
50. *Wilcox:* Sound Propagation in Stratified Fluids.
51. *Golubitsky/Schaeffer:* Bifurcation and Groups in Bifurcation Theory, Vol. I.
52. *Chipot:* Variational Inequalities and Flow in Porous Media.
53. *Majda:* Compressible Fluid Flow and Systems of Conservation Laws in Several Space Variables.
54. *Wasow:* Linear Turning Point Theory.
55. *Yosida:* Operational Calculus: A Theory of Hyperfunctions.
56. *Chang/Howes:* Nonlinear Singular Perturbation Phenomena: Theory and Applications.
57. *Reinhardt:* Analysis of Approximation Methods for Differential and Integral Equations.
58. *Dwoyer/Hussaini/Voigt (eds):* Theoretical Approaches to Turbulence.
59. *Sanders/Verhulst:* Averaging Methods in Nonlinear Dynamical Systems.

(continued following index)

Jari Kaipio   Erkki Somersalo

Statistical and Computational Inverse Problems

With 102 Figures

 Springer

Jari Kaipio
Department of Applied Physics
University of Kuopio
70211 Kuopio
Finland
Jari.Kaipio@uku.fi

Erkki Somersalo
Institute of Mathematics
Helsinki University of Technology
02015 HUT
Finland
Erkki.Somersalo@hut.fi

Editors:

S.S. Antman
Department of Mathematics
and
Institute for Physical Science
 and Technology
University of Maryland
College Park, MD 20742-4015
USA
ssa@math.umd.edu

J.E. Marsden
Control and Dynamical
 Systems, 107-81
California Institute of
 Technology
Pasadena, CA 91125
USA
marsden@cds.caltech.edu

L. Sirovich
Laboratory of Applied
 Mathematics
Department of
 Biomathematical Sciences
Mount Sinai School
 of Medicine
New York, NY 10029-6574
USA
chico@camelot.mssm.edu

Mathematics Subject Classification (2000): 15A29, 35R25, 35R30, 62F15, 62P30, 62P35, 62K05, 65C05, 65C40, 65J20, 65J22, 65F10, 65F22, 65N21, 65R30, 65R32, 78A70

Library of Congress Cataloging-in-Publication Data
Kaipio, Jari.
  Statistical and computational inverse problems / Jari Kaipio and Erkki Somersalo.
    p. cm. — (Applied mathematical sciences)
  "April 1, 2004."
  Includes bibliographical references and index.
  ISBN 0-387-22073-9 (alk. paper)
  1. Inverse problems (Differential equations) 2. Inverse problems (Differential equations)—Numerical solutions. I. Somersalo, Erkki. II. Title. III. Applied mathematical sciences (Springer-Verlag New York, Inc.)
  QA377.K33   2004
  515'.357—dc22   2004049181

ISBN 0-387-22073-9   Printed on acid-free paper.

© 2005 Springer Science+Business Media, Inc.
All rights reserved. This work may not be translated or copied in whole or in part without the written permission of the publisher (Springer Science+Business Media, Inc., 233 Spring Street, New York, NY 10013, USA), except for brief excerpts in connection with reviews or scholarly analysis. Use in connection with any form of information storage and retrieval, electronic adaptation, computer software, or by similar or dissimilar methodology now known or hereafter developed is forbidden.
The use of general descriptive names, trade names, trademarks, etc., in this publication, even if the former are not especially identified, is not to be taken as a sign that such names, as understood by the Trade Marks and Merchandise Marks Act, may accordingly be used freely by anyone.

Printed in the United States of America.   (MP)

9 8 7 6 5 4 3 2 1   SPIN 10980149

springeronline.com

To Eerika, Paula, Maija and Noora

Between the idea
And the reality
Between the motion
And the act
Falls the Shadow

T.S. Eliot

Preface

This book is aimed at postgraduate students in applied mathematics as well as at engineering and physics students with a firm background in mathematics. The first four chapters can be used as the material for a first course on inverse problems with a focus on computational and statistical aspects. On the other hand, Chapters 3 and 4, which discuss statistical and nonstationary inversion methods, can be used by students already having knowldege of classical inversion methods.

There is rich literature, including numerous textbooks, on the classical aspects of inverse problems. From the numerical point of view, these books concentrate on problems in which the measurement errors are either very small or in which the error properties are known exactly. In real-world problems, however, the errors are seldom very small and their properties in the deterministic sense are not well known. For example, in classical literature the error norm is usually assumed to be a known real number. In reality, the error norm is a random variable whose mean might be known.

Furthermore, the classical literature usually assumes that the operator equations that describe the observations are exactly known. Again, usually when computational solutions based on real-world measurements are required, one should take into account that the mathematical models are themselves only approximations of real-world phenomena. Moreover, for computational treatment of the problem, the models must be discretized, and this introduces additional errors. Thus, the discrepancy between the measurements and the predictions by the observation model are not only due to the "noise that has been added to the measurements." One of the central topics in this book is the statistical analysis of errors generated by modelling.

There is rich literature also in statistics, especially concerning Bayesian statistics, that is fully relevant in inverse problems. This literature has been fairly little known to the inverse problems community, and thus the main aim of this book is to introduce the statistical concepts to this community. As for statisticians, the book contains probably little new information regarding, for example, sampling methods. However, the development of realistic observation models based, for example, on partial differential equations and the analysis of the associated modelling errors might be useful.

As for citations, in Chapters 1–6 we mainly refer to books for further reading and do not discuss historical development of the topics. Chapter 7, which discusses our previous and some new research topics, also does not contain reviews of the applications. Here we refer mainly to the original publications as well as to sources that contain modifications and extensions which serve to illustrate the potential of the statistical approach.

Chapters 5–7, which form the second part of the book, focus on problems for which the models for measurement errors, errorless observations and the unknown are really taken as *models*, which themselves may contain uncertainties. For example, several observation models are based on partial differential equations and boundary value problems. It might be that part of the boundary value data are inherently unknown. We would then attempt to model these boundary data as random variables that could either be treated as secondary unknowns or taken as a further source of uncertainty and compute its contribution to the discrepancy between the observation model and the predictions given by the observation model.

In the examples, especially in Chapter 7 that discusses nontrivial problems, we concentrate on research that we have carried out earlier. However, we also treat topics that either have not yet been published or are discussed here with more rigor than in the original publications.

We have tried to enhance the readability of the book by avoiding citations in the main text. Every chapter has a section called "Notes and Comments" where the citations and further reading, as well as brief comments on more advanced topics, are given.

We are grateful to our colleague and friend, Markku Lehtinen, who has advocated the statistical approach to inverse problems for decades and brought this topic to our attention. Much of the results in Chapter 7 have been done in collaboration with our present and former graduate students - as well as other scientists. We have been privileged to work with them and thank them all. We mention here only the people who have contributed directly to this book by making modifications to their computational implementations or otherwise: Dr. Ville Kolehmainen for Sections 7.2 and 7.9, Dr. Arto Voutilainen for Section 7.4, Mr. Aku Seppänen for Sections 7.5 and 7.7 and Ms. Jenni Heino for Section 7.8. We are also much obliged to Daniela Calvetti for carefully reading and commenting the whole manuscript and to the above-mentioned people for reading some parts of the book. For possible errors that remain we assume full responsibility.

This work was financially supported by the Academy of Finland and the Finnish Academy of Science and Letters (JPK) to whom thanks are due. Thanks are also due to the inverse problems group at the University of Kuopio and to vice head of the Applied Physics department, Dr. Ari Laaksonen, who saw to the other author's duties during his leave. Thanks are also due to Dr. Geoff Nicholls and Dr. Colin Fox from the University of Auckland, NZ, where much of the novel material in this book was conceived during the authors' visits there.

Helsinki and Kuopio *Jari P. Kaipio*
June 2004 *Erkki Somersalo*

Contents

Preface IX

1 Inverse Problems and Interpretation of Measurements 1
   1.1 Introductory Examples 3
   1.2 Inverse Crimes 5

2 Classical Regularization Methods 7
   2.1 Introduction: Fredholm Equation 7
   2.2 Truncated Singular Value Decomposition 10
   2.3 Tikhonov Regularization 16
      2.3.1 Generalizations of the Tikhonov Regularization 24
   2.4 Regularization by Truncated Iterative Methods 27
      2.4.1 Landweber–Fridman Iteration 27
      2.4.2 Kaczmarz Iteration and ART 31
      2.4.3 Krylov Subspace Methods 39
   2.5 Notes and Comments 46

3 Statistical Inversion Theory 49
   3.1 Inverse Problems and Bayes' Formula 50
      3.1.1 Estimators 52
   3.2 Construction of the Likelihood Function 55
      3.2.1 Additive Noise 56
      3.2.2 Other Explicit Noise Models 58
      3.2.3 Counting Process Data 60
   3.3 Prior Models 62
      3.3.1 Gaussian Priors 62
      3.3.2 Impulse Prior Densities 62
      3.3.3 Discontinuities 65
      3.3.4 Markov Random Fields 66
      3.3.5 Sample-based Densities 70
   3.4 Gaussian Densities 72

		3.4.1 Gaussian Smoothness Priors 79
	3.5	Interpreting the Posterior Distribution 90
	3.6	Markov Chain Monte Carlo Methods...................... 91
		3.6.1 The Basic Idea.................................... 91
		3.6.2 Metropolis–Hastings Construction of the Kernel 94
		3.6.3 Gibbs Sampler 98
		3.6.4 Convergence 106
	3.7	Hierarcical Models..108
	3.8	Notes and Comments 112

4 Nonstationary Inverse Problems 115
- 4.1 Bayesian Filtering 115
  - 4.1.1 A Nonstationary Inverse Problem 116
  - 4.1.2 Evolution and Observation Models 118
- 4.2 Kalman Filters 123
  - 4.2.1 Linear Gaussian Problems 123
  - 4.2.2 Extended Kalman Filters 126
- 4.3 Particle Filters 129
- 4.4 Spatial Priors 133
- 4.5 Fixed-lag and Fixed-interval Smoothing 138
- 4.6 Higher-order Markov Models...............................140
- 4.7 Notes and Comments 143

5 Classical Methods Revisited 145
- 5.1 Estimation Theory 146
  - 5.1.1 Maximum Likelihood Estimation 146
  - 5.1.2 Estimators Induced by Bayes Costs 147
  - 5.1.3 Estimation Error with Affine Estimators 149
- 5.2 Test Cases...150
  - 5.2.1 Prior Distributions 150
  - 5.2.2 Observation Operators 152
  - 5.2.3 The Additive Noise Models 155
  - 5.2.4 Test Problems 157
- 5.3 Sample-Based Error Analysis 158
- 5.4 Truncated Singular Value Decomposition 159
- 5.5 Conjugate Gradient Iteration 162
- 5.6 Tikhonov Regularization 164
  - 5.6.1 Prior Structure and Regularization Level 166
  - 5.6.2 Misspecification of the Gaussian Observation Error Model 170
  - 5.6.3 Additive Cauchy Errors 173
- 5.7 Discretization and Prior Models 175
- 5.8 Statistical Model Reduction, Approximation Errors and Inverse Crimes 181
  - 5.8.1 An Example: Full Angle Tomography and CGNE......184

	5.9	Notes and Comments 186

6 Model Problems 189
6.1 X-ray Tomography 189
6.1.1 Radon Transform 190
6.1.2 Discrete Model 192
6.2 Inverse Source Problems 194
6.2.1 Quasi-static Maxwell's Equations 194
6.2.2 Electric Inverse Source Problems 197
6.2.3 Magnetic Inverse Source Problems 198
6.3 Impedance Tomography 202
6.4 Optical Tomography 208
6.4.1 The Radiation Transfer Equation 208
6.4.2 Diffusion Approximation 211
6.4.3 Time-harmonic Measurement 219
6.5 Notes and Comments 219

7 Case Studies 223
7.1 Image Deblurring and Recovery of Anomalies 223
7.1.1 The Model Problem 223
7.1.2 Reduced and Approximation Error Models 225
7.1.3 Sampling the Posterior Distribution 229
7.1.4 Effects of Modelling Errors 234
7.2 Limited Angle Tomography: Dental X-ray Imaging 236
7.2.1 The Layer Estimation 239
7.2.2 MAP Estimates 240
7.2.3 Sampling: Gibbs Sampler 241
7.3 Biomagnetic Inverse Problem: Source Localization 242
7.3.1 Reconstruction with Gaussian White Noise Prior Model 243
7.3.2 Reconstruction of Dipole Strengths with the ℓ^1-prior Model 245
7.4 Dynamic MEG by Bayes Filtering 249
7.4.1 A Single Dipole Model 250
7.4.2 More Realistic Geometry 253
7.4.3 Multiple Dipole Models 254
7.5 Electrical Impedance Tomography: Optimal Current Patterns . 260
7.5.1 A Posteriori Synthesized Current Patterns 260
7.5.2 Optimization Criterion 262
7.5.3 Numerical Examples 265
7.6 Electrical Impedance Tomography: Handling Approximation Errors 269
7.6.1 Meshes and Projectors 270
7.6.2 The Prior Distribution and the Prior Model 272
7.6.3 The Enhanced Error Model 273

		7.6.4 The MAP Estimates 275
	7.7	Electrical Impedance Process Tomography 278
		7.7.1 The Evolution Model 280
		7.7.2 The Observation Model and the Computational Scheme 283
		7.7.3 The Fixed-lag State Estimate 285
		7.7.4 Estimation of the Flow Profile 286
	7.8	Optical Tomography in Anisotropic Media 291
		7.8.1 The Anisotropy Model 292
		7.8.2 Linearized Model 296
	7.9	Optical Tomography: Boundary Recovery 299
		7.9.1 The General Elliptic Case 300
		7.9.2 Application to Optical Diffusion Tomography 303
	7.10	Notes and Comments 305
A	**Appendix: Linear Algebra and Functional Analysis** 311	
	A.1	Linear Algebra 311
	A.2	Functional Analysis 314
	A.3	Sobolev Spaces 316
B	**Appendix 2: Basics on Probability** 319	
	B.1	Basic Concepts 319
	B.2	Conditional Probabilities 323
References 329		
Index 337		

1

Inverse Problems and Interpretation of Measurements

Inverse problems are defined, as the term itself indicates, as the inverse of direct or forward problems. Clearly, such a definition is empty unless we define the concept of direct problems. Inverse problems are encountered typically in situations where one makes indirect observations of a quantity of interest. Let us consider an example: one is interested in the air temperature. Temperature itself is a quantity defined in statistical physics, and despite its usefulness and intuitive clarity it is not directly observable. A ubiquitous thermometer that gives us information of the air temperature relies on the fact that materials such as quicksilver expand in a very predictable way in normal conditions as the temperature increases. Here the forward model is the function relating the volume of the quicksilver as a function of the temperature. The inverse problem in this case is trivial, and therefore it is not usually considered as a separate inverse problem at all, namely the problem of determining the temperature from the volume measured. A more challenging inverse problem arises if we try to measure the temperature in a furnace. Due to the high temperature, the traditional thermometer is useless and we have to use more advanced methods. One possibility is to use ultrasound. The high temperature renders the gases in the furnace turbulent, thus changing their acoustic properties which in turn is reflected in the acoustic echoes. Now the forward model consists of the challenging problem of describing the turbulence as a function of temperature plus acoustic wave propagation in the medium, and its even more challenging inverse counterpart of determining the temperature from acoustic observations.

It is the legacy of Newton, Leibniz and others that laws of nature are often expressed as systems of differential equations. These equations are *local* in the sense that at a given point they express the dependence of the function and its derivatives on physical conditions at that location. Another typical feature of the laws is *causality*: later conditions depend on the previous ones. Locality and causality are features typically associated with direct models. Inverse problems on the other hand are most often *nonlocal* and/or *noncausal*. In our example concerning the furnace temperature measurement, the acoustic echo observed outside depends on the turbulence everywhere, and due to the finite signal speed, we can hope to reconstruct the temperature distribution in a time span prior to the measurement, i.e., computationally we try to go upstream in time.

The nonlocality and noncausality of inverse problems greatly contribute to their instability. To understand this, consider heat diffusion in materials. Small changes in the initial temperature distributions smear out in time, leaving the final temperature distribution practically unaltered. The forward problem is then stable as the result is little affected by changes in the initial data.

Going in the noncausal direction, if we try to estimate the initial temperature distribution based on the observed temperature distribution at the final time, we find that vastly different initial conditions may have produced the final condition, at least within the accuracy limit of our measurement. On the one hand, this is a serious problem that requires a careful analysis of the data; on the other hand we need to incorporate all possible information about the initial data that we may have had *prior* to the measurement. The *statistical inversion theory*, which is the main topic of this book, solves the inverse problems systematically in such a way that all the information available is properly incorporated in the model.

Statistical inversion theory reformulates inverse problems as problems of statistical inference by means of Bayesian statistics. In Bayesian statistics all quantities are modeled as random variables. The randomness, which reflects the observer's uncertainty concerning their values, is coded in the probability distributions of the quantities. From the perspective of statistical inversion theory, the solution to an inverse problem is the probability distribution of the quantity of interest when all information available has been incorporated in the model. This distribution, called the *posterior distribution*, describes the degree of confidence about the quantity after the measurement has been performed.

This book, unlike many of the inverse problems textbooks, is not concerned with analytic results such as questions of uniqueness of the solution of inverse problems or their a priori stability. This does not mean that we do not recognize the value of such results; to the contrary, we believe that uniqueness and stability results are very helpful when analyzing what complementary information is needed in addition to the actual measurement. In fact, designing methods that incorporate all prior information is one of the big challenges in statistical inversion theory.

There is another line of textbooks on inverse problems, which emphasize the numerical solution of *ill-posed problems* focusing on regularization techniques. Their point of view is likewise different from ours. Regularization techniques are typically aimed at producing *a reasonable estimate* of the quantities of interest based on the data available. In statistical inversion theory, the solution to an inverse problem is not a single estimate but a probability distribution that can be used to produce estimates. But it gives more than just a single estimate: it can produce very different estimates and evaluate their reliability. This book contains a chapter discussing the most commonly used regularization schemes, not only because they are useful tools for their own right but also since it is informative to interpret and analyze those methods from the Bayesian point of view. This, we believe, helps to reveal what sort of implicit assumptions these schemes are based on.

1.1 Introductory Examples

In this section, we illustrate the issues discussed above with characteristic examples. The first example concerns the problems arising from the noncausal nature of inverse problems.

Example 1: Assume that we have a rod of unit length and unit thermal conductivity with ends set at a fixed temperature, say 0. According to the standard model, the temperature distribution $u(x,t)$ satisfies the heat equation

$$\frac{\partial^2 u}{\partial t^2} - \frac{\partial u}{\partial t} = 0, \quad 0 < x < 1, \ t > 0,$$

with the boundary conditions

$$u(0,t) = u(1,t) = 0$$

and with given intial condition

$$u(x,0) = u_0(x).$$

The inverse problem that we consider is the following: Given the temperature distribution at time $T > 0$, what was the initial temperature distribution?

Let us write first the solution in terms of its Fourier components,

$$u(x,t) = \sum_{n=1}^{\infty} c_n e^{-(n\pi)^2 t} \sin n\pi x.$$

The coefficients c_n are the Fourier sine coefficients of the initial state u_0, i.e.,

$$u_0(x) = \sum_{n=1}^{\infty} c_n \sin n\pi x.$$

Thus, to determine u_0, one has only to find the coefficients c_n from the final data. Assume that we have two initial states $u_0^{(j)}$, $j = 1, 2$, that differ only by a single high-frequency component, i.e.,

$$u_0^{(1)}(x) - u_0^{(2)}(x) = c_N \sin N\pi x,$$

for N large. The corresponding solutions at the final time will differ by

$$u^{(1)}(x,T) - u^{(2)}(x,T) = c_N e^{-(N\pi)^2 T} \sin N\pi x,$$

i.e., the difference in the final data for the two initial states is exponentially small; thus any information about high-frequency components will be lost in the presence of measurement errors. ◇

Example 2: Consider the scattering of a time harmonic acoustic wave by an inhomogeneity. The acoustic pressure field u satisfies, within the framework of linear acoustic, the wave equation

$$\Delta u + \frac{\omega^2}{c^2} u = 0 \text{ in } \mathbb{R}^3, \tag{1.1}$$

where $\omega > 0$ is the angular frequency of the harmonic time dependence and $c = c(x)$ is the propagation speed. Assume that $c = c_0 =$ constant outside a bounded set $D \subset \mathbb{R}^3$. We shall denote

$$\frac{\omega^2}{c(x)^2} = k^2(1 + q(x))$$

where $k = \omega/c_0$ is the wave number and q is a compactly supported perturbation defined as

$$q(x) = \frac{c_0^2}{c(x)^2} - 1.$$

Assume that we send in a plane wave u_0 traveling in the direction $\omega \in S^2$. Then the total field is decomposed as

$$u(x) = u_0(x) + u_{sc}(x) = e^{ik\omega \cdot x} + u_{sc}(x),$$

where the scattered field u_{sc} satisfies the Sommerfeld radiation condition at infinity,

$$\lim_{r \to \infty} r \left(\frac{\partial u_{sc}}{\partial r} - ik u_{sc} \right) = 0, \quad r = |x|. \tag{1.2}$$

The field u satisfies the Lippmann–Schwinger integral equation

$$u(x) = u_0(x) - \frac{k^2}{4\pi} \int_D \frac{e^{ik|x-y|}}{|x-y|} q(y) u(y) dy. \tag{1.3}$$

Expanding the integral kernel in Taylor series with respect to $1/r$, we find that asymptotically, the scattered part is of the form

$$u_{sc} = \frac{e^{ikr}}{4\pi r} \left(u_\infty(\hat{x}) + \mathcal{O}\left(\frac{1}{r}\right) \right), \quad \hat{x} = \frac{x}{r},$$

where the function u_∞, called the *far field pattern*, is obtained as

$$u_\infty(\hat{x}) = -k^2 \int_D e^{-i\hat{x} \cdot y} q(y) u(y) dy. \tag{1.4}$$

The forward scattering problem is to determine the pressure field u when the wave speed c is known.

The inverse scattering problem wants to determine the unknown wave speed from the knowledge of the far field patterns with different incoming plane wave directions.

We observe the fundamental difference between the direct and inverse problem. The direct problem requires the solution of one linear differential equation (1.1) with the radiation condition (1.2) or equivalently the Lippmann–Schwinger equation (1.3) which are linear problems. The inverse problem on the other hand is highly nonlinear since u in the formula (1.4) depends on q. Quite advanced techniques are needed to investigate the solvability of this problem as well as to implement a numerical solution. ◇

1.2 Inverse Crimes

Throughout this book, we shall use the term *inverse crime*.[1] By inverse crimes we mean that the numerical methods contain features that effectively render the inverse problem less ill-posed than it actually is, thus yielding unrealistically optimistic results. Inverse crimes can be summarized concisely by saying that the *model and the reality are identified*, i.e., the researcher believes that the computational model is exact. In practice, inverse crimes arise when

1. the numerically produced simulated data is produced by the same model that is used to invert the data, and
2. the discretization in the numerical simulation is the same as the one used in the inversion.

Throughout this book, these obvious versions of inverse crimes are avoided. Moreover, we show that the statistical inversion theory allows us to analyze the effects of modelling errors. We shall illustrate with examples what a difference the inverse crimes can make in simulated examples and, more importantly, how proper statistical error modelling effectively can remove problems related to discretization.

---

[1] To the knowledge of the authors, this concept was introduced by Rainer Kress in one of his survey talks on inverse problems.

2
Classical Regularization Methods

In this section we review some of the most commonly used methods used when ill-posed inverse problems are treated. These methods are called regularization methods. Although the emphasis in this book is not on regularization techniques, it is important to understand the philosophy behind them and how the methods work. Later we analyze these methods also from the point of view of statistics which is one of the main themes in this book.

2.1 Introduction: Fredholm Equation

To explain the basic ideas of regularization, we consider a simple linear inverse problem. Following the traditions, the discussion in this chapter is formulated in terms of Hilbert spaces. A brief review of some of the functional analytic results can be found in Appendix A of the book.

Let H_1 and H_2 be separable Hilbert spaces of finite or infinite dimensions and $A : H_1 \to H_2$ a compact operator. Consider first the problem of finding $x \in H_1$ satisfying the equation

$$Ax = y, \qquad (2.1)$$

where $y \in H_2$ is given. This equation is said to be a *Fredholm equation of the first kind*. Since, clearly

1. the solution *exists* if and only if $y \in \text{Ran}(A)$, and
2. the solution is *unique* if and only if $\text{Ker}(A) = \{0\}$,

both conditions must be satisfied to ensure that the problem has a unique solution. From the practical point of view, there is a third obstacle for finding a useful solution. The vector y typically represents measured data which is therefore contaminated by errors, i.e., instead of the exact equation (2.1), we have an approximate equation

$$Ax \approx y.$$

It is well known that even when the inverse of A exists, it cannot be continuous unless the spaces H_j are finite-dimensional. Thus, small errors in y may cause errors of arbitrary size in x.

Example 1: A classical ill-posed inverse problem is the deconvolution problem. Let $H_1 = H_2 = L^2(\mathbb{R})$ and define

$$A: L^2(\mathbb{R}) \to L^2(\mathbb{R}), \quad (Af)(x) = \phi * f(x) = \int_{-\infty}^{\infty} \phi(x-y)f(y)dy,$$

where ϕ is a Gaussian convolution kernel,

$$\phi(x) = \frac{1}{\sqrt{2\pi}} e^{-x^2/2}.$$

The operator A is injective, which is seen by applying the Fourier transform on Af, yielding

$$\mathcal{F}(Af)(\xi) = \int_{-\infty}^{\infty} e^{-i\xi x} Af(x)dx = \hat{\phi}(\xi)\hat{f}(\xi)$$

with

$$\hat{\phi}(\xi) = \frac{1}{\sqrt{2\pi}} e^{-\xi^2/2} > 0.$$

Therefore, if $Af = 0$, we have $\hat{f} = 0$, hence $f = 0$. Formally, the solution to the equation $Af = g$ is

$$f(x) = \mathcal{F}^{-1}(\hat{\phi}^{-1}\hat{g})(x).$$

However, the above formula is not well defined for general $g \in L^2(\mathbb{R})$ (or even in the space of tempered distributions) since the inverse of $\hat{\phi}$ grows exponentially. Measurement errors of arbitrarily small L^2-norm in g can cause g to be not in $\text{Ran}(A)$ and the integral not to converge, thus making the inversion formula practically useless. ◇

The following example shows that even when the Hilbert spaces are finite-dimensional, serious practical problems may occur.

Example 2: Let f be a real function defined over the interval $[0, \infty)$. The *Laplace transform* $\mathcal{L}f$ of f is defined as the integral

$$\mathcal{L}f(s) = \int_0^{\infty} e^{-st} f(t)dt,$$

provided that the integral is convergent. We consider the following problem: Given the values of the Laplace transform at points s_j, $0 < s_1 < \cdots < s_n < \infty$, we want to estimate the function f. To this end, we approximate first the integral defining the Laplace transform by a finite sum,

$$\int_0^{\infty} e^{-s_j t} f(t)dt \approx \sum_{k=1}^{n} w_k e^{-s_j t_k} f(t_k),$$

where, w_k's are the weights and t_k's are the nodes of the quadrature rule, e.g., Gauss quadrature, Simpson's rule or the trapezoid rule. Let $x_k = f(t_k)$, $y_j = \mathcal{L}f(s_j)$ and $a_{jk} = w_k e^{-s_j t_k}$, and write the numerical approximation of the Laplace transform in the form (2.1), where A is an $n \times n$ square matrix. Here, $H_1 = H_2 = \mathbb{R}^n$. In this example, we choose the data points logarithmically distributed, e.g.,

$$\log(s_j) = \left(-1 + \frac{j-1}{20}\right) \log 10, \quad 1 \leq j \leq 40,$$

to guarantee denser sampling near the origin. The quadrature rule is the 40-point Gauss–Legendre rule and the truncated interval of integration $(0,5)$. Hence, $A \in \mathbb{R}^{40 \times 40}$.

Let the function f be

$$f(t) = \begin{cases} t, & \text{if } 0 \leq t < 1, \\ \frac{3}{2} - \frac{1}{2}t, & \text{if } 1 \leq t < 3, \\ 0, & \text{if } t \geq 3, \end{cases}$$

The Laplace transform can then be calculated analytically. We have

$$\mathcal{L}f(s) = \frac{1}{2s^2}(2 - 3e^{-s} + e^{-3s}).$$

The function f and its Laplace transform are depicted in Figure 2.1.

An attempt to estimate the values $x_j = f(t_j)$ by direct solution of the system (2.1) even without adding any error leads to the catastrophic results shown also in Figure 2.1. The reason for the bad behaviour of this solution is that in this example, the condition number of the matrix A, defined as

$$\kappa(A) = \|A\| \, \|A^{-1}\|$$

is very large, i.e., $\kappa(A) \approx 8.5 \times 10^{20}$. Hence, even roundoff errors that in double precision are numerical zeroes are negatively affecting the solution. ◇

The above example demonstrates that the conditions 1 and 2 that guarantee the unique existence of a solution of equation (2.1) are not sufficient in practical applications. Even in the finite-dimensional problems, we must require further that the condition number is not excessively large. This can be formulated more precisely using the singular value decomposition of operators discussed in the following section.

Classical regularization methods are designed to overcome the obstacles illustrated in the examples above. To summarize, the basic idea of regularization methods is that, instead of trying to solve equation (2.1) exactly, one seeks to find a nearby problem that is uniquely solvable and that is robust in the sense that small errors in the data do not corrupt excessively this approximate solution.

In this chapter, we review three families of classical methods. These methods are (1) regularization by singular value truncation, (2) the Tikhonov regularization and (3) regularization by truncated iterative methods.

10   2 Classical Regularization Methods

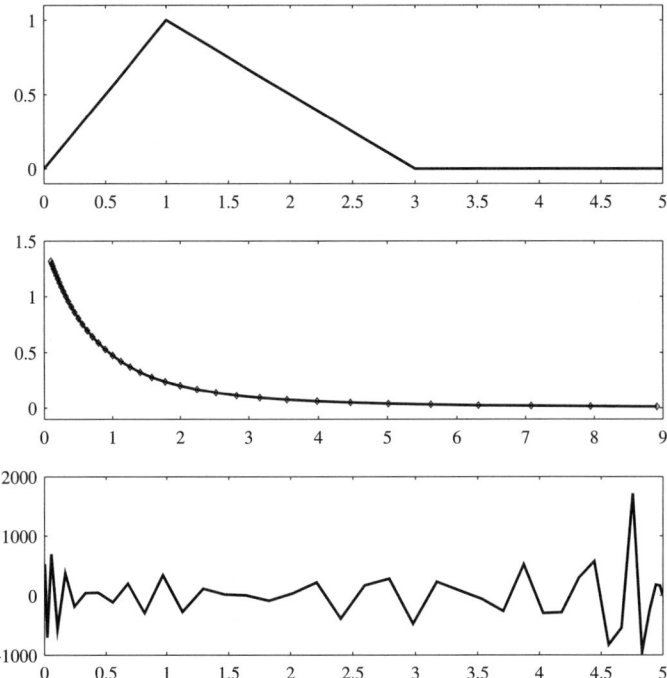

Figure 2.1. The original function (top), its Laplace transform (center) and the estimator obtained by solving the linear system (bottom).

2.2 Truncated Singular Value Decomposition

In this section, H_1 and H_2 are Hilbert spaces of finite or infinite dimension, equipped with the inner products $\langle x, y \rangle_j$, $x, y \in H_j$, $j = 1, 2$, and $A : H_1 \to H_2$ is a compact operator. When there is no risk of confusion, the subindices in the inner products are suppressed. For the sake of keeping the notation fairly straightforward, we assume that both H_1 and H_2 are infinite-dimensional.

The starting point in this section is the following proposition.

Proposition 2.1. *Let H_1, H_2 and A be as above, and let A^* be the adjoint operator of A. Then*

1. *The spaces H_j, $j = 1, 2$, allow orthogonal decompositions*

$$H_1 = \operatorname{Ker}(A) \oplus \left(\operatorname{Ker}(A)\right)^\perp = \operatorname{Ker}(A) \oplus \overline{\operatorname{Ran}(A^*)},$$

$$H_2 = \overline{\operatorname{Ran}(A)} \oplus \left(\operatorname{Ran}(A)\right)^\perp = \overline{\operatorname{Ran}(A)} \oplus \operatorname{Ker}(A^*).$$

2. *There exists orthonormal sets of vectors $(v_n) \in H_1$, $(u_n) \in H_2$ and a sequence (λ_j) of positive numbers, $\lambda \searrow 0+$ such that*

$$\overline{\mathrm{Ran}(A)} = \overline{\mathrm{span}}\{u_n \mid n \in \mathbb{N}\}, \quad (\mathrm{Ker}(A))^\perp = \overline{\mathrm{span}}\{v_n \mid n \in \mathbb{N}\},$$

and the operator A can be represented as

$$Ax = \sum_n \lambda_j \langle x, v_n \rangle u_n.$$

The system (v_n, u_n, λ_n) is called the *singular system* of the operator A.

3. The equation $Ax = y$ has a solution if and only if

$$y = \sum_n \langle y, u_n \rangle u_n, \quad \sum_n \frac{1}{\lambda_n^2} |\langle y, u_n \rangle|^2 < \infty.$$

In this case a solution is of the form

$$x = x_0 + \sum_n \frac{1}{\lambda_j} \langle y, u_n \rangle v_n,$$

where $x_0 \in \mathrm{Ker}(A)$ can be chosen arbitrarily.

The proofs of these results, with proper references, are briefly outlined in Appendix A.

The representation of the operator A in terms of its singular system is called the *singular value decomposition* of A, abbreviated as SVD of A. The above proposition gives a good picture of the possible difficulties in solving the equation $Ax = y$. First of all, let P denote the orthogonal projection on the closure of the range of A. By the above proposition, we see that P is given as

$$P : H_2 \to \overline{\mathrm{Ran}(A)}, \quad y \mapsto \sum_n \langle y, u_n \rangle u_n. \tag{2.2}$$

It follows that for any $x \in H_1$, we have

$$\|Ax - y\|^2 = \|Ax - Py\|^2 + \|(1-P)y\|^2 \geq \|(1-P)y\|^2.$$

Hence, if y has a nonzero component in the subspace orthogonal to the range of A, the equation $Ax = y$ cannot be satisfied exactly. Thus, the best we can do is to solve the projected equation,

$$Ax = PAx = Py. \tag{2.3}$$

This projection removes the most obvious obstruction of the solvability of the equation by replacing it with another substitute equation. However, given a noisy data vector y, there is in general no guarantee that the components $\langle y, u_n \rangle$ tend to zero rapidly enough to guarantee convergence of the quadratic sum in the solvability condition 3 of Proposition 2.1.

Let P_k denote the finite-dimensional orthogonal projection

12   2 Classical Regularization Methods

$$P_k : H_2 \to \text{span}\{u_1, \ldots, u_k\}, \quad y \mapsto \sum_{n=1}^{k} \langle y, u_n \rangle u_n. \qquad (2.4)$$

Since P_k is finite dimensional, we have $P_k y \in \text{Ran}(A)$ for all $k \in \mathbb{N}$, and more importantly, $P_k y \to Py$ in H_2 as $k \to \infty$. Thus, instead of equation (2.3), we consider the projected equation

$$Ax = P_k y, \quad k \in \mathbb{N}. \qquad (2.5)$$

This equation is always solvable. Taking on both sides the inner product with u_n, we find that

$$\lambda_n \langle x, v_n \rangle = \begin{cases} \langle y, u_n \rangle, & 1 \leq n \leq k, \\ 0, & n > k. \end{cases}$$

Hence, the solution to equation (2.5) is

$$x_k = x_0 + \sum_{n=1}^{k} \frac{1}{\lambda_j} \langle y, u_n \rangle,$$

for some $x_0 \in \text{Ker}(A)$. Observe that since for increasing k,

$$\|Ax_k - Py\|^2 = \|(P - P_k)y\|^2 \to 0,$$

the residual of the projected equation can be made arbitrarily small.

Finally, to remove the ambiguity of the sought solution due to the possible noninjectivity of A, we select $x_0 = 0$. This choice minimizes the norm of x_k, since by orthogonality,

$$\|x_k\|^2 = \|x_0\|^2 + \sum_{j=1}^{k} \frac{1}{\lambda_j^2} |\langle y, u_j \rangle|^2.$$

These considerations lead us to the following definition.

Definition 2.2. *let $A : H_1 \to H_2$ be a compact operator with the singular system (λ_n, v_n, u_n). By the* truncated SVD approximation (TSVD) *of the problem $Ax = y$ we mean the problem of finding $x \in H_1$ such that*

$$Ax = P_k y, \quad x \perp \text{Ker}(A)$$

for some $k \geq 1$.

We are now ready to state the following result.

Theorem 2.3. *The problem given in Definition 2.2 has a unique solution x_k, called the* truncated SVD (or TSVD) solution, *which is*

$$x_k = \sum_{n=1}^{k} \frac{1}{\lambda_j} \langle y, u_n \rangle v_n.$$

2.2 Truncated Singular Value Decomposition

Furthermore, the TSVD *solution satisfies*

$$\|Ax_k - y\|^2 = \|(1-P)y\|^2 + \|(P-P_k)y\|^2 \to \|(1-P)y\|^2$$

as $k \to \infty$, *where the projections* P *and* P_k *are given by formulas (2.2) and (2.4), respectively.*

Before presenting numerical examples, we briefly discuss the above regularization scheme in the finite-dimensional case. Therefore, let $A \in \mathbb{R}^{m \times n}$, $A \neq 0$, be a matrix defining a linear mapping $\mathbb{R}^n \to \mathbb{R}^m$, and consider the matrix equation

$$Ax = y.$$

In Appendix A, it is shown that the matrix A has a singular value decomposition

$$A = U\Lambda V^{\mathrm{T}},$$

where $U \in \mathbb{R}^{m \times m}$ and $V \in \mathbb{R}^{n \times n}$ are orthogonal matrices, i.e.,

$$U^{\mathrm{T}} = U^{-1}, \quad V^{\mathrm{T}} = V^{-1},$$

and $\Lambda \in \mathbb{R}^{m \times n}$ is a diagonal matrix with diagonal elements

$$\lambda_1 \geq \lambda_2 \geq \cdots \lambda_{\min(m,n)} \geq 0.$$

Let us denote by p, $1 \leq p \leq \min(m,n)$, the largest index for which $\lambda_p > 0$, and let us think of $U = [u_1, u_2, \ldots, u_m]$ and $V = [v_1, v_2, \ldots, v_n]$ as arrays of column vectors. The orthogonality of the matrices U and V is equivalent to saying that the vectors v_j and u_j form orthonormal base for $\mathbb{R}^n$ and $\mathbb{R}^m$, respectively. Hence, the singular system of the mapping A is $(v_j, u_j, \lambda_j)_{1 \leq j \leq p}$.

We observe that If $p = n$,

$$\mathbb{R}^n = \mathrm{span}\{v_1, \ldots, v_n\} = \mathrm{Ran}(A^{\mathrm{T}}),$$

and consequently, $\mathrm{Ker}(A) = \{0\}$. If $p < n$, then we have

$$\mathrm{Ker}(A) = \mathrm{span}\{v_{p+1}, \ldots, v_n\}.$$

Hence, any vector x_0 in the kernel of A is of the form

$$x_0 = V_0 c, \quad V_0 = [v_{p+1}, \ldots, v_n] \in \mathbb{R}^{n \times (n-p)}$$

for some $c \in \mathbb{R}^{n-p}$.

In the finite-dimensional case, we need not to worry about the convergence condition 3 of Proposition 2.1; hence the projected equation (2.3) always has a solution,

$$x = x_0 + A^{\dagger} y,$$

where x_0 is an arbitrary vector in the kernel of A. The matrix $A^{\dagger}$ is called the *pseudoinverse* or *Moore–Penrose inverse* of A, and it is defined as

2 Classical Regularization Methods

$$A^\dagger = V\Lambda^\dagger U^\mathrm{T},$$

where

$$\Lambda^\dagger = \begin{bmatrix} 1/\lambda_1 & 0 & \cdots & & & 0 \\ 0 & 1/\lambda_2 & & & & \\ & & \ddots & & & \\ \vdots & & & 1/\lambda_p & & \vdots \\ & & & & 0 & \\ & & & & & \ddots \\ 0 & & & \cdots & & 0 \end{bmatrix} \in \mathbb{R}^{n \times m}.$$

Properties of the pseudoinverse are listed in the "Notes and Comments" at the end of this chapter.

When $x_0 = 0$, the solution $x = A^\dagger y$ is called simply the *minimum norm solution* of the problem $Ax = y$, since

$$\|A^\dagger y\| = \min\{\|x\| \mid \|Ax - y\| = \|(1-P)y\|\},$$

where P is the projection onto the range of A. Thus, the minimum norm solution is the solution that minimizes the residual error and that has the minimum norm. Observe that in this definition, there is no truncation since we keep all the nonzero singular values.

In the case of inverse problems, the minimum norm solution is often useless due to the ill-conditioning of the matrix A. The smallest positive singular values are very close to zero and the minimum norm solution is sensitive to errors in the vector y. Therefore, in practice we need to choose the truncation index $k < p$ in Definition 2.2. The question that arises is: what is a judicious choice for the value of the for the truncation level k? There is a rule of thumb that is often referred to as the *discrepancy principle*. Assume that the data vector y is a noisy approximation of a noiseless vector y_0. While y_0 is unknown to us, we may have an estimate of the noise level, e.g., we may have

$$\|y - y_0\| \simeq \varepsilon \qquad (2.6)$$

for some $\varepsilon > 0$. The discrepancy principle states that we cannot expect the approximate solution to yield a smaller residual error than the measurement error, since otherwise we would be fitting the solution to the noise. This principle leads to the following selection criterion for the truncation parameter k: choose k, $1 \leq k \leq m$ the largest index that satisfies

$$\|y - Ax_k\| = \|y - P_k y\| \leq \varepsilon.$$

In the following example, the use of the minimum norm solution and the TSVD solution are demonstrated.

Example 3: We return to the Laplace inversion problem of Example 2. Let A be the same matrix as before. A plot of the logarthms of its singular values is shown in Figure 2.2.

2.2 Truncated Singular Value Decomposition

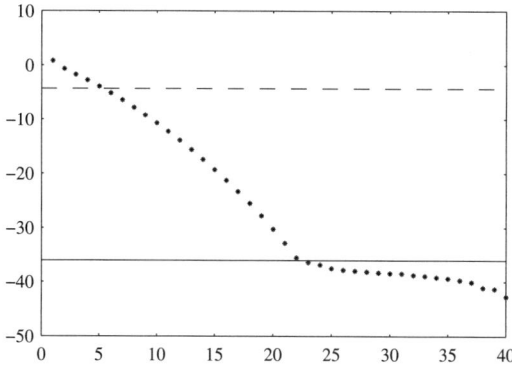

Figure 2.2. The singular values of the discretized Laplace transform on a logarithmic scale. The solid line indicates the level of the machine epsilon.

Let ε_0 denote the *machine epsilon*, i.e., the smallest floating point number that the machine recognizes to be nonzero. In IEEE double precision arithmetic, this number is of the order 10^{-16}. In Figure 2.2, we have marked this level by a solid horizontal line. The plot clearly demonstrates that the matrix is numerically singular: Singular values smaller than ε_0 represent roundoff errors and should be treated as zeros.

First, we consider the case where only the roundoff error is present and the data is precise within the arithmetic. We denote in this case $y = y_0$. Here, the minimum norm solution $x = A^\dagger y_0$ should give a reasonable estimate for the discrete values of f. It is also clear that although 22 of the singular values are larger than ε_0, the smallest ones above this level are quite close to ε_0.

In Figure 2.3 we have plotted the reconstruction of f with $x = A^\dagger y_0$ computed with $p = 20, 21$ and 22 singular values retained.

For comparison, let us add artificial noise, i.e., the data vector is

$$y = y_0 + e,$$

where the noise vector e is normally distributed zero mean noise with the standard deviation (STD) σ being 1% of the maximal data component, i.e., $\sigma = 0.01 \|y_0\|_\infty$. The logarithm of this level is marked in Figure 2.2 by a dashed horizontal line. In this case only five singular values remain above σ.

When the standard deviation of the noise is given, it is not clear without further analysis how one should select the parameter ε in the dsicrepancy principle. In this example, expect somewhat arbitrarily the norm of the noise to be of the order of σ. Figure 2.3 depicts the reconstructions of f obtained from the TSVD solutions x_k with $k = 4, 5$ and 6. We observe that for $k = 6$, the solution is oscillatory.

Let us remark here that the noise level criterion in the discrepancy principle does not take into account the stochastic properties of the noise. Later in this chapter, we discuss in more detail how to choose the cutoff level.

Let us further remark that single reconstructions such as those displayed in Figure 2.3 are far from giving a complete picture of the stability of the reconstruction. Instead, one should analyze the variance of the solutions by performing several runs from independently generated data. This issue will be discussed in Chapter 5, where the classical methods are revisited and analyzed from the statistical point of view. ◇

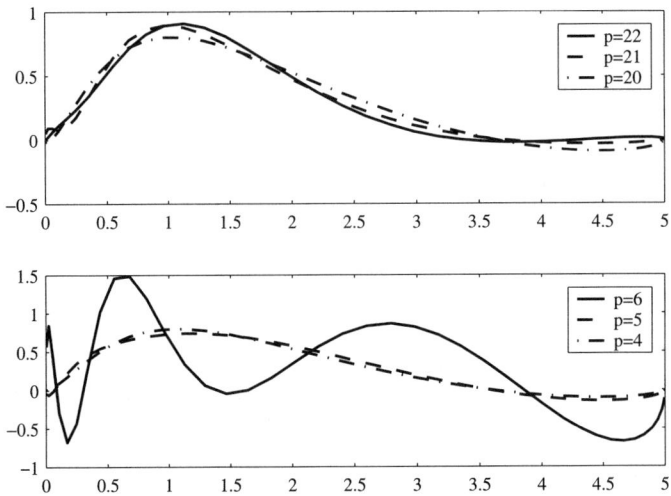

Figure 2.3. The inverse Laplace transform by using the singular value truncation. The top figure corresponds to no artificial noise in the data, the bottom one with 1% additive artificial noise.

2.3 Tikhonov Regularization

The discussion in Section 2.2 demonstrates that when solving the equation $Ax = y$, problems occur when the singular values of the operator A tend to zero rapidly, causing the norm of the approximate solution x_k to go to infinity when $k \to \infty$. The idea in the basic regularization scheme discussed in this section is to control simultaneously the norm of the residual $r = Ax - y$ and the norm of the approximate solution x. We start with the following definition.

Definition 2.4. *Let $\delta > 0$ be a given constant. The* Tikhonov *regularized solution $x_\delta \in H_1$ is the minimizer of the functional*

$$F_\delta(x) = \|Ax - y\|^2 + \delta \|x\|^2,$$

provided that a minimizer exists. The parameter $\delta > 0$ is called the regularization parameter.

2.3 Tikhonov Regularization

Observe that the regularization parameter plays essentially the role of a Lagrange multiplier, i.e., we may think that we are solving a minimization problem with the constraint $\|x\| = R$, for some $R > 0$.

The following theorem shows that Definition 2.4 is reasonable.

Theorem 2.5. *Let $A : H_1 \to H_2$ be a compact operator with the singular system (λ_n, v_n, u_n). Then the Tikhonov regularized solution exists, is unique, and is given by the formula*

$$x_\delta = (A^*A + \delta I)^{-1} A^* y = \sum_n \frac{\lambda_n}{\lambda_n^2 + \delta} \langle y, u_n \rangle v_n. \qquad (2.7)$$

Proof: We have
$$\langle x, (A^*A + \delta I)x \rangle \geq \delta \|x\|^2,$$
i.e., the operator $(A^*A + \delta I)$ is bounded from below. It follows from the Riesz representation theorem (see Appendix A) that the inverse of this operator exists and
$$\|(A^*A + \delta I)^{-1}\| \leq \frac{1}{\delta}. \qquad (2.8)$$

Hence, x_δ in (2.7) is well defined. Furthermore, expressing the equation
$$(A^*A + \delta I)x = A^* y$$
in terms of the singular system of A, we have
$$\sum_n (\lambda_n^2 + \delta)\langle x, v_n \rangle v_n + Px = \sum \lambda_n \langle y, u_n \rangle v_n,$$
where $P : H_1 \to \mathrm{Ker}(A)$ is the orthogonal projector. By projecting onto the eigenspaces $\mathrm{sp}\{v_n\}$, we find that $Px = 0$ and $(\lambda_n^2 + \delta)\langle x, v_n \rangle = \lambda_n \langle y, u_n \rangle$.

To show that x_δ minimizes the quadratic functional F_δ, let x be any vector in H_1. By decomposing x as
$$x = x_\delta + z, \quad z = x - x_\delta,$$
and arranging the terms in $F_\delta(x)$ according to the degree with respect to z, we obtain
$$F_\delta(x_\delta + z) = F_\delta(x_\delta) + \langle z, (A^*A + \delta I)x_\delta - A^* y \rangle + \langle z, (A^*A + \delta I)z \rangle$$
$$= F_\delta(x_\delta) + \langle z, (A^*A + \delta I)z \rangle$$
by definition of x_δ. The last term is nonnegative and vanishes only if $z = 0$. This proves the claim. $\square$

Remark: When the spaces H_j are finite-dimensional and A is a matrix, we may write
$$F_\delta(x) = \left\| \begin{bmatrix} A \\ \sqrt{\delta} I \end{bmatrix} x - \begin{bmatrix} y \\ 0 \end{bmatrix} \right\|^2.$$

From the inequality (2.8) it follows that the singular values of the matrix

$$K_\delta = \begin{bmatrix} A \\ \sqrt{\delta} I \end{bmatrix}$$

are bounded from below by $\sqrt{\delta}$, so the minimizer of the functional F_δ is simply

$$x_\delta = K_\delta^\dagger \begin{bmatrix} y \\ 0 \end{bmatrix}.$$

This formula is particularly handy in numerical implementation of the Tikhonov regularization method.

The choice of the value of the regularization parameter δ based on the noise level of the measurement y is a central issue in the literature discussing Tikhonov regularization. Several methods for choosing δ have been proposed. Here, we discuss briefly only one of them, known as the *Morozov discrepancy principle*. This principle is essentially the same as the discrepancy principle discussed in connection with the singular value truncation method.

Let us assume that we have an estimate $\varepsilon > 0$ of the norm of the error in the data vector as in (2.6). Then any $x \in H_1$ such that

$$\|Ax - y\| \leq \varepsilon$$

should be considered an acceptable approximate solution. Let x_δ be defined by (2.7), and

$$f : \mathbb{R}_+ \to \mathbb{R}_+, \quad f(\delta) = \|Ax_\delta - y\| \tag{2.9}$$

the discrepancy related to the parameter δ. The Morozov discrepancy principle says that the regularization parameter δ should be chosen from the condition

$$f(\delta) = \|Ax_\delta - y\| = \varepsilon, \tag{2.10}$$

if possible, i.e., the regularized solution should not try to satisfy the data more accurately than up to the noise level.

The following theorem gives a condition when the discrepancy principle can be used.

Theorem 2.6. *The discrepancy function (2.9) is strictly increasing and*

$$\|Py\| \leq f(\delta) \leq \|y\|, \tag{2.11}$$

where $P : H_2 \to \mathrm{Ker}(A^*) = \mathrm{Ran}(A)^\perp$ is the orthogonal projector. Hence, the equation (2.10) has a unique solution $\delta = \delta(\varepsilon)$ if and only if $\|Py\| \leq \varepsilon \leq \|y\|$.

Proof: By using the singular system representation of the vector x_δ, we have

$$\|Ax_\delta - y\|^2 = \sum \left(\frac{\lambda_n^2}{\lambda_n^2 + \delta} - 1\right)^2 \langle y, u_n \rangle^2 + \|Py\|^2$$

$$= \sum \left(\frac{\delta}{\lambda_n^2 + \delta}\right)^2 \langle y, u_n \rangle^2 + \|Py\|^2.$$

Since, for each term of the sum,

$$\frac{d}{d\delta}\left(\frac{\delta}{\lambda_n^2+\delta}\right)^2 = \frac{2\delta\lambda_n^2}{(\lambda_n^2+\delta)^3} > 0, \tag{2.12}$$

the mapping $\delta \mapsto \|Ax_\delta - y\|^2$ is strictly increasing, and

$$\|Py\|^2 = \lim_{\delta\to 0+} \|Ax_\delta - y\|^2 \leq \|Ax_\delta - y\|^2 \leq \lim_{\delta\to\infty} \|Ax_\delta - y\|^2 = \|y\|^2,$$

as claimed. □

Remark The condition $\|Py\| \leq \varepsilon$ is natural in the sense that any component in the data y that is orthogonal to the range of A must be due to noise. On the other hand, the condition $\varepsilon < \|y\|$ can be understood in the sense that the error level should not exceed the signal level. Indeed, if $\|y\| < \varepsilon$, we might argue that, from the viewpoint of the discrepancy principle, $x = 0$ is an acceptable solution.

The Morozov discrepancy principle is rather straightforward to implement numerically, apart of problems that arise from the size of the matrices. Indeed, if A is a matrix with nonzero singular values $\lambda_1 \geq \cdots \geq \lambda_r$, one can employ e.g., Newton's method to find the unique zero of the function

$$f(\delta) = \sum_{j=1}^{r} \left(\frac{\delta}{\lambda_n^2+\delta}\right)^2 \langle y, u_n\rangle^2 + \|Py\|^2 - \varepsilon^2.$$

The derivative of this function with respect to the parameter δ can be expressed without a reference to the singular value decomposition. Indeed, from formula (2.12), we find that

$$f'(\delta) = \sum \frac{2\delta\lambda_n^2}{(\lambda_n^2+\delta)^3} \langle u_n, y\rangle^2 = \langle x_\delta, \delta(A^*A+\delta I)^{-1}x_\delta\rangle.$$

This formula is valuable in particular when A is a large sparse matrix and the linear system with the matrix $A^*A+\delta I$ is easier to calculate than the singular value decomposition.

Example 4: Anticipating the statistical analysis of the inverse problems, we consider the problem of how to set the noise level ε appearing in the discrepancy principle. Assume that we have a linear inverse problem with additive noise model, i.e., $A \in \mathbb{R}^{k\times m}$ is a known matrix and the model is

$$y = Ax + e = y_0 + e.$$

Furthermore, assume that we have information about the statistics of the noise vector $e \in \mathbb{R}^k$. The problem is, how does one determine a reasonable noise level based on the probability distribution of the noise. In principle, there are several possible candidates. Remembering that e is a random variable, we might in fact define

$$\varepsilon = \mathrm{E}\{\|e\|\}, \tag{2.13}$$

where E is the expectation (see Appendix B). Equally well, one could argue that another judicious choice is to set

$$\varepsilon^2 = \mathrm{E}\{\|e\|^2\}, \tag{2.14}$$

leading to a slightly different value of ε. In general, these levels can be computed either numerically by generating randomly a sample of noise vectors and averaging, or analytically, if the explicit integrals of the probability densities are available.

For a simple illustration of how (2.13) and (2.14) differ from each other, assume that $k = 1$, i.e., the data y is a real number and $e \sim \mathcal{U}(0,1)$, i.e., e has a uniform probability distribution on the interval $[0,1]$. The criterion (2.13) would give

$$\varepsilon = \int_0^1 t\,dt = \frac{1}{2},$$

while the second criterion leads to

$$\varepsilon = \left(\int_0^1 t^2\,dt\right)^{1/2} = \frac{1}{\sqrt{3}}.$$

for another, more frequently encountered example, consider k-variate zero mean Gaussian noise with independent components, i.e., $e \sim \mathcal{N}(0, \sigma^2 I)$, where σ^2 is the variance and I is the unit matrix of dimension k. In this case, the criterion (2.14) immediately yields

$$\varepsilon^2 = \mathrm{E}\{\|e\|^2\} = k\sigma^2.$$

The first criterion requires more work. We have

$$\varepsilon = \frac{1}{(2\pi\sigma^2)^{k/2}} \int_{\mathbb{R}^k} \|t\| \exp\left(-\frac{1}{2\sigma^2}\|t\|^2\right) dt,$$

which, after passing to polar coordinates and properly scaling the variables, yields

$$\varepsilon = \frac{|\mathbb{S}^{k-1}|}{(2\pi)^{k/2}} \sigma \int_0^\infty t^k \exp\left(-\frac{1}{2}\|t\|^2\right) dt = \gamma_k \sigma.$$

Here, $|\mathbb{S}^{k-1}|$ denotes the surface area of the unit ball. It is left as an excercise to evaluate the scaling factor γ_k. The important thing to notice is that both results scale linearly with σ.

The important thing to notice here that $\|e\|$ is a random variable. For example, taking above $k = 100$ and $\sigma = 1$, the probability of $9 < \|e\| \leq 11$ is approximately 0.84.

Often, in classical regularization literature, the noise level used in the Morozov discrepancy principle is adjusted by an extra parameter $\tau > 1$ to avoid underregularization. Using the k-variate Gaussian white noise model, the discrepancy condition would give

$$\|Ax_\alpha - y\| = \tau\sqrt{k}\sigma.$$

A common choice is $\tau = 1.1$. ◇

Example 5: As an example of the use of the Tikhonov regularization method, consider the image deblurring problem, i.e., a deconvolution problem in the plane, introduced in Example 1. It is instructive to express the Tikhonov regularized solution using Fourier analysis. Therefore, let $H_1 = H_2 = L^2(\mathbb{R}^2)$. To guarantee the integrabilty of the image, we assume that f is compactly supported. With respect to the inner product of $L^2(\mathbb{R}^2)$, the adjoint of the convolution operator with a real valued kernel is

$$A^*f(x) = \int_{\mathbb{R}^2} \phi(y - x) f(y) dy.$$

Moreover, if the kernel is even, i.e., $\phi(-x) = \phi(x)$, A is self-adjoint. Since $\widehat{Af} = \widehat{\phi}\widehat{f}$, the operator A allows a Fourier representation

$$Af(x) = \mathcal{F}^{-1}(\widehat{\phi}\,\widehat{f})(\xi) = \frac{1}{(2\pi)^2} \int_{\mathbb{R}^2} e^{i\langle \xi, x\rangle} \widehat{\phi}(\xi) \widehat{f}(\xi) d\xi.$$

Similarly, the adjoint operator can be written as

$$A^*f(x) = \mathcal{F}^{-1}(\overline{\widehat{\phi}}\,\widehat{f})(\xi).$$

Based on this representation, we have

$$(A^*A + \delta I)f(x) = \mathcal{F}^{-1}\big((|\widehat{\phi}|^2 + \delta)\widehat{f}\big)(x)$$

and further, the operator defining the Tikhonov regularized solution is simply

$$(A^*A + \delta I)^{-1} A^* g(x) = \mathcal{F}^{-1}\left(\frac{\overline{\widehat{\phi}}}{|\widehat{\phi}|^2 + \delta} \widehat{g}\right)(x).$$

We have

$$\left|\frac{\overline{\widehat{\phi}(\xi)}}{|\widehat{\phi}(\xi)|^2 + \delta}\right| \leq \frac{1}{\delta},$$

so the operator is well defined in $L^2(\mathbb{R}^2)$ as the theory predicts.

Although is possible to determine a numerical solution based on the above formula, we represent the numerical solution here by using direct matrix discretization.

Let the image area be the unit rectangle $[0,1] \times [0,1]$, the true image f being identically zero outside this area. Assume that the image area is divided into pixels of equal size, $P_{jk} = [(j-1)\Delta, j\Delta] \times [(k-1)\Delta, k\Delta]$, $1 \leq j,k \leq N$, where $\Delta = 1/N$. If p_{jk} denotes the centerpoint of P_{jk}, we write the approximation

$$g(p_{jk}) \approx \sum_{\ell,m=1}^{N} \Delta^2 \phi(p_{jk} - p_{\ell m}) f(p_{\ell m}),$$

or, in the matrix form,

$$y = Ax,$$

where $x, y \in \mathbb{R}^{N^2}$ are vectors with the pixel values stacked in a long vector and $A \in \mathbb{R}^{N^2 \times N^2}$ a convolution matrix arranged accordingly.

In our numerical example, we consider the convolution kernel

$$\phi(x) = e^{-\alpha |x|}$$

with $\alpha = 20$. The convolution kernel is plotted in Figure 2.4.

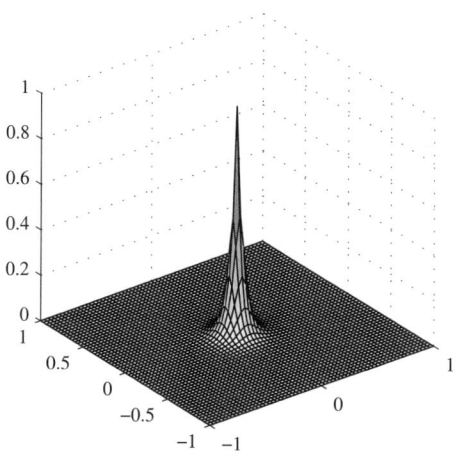

Figure 2.4. The convolution kernel used for image blurring.

To avoid the infamous inverse crime - or at least the most evident version of it - we have computed the blurred data using a finer mesh (size 50×50 pixels) than the one in which the blurred image is given (size 32×32). The true and the blurred images are shown in Figure 2.5.

The noise model we use here is Gaussian additive noise,

$$y = Ax + e,$$

where e is a random vector with independent components, each component being zero mean normally distributed. The standard deviation of each component of e is $\sigma = 0.01 \max(Ax)$, i.e., 1% of the maximum pixel value in the blurred noiseless image. To fix the noise level used in the Morozov discrepancy principle, we use the criterion (2.14) of the previous example, i.e., in this example we set

2.3 Tikhonov Regularization 23

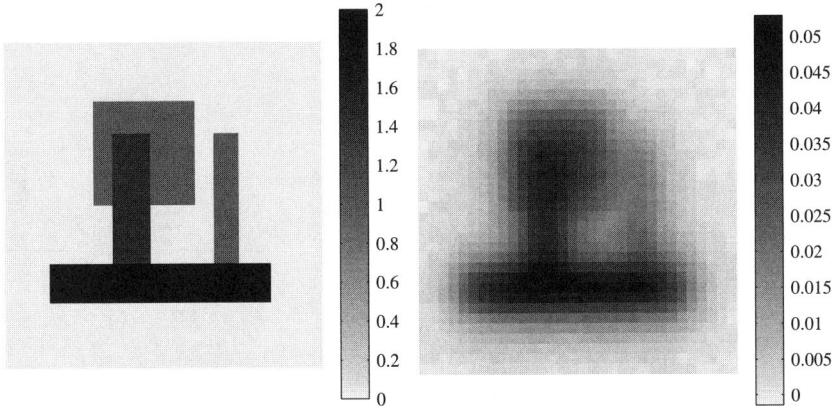

Figure 2.5. Original image and the noisy blurred image.

$$\varepsilon = N\sigma.$$

In Figure 2.6, we have plotted a piece of the curve $\delta \mapsto \|Ax_\delta - y\|$. The noise level is marked with a dashed line. Evidently, in this case, the condition (2.11) of the Theorem 2.6 is satisfied, so the Morozov discrepancy principle is applicable.

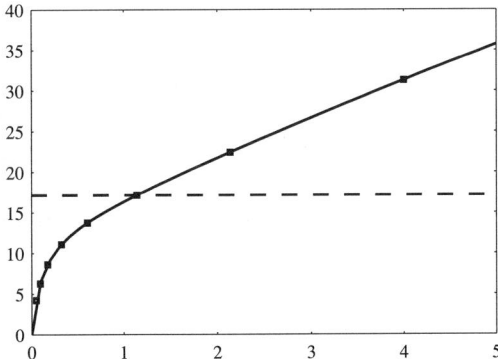

Figure 2.6. The discrepancy versus the regularization parameter δ. The estimated noise level is marked by a dashed line. The asterisks mark the values of the regularization parameters corresponding to the regularised solutions of the Figure 2.7

The value $\delta = \delta(\varepsilon)$ in this example is calculated using a bisection method. To illustrate the effect on the solution of the regularization parameter, we calculate Tikhonov regularized solutions with nine different values of the regularization parameter. These values of δ are marked in Figure 2.6 by an asterisk. The outcomes are shown in Figure 2.7. When the regularization parameter is significantly below $\delta(\varepsilon)$, the outcome is noisy, i.e., the solution is *underregularized*, while for large values, the results get again blurred. These solutions are often said to be *overregularized*. ◇

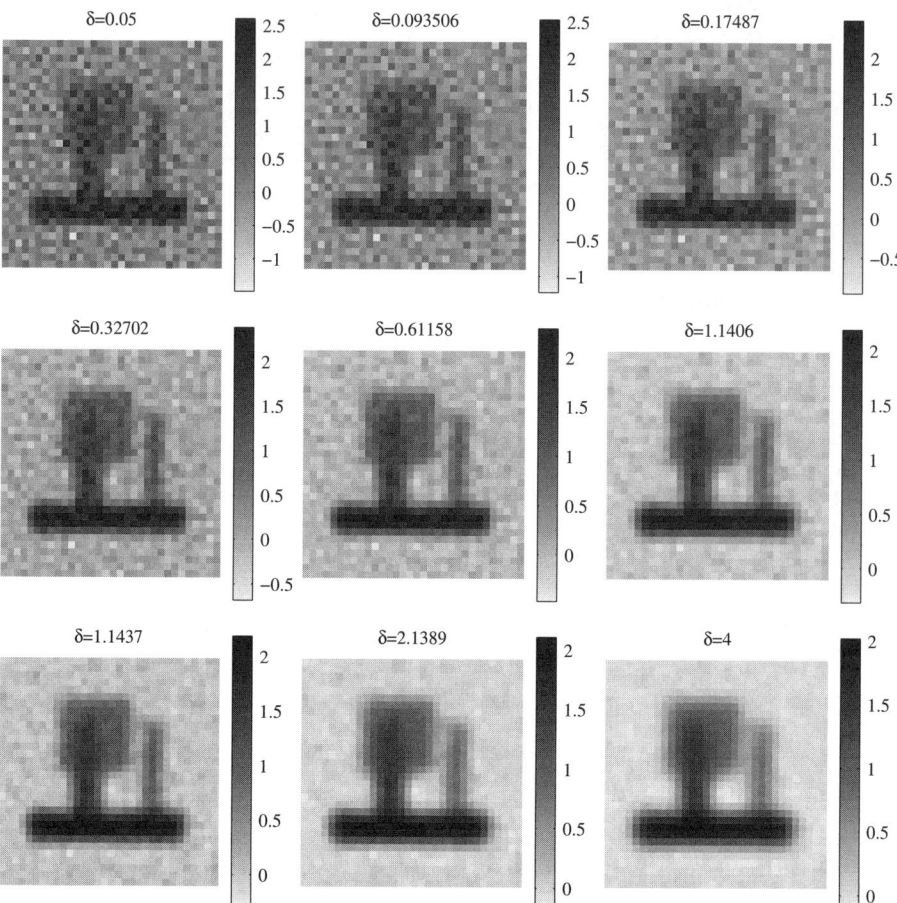

Figure 2.7. Nine reconstructions from the same noisy data with various values of the regularization parameter δ. The reconstruction that corresponds to the Morozov discrepancy principle is in the second row at right.

2.3.1 Generalizations of the Tikhonov Regularization

The Tikhonov regularization method is sometimes applicable also when nonlinear problems are considered, i.e., to find $x \in H_1$ satisfying

$$y = A(x) + e,$$

2.3 Tikhonov Regularization

where $A : H_1 \to H_2$ is a nonlinear mapping and e is observation noise. If the mapping A is such that large changes in the vector x may produce small changes in $A(x)$, the problem is ill-posed and numerical methods, typically, iterative ones, may fail to find a satisfactory estimate of x. The nonlinear Tikhonov regularization scheme amounts to searching for an x that minimizes the functional

$$F_\delta(x) = \|A(x) - y\|^2 + \delta \|x\|^2.$$

As this functional is no longer a quadratic one, it is not clear whether a minimizer exists, it is unique or how to determine it. The most common method to search for a feasible solution is to use an iterative scheme based on successive linearizations of A.

Definition 2.7. *The operator $A : H_1 \to H_2$ is Fréchet differentiable at a point x_0 if it allows an expansion*

$$A(x_0 + z) = A(x_0) + R_{x_0} z + W(x_0, z).$$

Here $R_{x_0} : H_1 \to H_2$ is a continuous linear operator and

$$\|W(x_0, z)\| \leq \|z\| \epsilon(x_0, z),$$

where the functional $z \mapsto \epsilon(x_0, z)$ tends to zero as $z \to 0$.

Let A be Fréchet differentiable. The linearization of A around a given point x_0 leads to the approximation of the functional F_δ,

$$\begin{aligned} F_\delta(x) \approx \tilde{F}_\delta(x; x_0) &= \|A(x_0) + R_{x_0}(x - x_0) - y\|^2 + \delta \|x\|^2 \\ &= \|R_{x_0} x - g(y, x_0)\|^2 + \delta \|x\|^2, \end{aligned}$$

where

$$g(y, x_0) = y - A(x_0) + R_{x_0} x_0.$$

From the previous section we know that the minimizer of the functional $\tilde{F}_\delta(x; x_0)$ is

$$x = (R_{x_0}^* R_{x_0} + \delta I)^{-1} R_{x_0}^* g(y, x_0).$$

While a straightforward approach would suggest to choose the new approximate solution as the base point for a new linearization, it may happen that the solution of the linearized problem does not reflect adequately the nonlinearities of the original function. Therefore a better strategy is to implement some form of stepsize control. This leads us to the following iterative method.

1. Pick an initial guess x_0 and set $k = 0$.
2. Calculate the Fréchet derivative R_{x_k}.
3. Determine

$$x = (R_{x_k}^* R_{x_k} + \delta I)^{-1} R_{x_k}^* g(y, x_k), \quad g(y, x_k) = y - A(x_k) + R_{x_k} x_k,$$

and define $\delta x = x - x_k$.

4. Find $s > 0$ by minimizing the function

$$f(s) = \|A(x_k + s\delta x) - y\|^2 + \|x_k + s\delta x\|^2.$$

5. Set $x_{k+1} = x_k + s\delta x$ and increase $k \leftarrow k+1$.
6. Repeat steps 2.–5. until the method converges.

In the generalization of Tikhonov regularization that we just described, the linear operator A has been replaced by a nonlinear one. Another way of generalizing Tikhonov regularization method is concerned with the choice of the penalty term.[1] Indeed, we may consider the following minimization problem: Find $x \in H_1$ that minimizes the functional

$$\|Ax - y\|^2 + \delta G(x),$$

where $G : H_1 \to \mathbb{R}$ is a nonnegative functional. The existence and uniqueness of the solution of this problem depends on the choice of the functional G.

The most common version of this generalization sets

$$G(x) = \|L(x - x_0)\|^2, \tag{2.15}$$

where $L : \mathcal{D}(L) \to H_2$, $\mathcal{D}(L) \subset H_1$ is a linear operator and $x_0 \in H_1$ is given. Typically, when H_1 is a function space, L will be a differential operator. This choice forces the solutions of the corresponding minimization problem to be smooth.

In the finite-dimensional case, the operator L is a matrix in $\mathbb{R}^{k \times n}$. The Tikhonov functional to be minimized can then be written as

$$\|Ax - y\|^2 + \delta \|L(x - x_0)\|^2 = \left\| \begin{bmatrix} A \\ \sqrt{\delta}L \end{bmatrix} x - \begin{bmatrix} y \\ \sqrt{\delta}Lx_0 \end{bmatrix} \right\|^2.$$

The minimizer of this functional is

$$x_\delta = K^\dagger \begin{bmatrix} y \\ \sqrt{\delta}Lx_0 \end{bmatrix}, \quad K = \begin{bmatrix} A \\ \sqrt{\delta}L \end{bmatrix},$$

provided that the singular values of K are all positive. If some of the singular values of K vanish, one may argue that the choice of L does not regularize the problem properly.

Finally, we may combine both the generalizations above and consider the problem of minimizing a functional of the type

$$F_\delta(x) = \|A(x) - y\|^2 + \delta G(x).$$

This problem leads to a general nonlinear optimization problem which is not discussed in detail here.

---

[1] This, in fact is the original form of Tikhonov regularization; see; "Notes and Comments."

2.4 Regularization by Truncated Iterative Methods

Consider again the simple linear matrix equation (2.1), $Ax = y$. Numerical analysis offers a rich selection of various iterative solvers for this equation. It turns out that these solvers, albeit not originally designed for regularization purposes, can often be used as regularizers when the data y are corrupted by noise. In this section, we discuss three different iterative methods and their regularizing properties.

2.4.1 Landweber–Fridman Iteration

The first iterative scheme discussed here is a method based on *fixed point iteration*. We start recalling a few concepts. Let H be a Hilbert space and $S \subset H$. Consider a mapping, not necessarily linear, $T: H \to H$. We say that S is an *invariant set* for T if $x \in S$ implies $T(x) \in S$, or briefly $T(S) \subset S$. The operator T is said to be a *contraction* on an invariant set S if there is $\kappa \in \mathbb{R}$, $0 \leq \kappa < 1$ such that for all $x, z \in S$,

$$\|T(x) - T(z)\| < \kappa \|x - z\|.$$

A vector $x \in H$ is called a *fixed point* of T if we have

$$T(x) = x.$$

The following elementary result, known as the *fixed point theorem*, is proved in Appendix A.

Proposition 2.8. *Let H be a Hilbert space and $S \subset H$ a closed invariant set for the mapping $T: H \to H$. Assume further that T is a contraction in S. Then there is a unique $x \in S$ such that $T(x) = x$. The fixed point x can be found by the fixed point iteration as*

$$x = \lim_{k \to \infty} x_k, \quad x_{k+1} = T(x_k),$$

where the initial value $x_0 \in S$ is arbitrary.

Consider now the linear equation (2.1). By using the notation of Section 2.2, we write first the right-hand side y as

$$y = Py + (1-P)y, \quad Py \in \overline{\mathrm{Ran}(A)}, \quad (1-P)y \in \mathrm{Ker}(A^*).$$

Since there is no way of matching Ax with the vector $(1-P)y$ that is orthogonal to the range of A, we filter it out by applying A^* to both sides of the equation. This leads to the *normal equations*

$$A^*Ax = A^*Py + A^*(1-P)y = A^*y. \tag{2.16}$$

We then seek to solve this normal equations by an iterative method. To this end, observe that when the normal equations are satisfied, we have

$$x = x + \beta(A^*y - A^*Ax) = T(x) \qquad (2.17)$$

for all $\beta \in \mathbb{R}$. Therefore the solution x of the equation is a fixed point for the affine map T. Our aim is to solve this equation by fixed point iterations. Hence, let $x_0 = 0$ and define

$$x_{k+1} = T(x_k).$$

In the following theorem, we assume that the dimension of $\text{Ran}(A)$ is finite. For finite-dimensional matrix equations, this is always true. More generally, this assumption means that there are only finitely many nonzero singular values in the singular system of A, and we can write Ax as

$$Ax = \sum_{j=1}^{N} \lambda_j \langle v_j, x \rangle u_j.$$

We are now ready to prove the following result.

Theorem 2.9. *Let* $\dim(\text{Ran}(A)) = N < \infty$ *and let* $0 < \beta < 2/\lambda_1^2$, *where* λ_1 *is the largest singular value of* A. *Then the fixed point iteration sequence* (x_k) *converges to an* $x \in \text{Ker}(A)^\perp$ *which satisfies the normal equations (2.16).*

Proof: Let $S = \text{Ker}(A)^\perp = \overline{\text{Ran}(A^*)}$. First we observe that S is an invariant set for the affine mapping T given by the formula (2.17), i.e., $T(S) \subset S$. We show that the mapping T is a contraction on S. Indeed, if (v_n, u_n, λ_n) is the singular system of A, then for any $x, z \in S = \text{sp}\{v_1, \ldots, v_N\}$, we have

$$\|T(x) - T(z)\|^2 = \|(1 - \beta A^*A)(x - z)\|^2$$
$$= \sum_{j=1}^{N} (1 - \beta\lambda_j^2)^2 \langle v_j, x - z \rangle^2 \leq \kappa^2 \|x - z\|^2,$$

where

$$\kappa^2 = \max_{1 \leq j \leq N} (1 - \beta\lambda_j^2)^2.$$

We observe that $\kappa < 1$ provided that for all j, $1 \leq j \leq N$,

$$0 < \beta\lambda_j^2 < 2,$$

which holds true when $0 < \beta < 2/\lambda_1^2$.

Let $x = \lim x_n$. We have

$$0 = T(x) - x = \beta(A^*y - A^*Ax),$$

i.e., the limit satisfies the normal equations (2.16). $\square$

2.4 Regularization by Truncated Iterative Methods

In general, when $\dim(\mathrm{Ran}(A)) = \infty$ and A is compact, we cannot hope that the Landweber–Fridman iteration converges because the normal equations do not, in general, have a solution. This does not prevent us from using the iteration provided that we truncate it after finitely many steps.

To understand the regularization aspect of this iterative scheme, let us introduce
$$R = 1 - \beta A^* A : S \to S.$$
Inductively, we see that the kth iterate x_k can be written simply as
$$x_k = \sum_{j=0}^{k} R^j \beta A^* y,$$
and in particular,
$$\langle x_k, v_n \rangle = \sum_{j=0}^{k} \beta \lambda_n (1 - \beta \lambda_n^2)^j \langle y, u_n \rangle = \frac{1}{\lambda_n}(1 - (1 - \beta \lambda_n^2)^{k+1})\langle y, u_n \rangle$$
by the geometric series sum formula. From this formula it is evident that when the singular value λ_n is small, the factor $(1 - (1 - \beta \lambda_n^2)^{k+1}) < 1$ in the numerator is also small. Therefore, one can expect that the sum $\sum \langle x_k, v_n \rangle$ is less sensitive to noise in y than the minimum norm solution.

When iterative methods are applied for regularization, the crucial issue is to equip the algorithm with a good stopping criterion. It should be pointed out that none of the criteria proposed in the literature has been proved to be failproof. Similar to the TSVD and Tikhonov regularization, one can try to apply also here the discrepancy principle and stop the iterations when
$$\|A x_k - y\| = \varepsilon, \tag{2.18}$$
where ε is the estimated noise level.

We illustrate the behavior of this method with the stopping criterion just described in the following simple example.

Example 6: Consider the one-dimensional deconvolution problem of finding $f(t)$, $0 \le t \le 1$ from noisy observations of the function
$$g(s) = \int_0^1 \phi(s-t) f(t) dt, \quad 0 \le s \le 1,$$
where the convolution kernel is
$$\phi(t) = e^{-a|t|}, \quad a = 20.$$
As a test function, we use
$$f(t) = t(1-t).$$
The function g can be computed analytically, yielding

$$g(s) = \frac{2}{a}s(1-s) + \frac{1}{a^2}(e^{-as} + e^{-a(1-s)}) + \frac{2}{a^3}(e^{-as} + e^{-a(1-s)} - 2).$$

The data is recorded on an even mesh with mesh size $1/100$. Random normally distributed zero mean noise is added to the exact data with independent components and standard deviation 5% of the maximum value of the noiseless data. The reconstruction mesh is also an equispaced mesh with mesh size $1/80$. The matrix A has entries

$$A_{ij} = \frac{1}{80}e^{-a|t_i - s_j|}, \quad t_i = \frac{i}{80}, \quad s_j = \frac{j}{100}, \quad 0 \le i \le 80,\ 0 \le j \le 100.$$

The condition number of the matrix A is $\kappa(A) \approx 110$, so it is possible to calculate directly the minimum norm solution $f^\dagger = K^\dagger g$. In Figure 2.8, the noiseless and noisy data are displayed, as well as the exact f and the minimum norm solution $f^\dagger$. The latter is essentially pure noise, showing that some form of regularization is required. We apply the Landweber–Fridman iteration. The relaxation parameter of the iterative scheme was chosen as $\beta = 0.1\beta_{\max}$, where $\beta_{\max} = 2/\|A\|^2$. We terminate the iteration according to the stopping criterion (2.18) with $\varepsilon = \sqrt{81}\sigma$, σ being the standard deviation of the added noise. With the simulated data, the requested discrepancy level is attained after 38 iterations. In Figure 2.9, the final solution is displayed against the true solution (left). To get an idea of the convergence, we also display the iterated solutions f_n (right). ◇

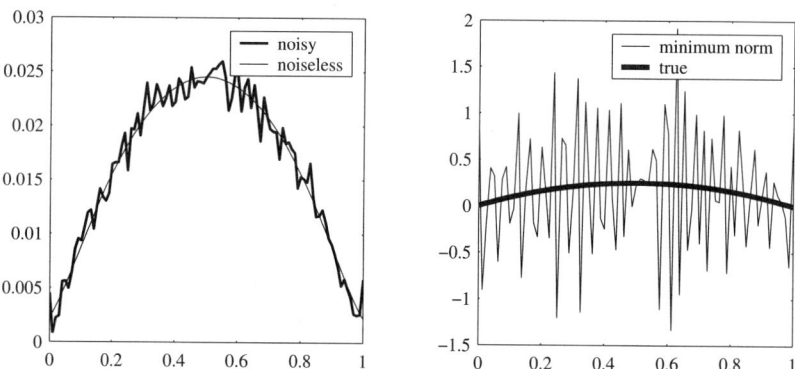

Figure 2.8. Noisy and noiseless data and the minimum norm solution.

Usually, the Landweber–Fridman iteration progresses much slower than several other iterative methods. The slow convergence of the method is sometimes argued to be a positive feature of the algorithm, since a fast progress would bring us quickly close to the minimum norm solution that is usually nonsense, as in the previous example.

2.4 Regularization by Truncated Iterative Methods 31

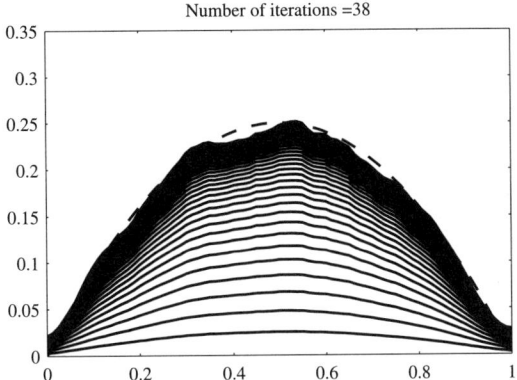

Figure 2.9. Iterated solutions. The final solution satisfies the discrepancy criterion.

2.4.2 Kaczmarz Iteration and ART

The idea of the Kaczmarz iteration to solve the matrix equation (2.1), $Ax = y$, is to partition the system rowwise, either into single rows or into blocks of rows. For the sake of definiteness, we consider first the single-row version. Writing

$$A = \begin{bmatrix} a_1^{\mathrm{T}} \\ \vdots \\ a_m^{\mathrm{T}} \end{bmatrix} \in \mathbb{R}^{m \times n}, \quad a_j \in \mathbb{R}^n,$$

where $a_j^{\mathrm{T}} \neq 0$ is the jth row of the matrix A, the equation $Ax = y$ can be thought of as a system of equations

$$A_j x = a_j^{\mathrm{T}} x = y_j, \quad 1 \leq j \leq m,$$

where $A_j : \mathbb{R}^n \to \mathbb{R}$. Each of these underdetermined equations define a hyperplane of dimension $n - 1$. The idea of the Kaczmarz iteration is to project the current approximate solution successively onto each one of these hyperplanes. It turns out that such a procedure converges to the solution of the system, provided that a solution exists.

More generally, we may write

$$A = \begin{bmatrix} A_1 \\ \vdots \\ A_\ell \end{bmatrix} \in \mathbb{R}^{m \times n}, \quad A_j \in \mathbb{R}^{k_j \times n},$$

where $k_1 + \cdots + k_\ell = m$. In this block decomposition of A, we must require that each row block A_j has full row rank and defines thus a surjective mapping.

In the following discussion, we generalize the setting slightly. Let us denote by H and H_j, $1 \leq j \leq m$ denote separable Hilbert spaces. We consider the system

2 Classical Regularization Methods

$$A_j x = y_j, \quad 1 \leq j \leq m,$$

where the operators

$$A_j : H \to H_j, \quad 1 \leq j \leq m$$

are given linear bounded operators and $y_j \in \mathrm{Ran}(A_j)$. Let

$$X_j = \{x \in H \mid A_j x = y_j\},$$

and denote by $P_j : H \to X_j$ the orthogonal projectors onto these affine subspaces. Furthermore, we define the sequential projection

$$P = P_m P_{m-1} \cdots P_2 P_1.$$

The following definition essentially defines the Kaczmarz iteration.

Definition 2.10. *With the notations introduced above, we define the Kaczmarz sequence $(x_k) \subset H$ recursively as*

$$x_{k+1} = P x_k, \quad x_0 = 0.$$

The following theorem is helpful in understanding the behavior of the Kaczmarz iteration.

Theorem 2.11. *Assume that $X = \cap_{j=1}^{m} X_j \neq \emptyset$. Then the Kaczmarz sequence converges to the minimum norm solution of the equation $Ax = y$, i.e.,*

$$\lim_{k \to \infty} x_k = x, \quad Ax = y, \quad x \perp \mathrm{Ker}(A).$$

To sketch the main idea of the proof, let us denote by $\mathcal{Q}$ the orthogonal projection

$$\mathcal{Q} : H \to \bigcap_{j=1}^{m} \mathrm{Ker}(A_j) = \mathrm{Ker}(A),$$

and let $z \in X$ be arbitrary. We shall prove that

$$x_k \longrightarrow x = z - \mathcal{Q}z, \text{ as } k \to \infty.$$

Clearly, this limit x satisfies

$$A_j x = A_j z - A_j \mathcal{Q} z = y_j, \quad 1 \leq j \leq m,$$

and furthermore, x is by definition perpendicular to $\mathrm{Ker}(A)$.

To relate the partial projections P_j to $\mathcal{Q}$, let us denote by Q_j the orthogonal projections

$$Q_j : H \to \mathrm{Ker}(A_j), \quad 1 \leq j \leq m,$$

and by Q the sequential projection

$$Q = Q_m Q_{m-1} \cdots Q_2 Q_1. \tag{2.19}$$

For any $z \in X$, we have
$$P_j x = z + Q_j(x - z).$$
Indeed,
$$A_j P_j x = A_j z + A_j Q_j(x - z) = y_j,$$
and for arbitrary $z_1, z_2 \in X_j$, the difference $\delta z = z_1 - z_2$ is in $\mathrm{Ker}(A_j)$. Therefore, it follows that
$$\langle x - (z + Q_j(x - z)), \delta z \rangle = \langle (1 - Q_j)(x - z), \delta x \rangle = 0.$$
Now we may write the sequential projection P in terms of Q as follows. For $z \in X$, and $x \in H$, we have
$$P_2 P_1 x = z + Q_2(P_1 x - z) = z + Q_2 Q_1(x - z),$$
and inductively,
$$Px = z + Q(x - z).$$
Similarly, it holds that
$$P^2 x = z + Q(Px - z) = z + Q^2(x - z),$$
and again inductively,
$$P^k x = z + Q^k(x - z),$$
i.e., by the definition of the Kaczmarz sequence, we have
$$x_k = z - Q^k z.$$
Hence, it suffices to show that for any $z \in H$, we have
$$\lim_{k \to \infty} Q^k z = Qz.$$
This result is the consequence of the following three technical lemmas.

Lemma 2.12. *Let $(x_k) \subset H$ be a sequence satisfying*
$$\|x_k\| \le 1, \quad \lim_{k \to \infty} \|Q x_k\| = 1,$$
where Q is given by (2.19). Then
$$\lim_{k \to \infty} (1 - Q) x_k = 0.$$

Proof. The proof is given by induction on the number of the projections Q_j. For Q_1, the claim is immediate since, by orthogonality,
$$\|(1 - Q_1) x_k\|^2 = \|x_k\|^2 - \|Q_1 x_k\|^2 \le 1 - \|Q_1 x_k\|^2 \to 0,$$

as k increases.

Next, assume that the claim holds for $Q_j \cdots Q_1$. We have

$$\|Q_{j+1}Q_j \cdots Q_1 x_k\| \leq \|Q_j \cdots Q_1 x_k\| \leq 1,$$

so $\lim_{k\to\infty} \|Q_{j+1}Q_j \cdots Q_1 x_k\| = 1$ implies $\lim_{k\to\infty} \|Q_j \cdots Q_1 x_k\| = 1$, and the induction assumption implies that

$$(1 - Q_j \cdots Q_1)x_k \to 0.$$

We write

$$(1 - Q_{j+1}Q_j \cdots Q_1)x_k = (1 - Q_j \cdots Q_1)x_k + (1 - Q_{j+1})Q_j \cdots Q_1 x_k.$$

Here, the first term on the right tends to zero as we have seen. Similarly, by denoting $z_k = Q_j \cdots Q_1 x_k$, it holds that

$$\|z_k\| = \|Q_j \cdots Q_1 x_k\| \leq 1$$

and

$$\|Q_{j+1} z_k\| = \|Q_{j+1} Q_j \cdots Q_1 x_k\| \to 1,$$

proving that the second term tends also to zero. □

Lemma 2.13. *We have*

$$\mathrm{Ker}(1 - Q) = \mathrm{Ker}(1 - \mathcal{Q}) = \bigcap_{j=1}^{m} \mathrm{Ker}(A_j).$$

Proof: Let $x \in \mathrm{Ker}(1 - \mathcal{Q})$. Then $x \in \mathrm{Ker}(A_j)$ for all j and so $x = Q_j x$, implying that $x = Q_m \cdots Q_1 x = Qx$, i.e., $x \in \mathrm{Ker}(1 - Q)$.

To prove the converse inclusion $\mathrm{Ker}(1 - Q) \subset \mathrm{Ker}(1 - \mathcal{Q})$, assume that $x = Qx$. We have

$$\|x\| = \|Q_m \cdots Q_2 Q_1 x\| \leq \|Q_1 x\| \leq \|x\|,$$

i.e., $\|Q_1 x\| = \|x\|$. By the orthogonality,

$$\|(1 - Q_1)x\|^2 = \|x\|^2 - \|Q_1 x\|^2 = 0,$$

i.e., $x = Q_1 x$. Hence, $x = Q_m \cdots Q_2 x$. Inductively, we show that $x = Q_j x$ for all j, i.e., $x \in \cap_{j=1}^{m} \mathrm{Ker}(A_j) = \mathrm{Ker}\mathcal{Q}$. □

We have the following decomposition result.

Lemma 2.14. *Assume that $Q : H \to H$ is linear and $\|Q\| \leq 1$. Then H can be decomposed into orthogonal subspaces,*

$$H = \mathrm{Ker}(1 - Q) \oplus \overline{\mathrm{Ran}(1 - Q)}.$$

2.4 Regularization by Truncated Iterative Methods

Proof: Since the decomposition claim 1 of Proposition 2.1 is valid for all continuous linear operators, not just for compact ones (see Appendix A), it suffices to show that $\operatorname{Ker}(1-Q) = \operatorname{Ker}(1-Q^*)$. Assume therefore that $Qx = x$. It follows that

$$\begin{aligned} \|x - Q^*x\|^2 &= \|x\|^2 - 2\langle x, Q^*x\rangle + \|Q^*x\|^2 \\ &= \|x\|^2 - 2\langle Qx, x\rangle + \|Q^*x\|^2 \\ &= -\|x\|^2 + \|Q^*x\|^2 \le -\|x\|^2 + \|x\|^2 = 0, \end{aligned}$$

implying that $x = Q^*x$.

The converse inclusion $\operatorname{Ker}(1 - Q^*) \subset \operatorname{Ker}(1 - Q)$ follows similarly. □

Now we are ready to prove Theorem 2.11.

Proof of Theorem 2.11: As we saw, it suffices to prove that

$$\lim_{j\to\infty} Q^j x = Qx.$$

Since $\|Q\| \le 1$, the decomposition result of the previous lemma holds. For any $x \in H$, it follows from Lemma 2.13 that $Qx \in \operatorname{Ker}(1 - Q) = \operatorname{Ker}(1 - Q)$, hence

$$x = Qx + (1 - Q)x \in \operatorname{Ker}(1 - Q) \oplus \overline{\operatorname{Ran}(1 - Q)},$$

and furthermore,

$$Q^k x = Qx + Q^k z, \quad z = (1 - Q)x \in \overline{\operatorname{Ran}(1 - Q)}.$$

Hence we need to show that $Q^k z \to 0$ for every $z \in \overline{\operatorname{Ran}(1 - Q)}$. Assume first that $z \in \operatorname{Ran}(1 - Q)$, or $z = (1 - Q)y$ for some $y \in H$. Consider the sequence $c_k = \|Q^k y\|$. This sequence is decreasing and positive. Let $c = \lim c_k$. If $c = 0$, then

$$Q^k z = Q^k y - Q^{k+1} y \to 0,$$

as claimed. Assume next that $c > 0$, and define the sequence

$$y_k = \frac{Q^k y}{c_k},$$

having the properties

$$\|y_k\| = 1, \quad \lim \|Qy_k\| = 1.$$

By Lemma 2.12, we have $\lim(1 - Q)y_k = 0$, or

$$Q^k z = Q^k y - Q^{k+1} y = c_k(1 - Q)y_k \to 0.$$

This result extends also to the closure of $\operatorname{Ran}(1 - Q)$. If $z \in \overline{\operatorname{Ran}(1 - Q)}$, we choose $z_0 \in \operatorname{Ran}(1 - Q)$ with $\|z - z_0\| < \varepsilon$, for arbitrary $\varepsilon > 0$. Then

$$\|Q^k z\| \le \|Q^k(z - z_0)\| + \|Q^k z_0\| < \varepsilon + \|Q^k z_0\| \to \varepsilon,$$

i.e., we must have $\lim_{k\to\infty} \|Qz\| = 0$. This completes the proof. □

Finally, we discuss the implementation of the Kaczmarz iteration in finite-dimensional spaces. The iterative algorithm that we present is commonly used especially in tomographic applications. The following lemma gives the explicit form of the projections P_j appearing in the algorithm.

Lemma 2.15. *Let $A_j \in \mathbb{R}^{k_j \times n}$ be a matrix such that the mapping $A_j : \mathbb{R}^n \to \mathbb{R}^{k_j}$ is surjective. For $y_j \in \mathbb{R}^{k_j}$, the orthogonal projection P_j to the affine subspace X_j is given by the formula*

$$P_j x = x + A_j^\mathrm{T}(A_j A_j^\mathrm{T})^{-1}(y_j - A_j x). \tag{2.20}$$

Proof: We observe first that the matrix $A_j A_j^\mathrm{T}$ is invertible. From the surjectivity of the mapping A_j,

$$\mathbb{R}^{k_j} = \mathrm{Ran}(A_j) = \mathrm{Ker}(A_j^\mathrm{T})^\perp,$$

i.e., the mapping defined by the matrix A_j^T and consequently $A_j A_j^\mathrm{T}$ are injective. Furthermore, if $z \perp \mathrm{Ran}(A_j A_j^\mathrm{T})$, we have in particular that

$$0 = z^\mathrm{T} A_j A_j^\mathrm{T} z = \|A_j^\mathrm{T} z\|^2,$$

so $z = 0$ by the injectivity of A_j^T.

As before, we may express P_j in terms of the projection Q_j as

$$P_j x = z_j + Q_j(x - z_j), \quad z_j \in X_j.$$

Since

$$x - P_j x = (1 - Q_j)(x - z) \in \mathrm{Ker}(A_j)^\perp = \mathrm{Ran}(A^\mathrm{T}),$$

there is a $u \in \mathbb{R}^{k_j}$ such that

$$x - P_j x = A_j^\mathrm{T} u. \tag{2.21}$$

Multiplying both sides by A_j, we obtain

$$A_j A_j^\mathrm{T} u = A_j x - y_j,$$

hence

$$u = (A_j A_j^\mathrm{T})^{-1}(A_j x - y_j).$$

Substituting this expression for u into formula (2.21) proves the claim. □

Remark: The Kaczmarz iteration allows a slightly more general form than the one given above. Instead of the projections P_j, one can use $P_{j\omega} = (1 - \omega)I + \omega P_j$, where ω is a relaxation parameter, $0 < \omega < 2$. The proofs above hold also in this more general setting, too. The formula (2.20) takes the form

$$P_{j\omega} x = x + \omega A_j^\mathrm{T}(A_j A_j^\mathrm{T})^{-1}(y_j - A_j x).$$

2.4 Regularization by Truncated Iterative Methods

Example 7: Probably the most typical application of the Kaczmarz iteration to inverse problems is in X-ray tomography.[2] Here we consider the two-dimensional discretized problem. The tomography data consists of projections, or shadow images, of the image into given directions. These projections can be described in terms of a linear operator that is approximated by a matrix. Thus, let $x \in \mathbb{R}^{N^2}$ be a vector containing the stacked pixel values of an $N \times N$ image, and $A \in \mathbb{R}^{M \times N^2}$ the sparse tomography matrix. We apply the Kaczmarz iteration row by row. Let

$$A = \begin{bmatrix} a_1^T \\ \vdots \\ a_M^T \end{bmatrix} \in \mathbb{R}^{M \times N^2}, \quad a_j \in \mathbb{R}^{N^2}.$$

The iterative scheme to solve the equation $Ax = y$, known in the context of X-ray tomography as the *algebraic reconstruction technique*, or ART for short, proceeds as follows:

Set $k = 0$, $x_0 = 0$;
Repeat until convergence:
  $z_0 = x_k$;
  for $j = 1 : M$ repeat
    $z_j = z_{j-1} + (1/\|a_j\|^2)(y_j - a_j^T z_{j-1})a_j$;
  end
  $x_{k+1} = z_M$; $k \leftarrow k + 1$;
end

To illustrate this algorithm, we apply it to both *full angle data* and *limited angle data*. The data is best understood by considering Figure 2.10. The original image is pixelized into 80×80 pixels. First, we compute the full angle data. The data consists of one-dimensional shadow images of the true image in different directions of projection. Let the projection angle vary over an interval of length π. We discretize the arc of a semicircle into 40 equal intervals and increase the look angle by $\pi/40$ each step, yielding to discrete angles ϕ_i, $1 \leq i \leq 40$. The line of projection is divided into 41 intervals. Hence, the projection data has the size 40×41. This data is the full angle data.

In Figure 2.10 this data without added noise is plotted with the angular variable on the horizontal axis. This representation is called the *sinogram* for rather obvious reasons. Now we apply the ART algorithm. To avoid an inverse crime, we use a different grid for the reconstruction. We seek to find an image in a pixel map of size 50×50. We add noise to the sinogram data by adding to each data entry independent uniformly distributed noise drawn from the interval $[0, \sigma]$, where σ is 2% of the maximum value of the data matrix.

In Figure 2.11 we display three ART reconstructions: The first one is after one iteration round; the second is the one that satisfies the discrepancy condition

---
[2] The X-ray tomography is discussed further in Chapter 6.

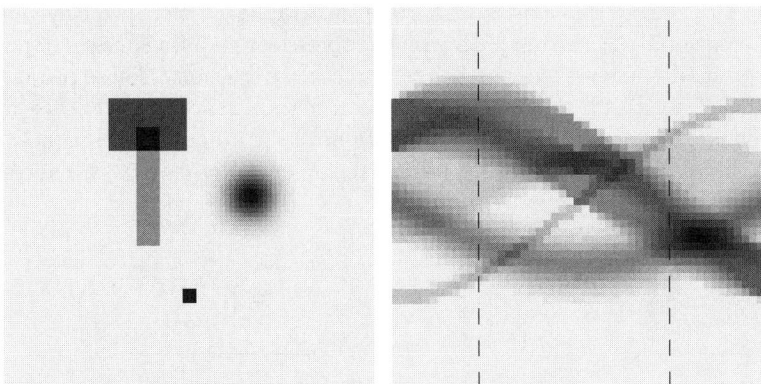

Figure 2.10. Original image and the sinogram data, the abscissa being the illumination angle. The limited angle data is the part of the sinogram between the vertical dashed lines.

$$\|Ax_j - y\| \leq \varepsilon = 50\sigma,$$

where σ is the standard deviation of the additive noise, and the factor 50 comes from the image size (see Example 4). Finally, the third reconstruction corresponds to 30 iterations. Evidently, the full angle data is so good that already after one single iteration the reconstruction is visually rather satisfactory. In fact, one can see some slight artifacts in the 30 iterations image.

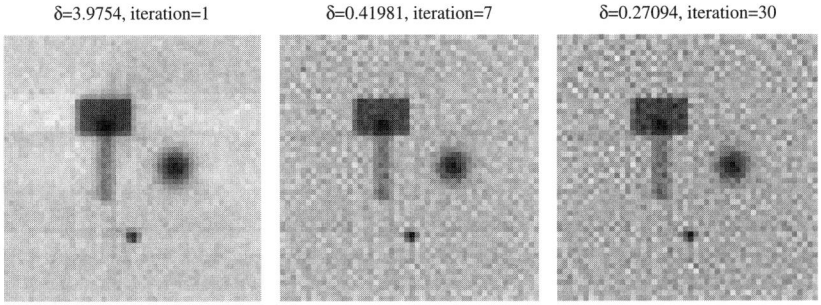

Figure 2.11. ART reconstruction from the full angle tomography data.

To get an idea of the convergence of the ART algorithm, we have also plotted the discrepancies in Figure 2.12

Now we repeat the computation starting with limited angle data. We assume that the look angle varies from $-\pi/4$ to $\pi/4$ around the vertical direction, i.e., the image is illuminated from below in an angle of $\pi/2$ opening. We do this by discarding from the full angle data those illumination lines where

2.4 Regularization by Truncated Iterative Methods

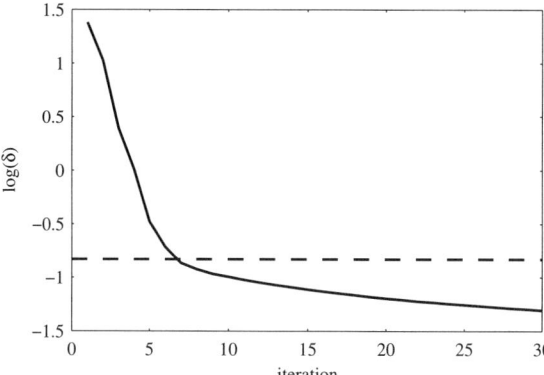

Figure 2.12. The full angle discrepancies. The estimated noise level is marked with a dashed line.

look angle differs from the vertical more than $\pi/4$, i.e., we use only the central part of the sinogram data. Figure 2.13 displays the ART reconstructions with limitid angle data analogous to those ones with full angle data. The fact that no horizontal or close to horizontal integration lines are included is reflected in the reconstructions that show long shadows in directions close to vertical. These reconstructions demonstrate the limitation of ART (or in fact, any inversion scheme) when data is scarce and no additional or prior information is used in the reconstruction. ◇

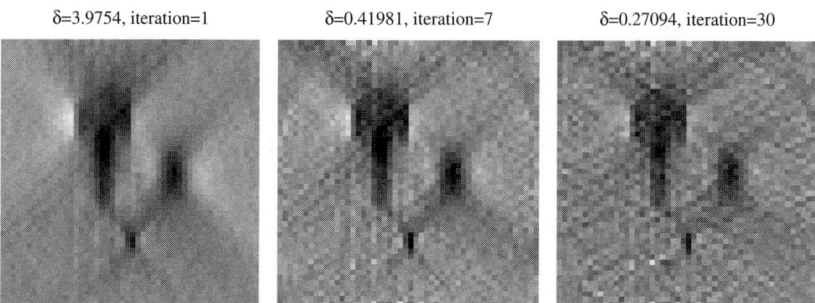

Figure 2.13. ART reconstruction from the limited angle tomography data.

2.4.3 Krylov Subspace Methods

The Krylov subspace methods refer to a class of iterative solvers of large linear equations of the form $Ax = y$. Roughly, the idea is to produce a sequence of approximate solutions as linear combinations of vectors of the type

$u, Au, A^2u, \ldots$. The best known of these methods when the matrix A is symmetric and positive definite is the *conjugate gradient method* (CG). Here, we restrict the discussion to that method.

In the sequel, we assume that $A \in \mathbb{R}^{n \times n}$ is a symmetric and strictly positive definite matrix, i.e.,

$$A^T = A, \quad u^T Au > 0 \text{ for } u \neq 0.$$

In particular, all the eigenvalues of A must be positive, and hence matrix A is invertible. The objective of the CG method is to find an approximating sequence (x_j) converging to the solution of the equation $Ax = y$ by solving a sequence of minimization problems. Let us denote by

$$x_* = A^{-1}y$$

the exact solution, and denote by e and r the error and the residual of a given approximation x,

$$e = x_* - x, \quad r = y - Ax = Ae.$$

Consider the quadratic functional

$$\phi(x) = e^T Ae = r^T A^{-1} r.$$

It is not possible to calculate the value of this functional for a given x without the knowledge of the exact solution x_* or, alternatively, A^{-1}. However, it is possible to consider the problem of minimizing this functional over a nested sequence of Krylov subspaces. First, let us observe that by the positive definiteness of A,

$$\phi(x) = 0 = \min_{x \in \mathbb{R}^n} \phi(x) \text{ if and only if } x = x_*.$$

Assume that we have an initial guess x_1 and an initial direction s_1, and we consider the problem of minimizing the function

$$\mathbb{R} \to \mathbb{R}, \quad \alpha \mapsto \phi(x_1 + \alpha s_1).$$

Interestingly, we can solve this minimization problem without knowing the value of ϕ.

Lemma 2.16. *The function $\alpha \mapsto \phi(x_1 + \alpha s_1)$ has a minimum at*

$$\alpha = \alpha_1 = \frac{s_1^T r_1}{s_1^T A s_1},$$

where r_1 is the residual of the initial guess,

$$r_1 = y - Ax_1.$$

2.4 Regularization by Truncated Iterative Methods

Proof: The residual corresponding to $x = x_1 + \alpha s_1$ is

$$y - Ax = y - Ax_1 - \alpha As_1 = r_1 - \alpha As_1,$$

and so

$$\begin{aligned}\phi(x) &= (r_1 - \alpha As_1)^T A^{-1}(r_1 - \alpha As_1) \\ &= \alpha^2 s_1^T As_1 - 2\alpha s_1^T r_1 + r_1^T A^{-1} r_1.\end{aligned}$$

The claim follows immediately from this formula. □

Hence, given a sequence (s_k) of directions, we may produce a sequence (x_k) of approximate solutions by setting

$$x_{k+1} = x_k + \alpha_k s_k, \quad \alpha_k = \frac{s_k^T r_k}{s_k^T As_k}, \tag{2.22}$$

where r_k is the residual of the previous iterate, i.e.,

$$r_k = y - Ax_k.$$

Note that the residuals in this scheme are updated according to the formula

$$r_{k+1} = y - A(x_k + \alpha_k s_k) = r_k - \alpha_k As_k.$$

This procedure can be carried out with any choice of the search directions s_k. The conjugate gradient method is characterized by a particular choice of the search directions. We give the following definition.

Definition 2.17. *We say that the linearly independent vectors $\{s_1, \ldots, s_k\}$ are A-conjugate, if*

$$s_i^T As_j = 0 \text{ for } i \neq j,$$

i.e., the vectors are orthogonal with respect to the inner product defined by the matrix A,

$$\langle u, v \rangle_A = u^T Av.$$

Observe that if a given set of vectors $\{u_1, \ldots, u_k\}$ are linearly independent, it is always possible to find A-conjugate vectors $v_j \in \text{sp}\{u_1, \ldots, u_k\}, 1 \leq j \leq k$ so that $\text{sp}\{u_1, \ldots, u_k\} = \text{sp}\{v_1, \ldots, v_k\}$. This can be done, e.g., by the Gram–Schmidt orthogonalization process with respect to the inner product $\langle \, \cdot \, , \, \cdot \, \rangle_A$.

Introduce the matrix $S_k = [v_1, \ldots, v_k] \in \mathbb{R}^{n \times k}$; then the A-conjugacy of the vectors $\{v_j\}$ is equivalent to

$$S_k^T AS_k = D_k = \text{diag}(d_1, \ldots, d_k) \in \mathbb{R}^{k \times k}$$

with $d_j \neq 0$, $1 \leq j \leq k$.

To understand the significance of using A-conjugate directions, consider the following *global* minimization problem: given the matrix $S = [s_1, \ldots, s_k]$ with linearly independent columns, find a minimum of the mapping

$$\mathbb{R}^k \to \mathbb{R}, \quad h \mapsto \phi(x_1 + S_k h),$$

i.e., seek to minimize $\phi(x)$ not sequentially over each given directions but over the whole subspace in one single step. The following result is analogous to Lemma 2.16.

Lemma 2.18. *The function $h \mapsto \phi(x_1 + S_k h)$ attains its minimum at*

$$h = (S_k^{\mathrm{T}} A S_k)^{-1} S_k^{\mathrm{T}} r_1. \tag{2.23}$$

Proof: We observe first that the matrix $S_k^{\mathrm{T}} A S_k$ is invertible. Indeed, if $S_k^{\mathrm{T}} A S_k x = 0$, then also $x^{\mathrm{T}} S_k^{\mathrm{T}} A S_k x = 0$, and the positive definiteness of A and the linear independence of the columns of S_k imply that $x = 0$.

Since
$$r = y - A(x_1 + S_k h) = r_1 - A S_k h,$$

we have
$$\phi(x_1 + S_k h) = (r_1 - A S_k h)^{\mathrm{T}} A^{-1} (r_1 - A S_k h)$$
$$= h^{\mathrm{T}} S_k^{\mathrm{T}} A S_k h - 2 r_1^{\mathrm{T}} S_k h + r_1^{\mathrm{T}} A^{-1} r_1.$$

The minimum of this quadratic functional satisfies
$$S_k^{\mathrm{T}} A S_k h - S_k^{\mathrm{T}} r_1 = 0,$$

so the claim follows. □

The computation of the minimizer h becomes trivial if the matrix $D_k = S_k^{\mathrm{T}} A S_k$ is diagonal. But this is not the only advantage of using A-conjugate directions. Assume that the sequential minimizers $x_1, \ldots, x_{k+1}$ have been calculated as given by (2.22). Since $x_{k+1} \in x_1 + \mathrm{sp}\{s_1, \ldots, s_k\}$, we have

$$\phi(x_{k+1}) \geq \phi(x_1 + S_k h),$$

with $h \in \mathbb{R}$ given by (2.23). We are now ready to establish the following result.

Theorem 2.19. *Assume that the vectors $\{s_1, \ldots, s_k\}$ are linearly independent and A-conjugate. Then*
$$x_{k+1} = x_1 + S_k h,$$

i.e., the $(k+1)$th sequential minimizer is also the minimizer over the subspace spanned by the directions s_j, $1 \leq j \leq k$.

Proof: Let $a_j = [\alpha_1, \ldots, \alpha_j]^{\mathrm{T}}$. With this notation, we have

$$x_j = x_1 + S_{j-1} a_{j-1},$$

and the corresponding residual is

$$r_j = y - A x_j = r_1 - A S_{j-1} a_{j-1}.$$

We observe that by the A-conjugacy,
$$s_j^T r_j = s_j^T r_1 - s_j^T A S_{j-1} a_{j-1} = s_j^T r_1.$$
Therefore,
$$\alpha_j = \frac{s_j^T r_j}{s_j^T A s_j} = \frac{s_j^T r_1}{s_j^T A s_j} = h_j,$$
i.e., we have $a_k = h$. $\square$

As a corollary, we get also the following orthogonality result.

Corollary 2.20. *If the vectors $\{s_1, \ldots, s_k\}$ are A-conjugate and linearly independent, then*
$$r_{k+1} \perp \mathrm{sp}\{s_1, \ldots, s_k\}.$$

Proof: We have
$$r_{k+1} = y - A x_{k+1} = r_1 - A S_k h,$$
and so
$$r_{k+1}^T S_k = r_1^T S_k - h^T S_k^T A S_k = 0$$
by formula (2.23).
$\square$

The results above say that if we are able to choose the next search direction s_{k+1} to be A-conjugate with the previous ones, the search for the sequential minimum gives also the global minimum over the subspace. So the question is how to efficiently determine A-conjugate directions. It is well known that orthogonal polynomials satisfying a three-term recurrence relation could be used effectively to this end. However, it is possible to build an algorithm with quite elementary methods.

Definition 2.21. *Let $r_1 = y - A x_1$. The kth Krylov subspace of A with the initial vector r_1 is defined as*
$$\mathcal{K}_k = \mathcal{K}_k(A, r_1) = \mathrm{sp}\{r_1, A r_1, \ldots, A^{k-1} r_1\}, \quad k \geq 1.$$

What is the dimension of $\mathcal{K}_k$? Evidently, if r_1 is an eigenvector of the matrix A, then $\dim(\mathcal{K}_k) = 1$ for all k. More generally, if $K \subset \mathbb{R}^n$ is an invariant subspace of A and $\dim(K) = m$, then $r_1 \in K$ implies that $\mathcal{K}_k \subset K$ and so $\dim(\mathcal{K}_k) \leq m$. The implications will be discussed later.

Our aim is to construct the sequence of the search directions inductively. Assume that $r_1 \neq 0$, since otherwise $x_1 = x_*$ and we would be done. Then let $s_1 = r_1$.

We proceed by induction on k. Assume that for some $k \geq 1$, we have constructed an A-conjugate set $\{s_1, \ldots, s_k\}$ of linearly independent search directions such that
$$\mathrm{sp}\{s_1, \ldots, s_k\} = \mathrm{sp}\{r_1, \ldots, r_k\} = \mathcal{K}_k.$$

With our choice of s_1, this is evidently true for $k = 1$. The goal is to choose s_{k+1} so that the above conditions remain valid also for $k+1$.

Let $r_{k+1} = y - Ax_{k+1} = r_k - \alpha_k As_k$. If $r_{k+1} = 0$, we have $x_k = x_*$ and the search has converged. Assume therefore that $r_{k+1} \neq 0$. Since $r_k, s_k \in \mathcal{K}_k$ by the induction assumption, it follows that $r_{k+1} \in \mathcal{K}_{k+1}$. On the other hand, by Corollary 2.20, $r_{k+1} \perp s_j$ for all j, $1 \leq j \leq k$, thus

$$\mathrm{sp}\{s_1, \ldots, s_k, r_{k+1}\} = \mathrm{sp}\{r_1, \ldots, r_{k+1}\} = \mathcal{K}_{k+1}.$$

To ensure that s_{k+1} is A-conjugate to the previous search direction, we express it in the form

$$s_{k+1} = r_{k+1} + S_k \beta \in \mathcal{K}_{k+1}, \quad \beta \in \mathbb{R}^k.$$

The coefficient vector β is determined by imposing the A-conjugacy condition

$$S_k^\mathrm{T} A s_{k+1} = 0,$$

that is,

$$D_k \beta = S_k^\mathrm{T} A S_k \beta = -S_k^\mathrm{T} A r_{k+1} = -(AS_k)^\mathrm{T} r_{k+1}.$$

Here, we have

$$AS_k = [AS_{k-1}, As_k].$$

The columns of the matrix AS_{k-1} belong all to $A(\mathrm{sp}\{s_1, \ldots, s_{k-1}\}) = A(\mathcal{K}_{k-1}) \subset \mathcal{K}_k = \mathrm{sp}\{s_1, \ldots, s_k\}$, and $r_{k+1} \perp \mathrm{sp}\{s_1, \ldots, s_k\}$. Therefore,

$$D_k \beta = \begin{bmatrix} 0 \\ \vdots \\ 0 \\ -s_k^\mathrm{T} A r_{k+1} \end{bmatrix},$$

i.e., $\beta_1 = \cdots = \beta_{k-1} = 0$, and we have

$$s_{k+1} = r_{k+1} + \beta_k s_k, \quad \beta_k = -\frac{s_k^\mathrm{T} A r_{k+1}}{s_k^\mathrm{T} A s_k}.$$

Now we have all the necessary ingredients for the minimization algorithm. However, it is customary to make a small modification to the updating formulas that improves the computational stability of the algorithm. Since $r_k \perp s_{k-1}$, we have

$$s_k^\mathrm{T} r_k = (r_k + \beta_{k-1} s_{k-1})^\mathrm{T} r_k = \|r_k\|^2,$$

i.e., the formula (2.22) can be written as

$$\alpha_k = \frac{\|r_k\|^2}{s_k^\mathrm{T} A s_k}.$$

Furthermore, since $r_k \in \mathrm{sp}\{s_1, \ldots, s_k\}$, we have $r_{k+1} \perp r_k$, implying that

2.4 Regularization by Truncated Iterative Methods

$$\|r_{k+1}\|^2 = r_{k+1}^T(r_k - \alpha_k A s_k) = -\frac{\|r_k\|^2}{s_k^T A s_k} r_{k+1}^T A s_k$$

$$= \|r_k\|^2 \beta_k,$$

i.e., the expression for β_k simplifies to

$$\beta_k = \frac{\|r_{k+1}\|^2}{\|r_k\|^2}.$$

Now we are ready to state the CG algorithm.

Pick x_1. Set $k = 1$, $r_1 = y - Ax_1$, $s_1 = r_1$;
Repeat until convergence:
    $\alpha_k = \|r_k\|^2 / s_k^T A s_k$;
    $x_{k+1} = x_k - \alpha_k s_k$;
    $r_{k+1} = r_k - \alpha_k A s_k$;
    $\beta_k = \|r_{k+1}\|^2 / \|r_k\|^2$;
    $s_{k+1} = r_{k+1} + \beta s_k$;
    $k \leftarrow k + 1$;
end

Since the conjugate directions are linearly independent, the conjugate gradient algorithm needs at most n steps to converge. If the initial residual is in an invariant subspace K of A with $\dim(K) = m < n$, then the algorithm converges at most m steps. However, when using the conjugate gradient method to solve ill-posed inverse problems, one should not iterate until the residual is zero. Instead, the iterations are terminated e.g., as soon as the norm of the residual is smaller or equal to the estimated norm of the noise.

Example 8: We illustrate the use of the conjugate gradient method with the inversion of the Laplace transform. Let the data y and the matrix A be as in Examples 2 and 3 of this chapter with 1% normally distributed random noise e added to the data, i.e., we have

$$y = Ax + e.$$

To write this inverse problem in a form where the conjugate gradient method is applicable, we consider the normal equation,

$$A^T y = A^T A x + A^T e = Bx + \tilde{e}.$$

Observe that although the matrix B is numerically singular, this does not prevent us from using the method since the iteration process is terminated prior to convergence.

The approximate solutions computed by the conjugate gradient method in the iterations 1–9 from the noisy Laplace transform data are plotted in Figure 2.14. After the seventh iteration, the approximations get very rapidly

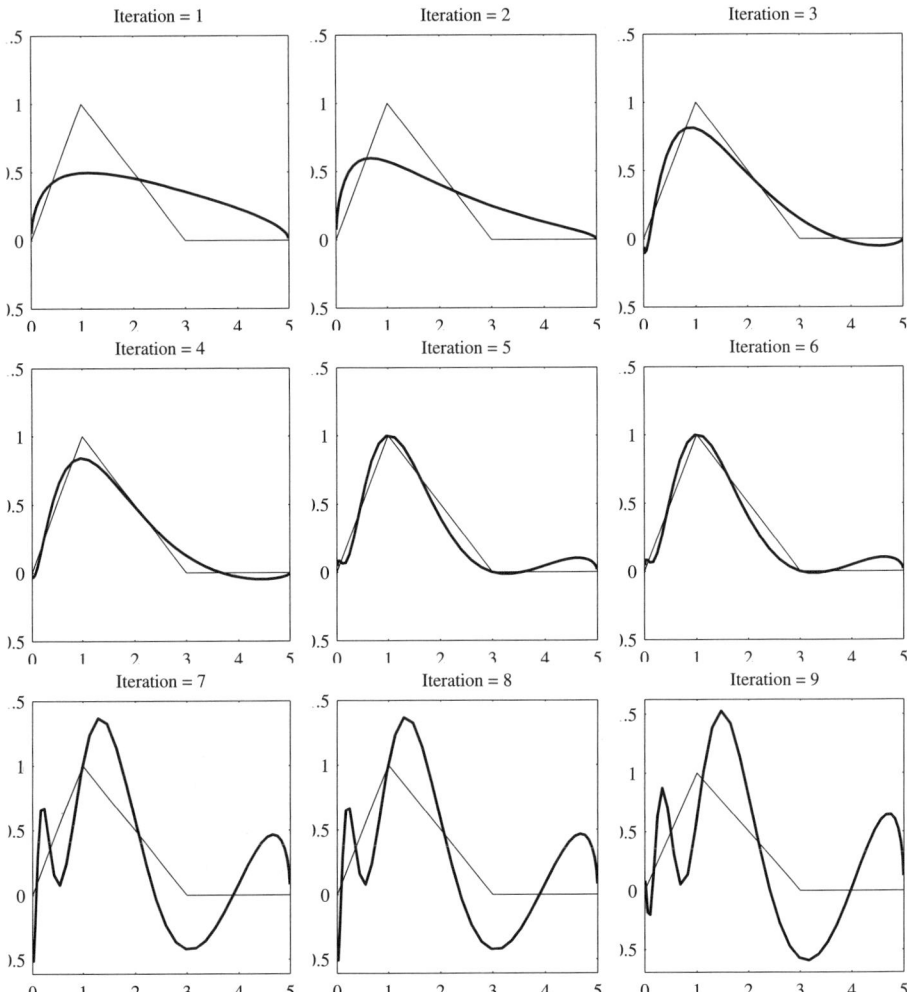

Figure 2.14. Conjugate gradient approximation after iterations 1–9.

out of control. By visual inspection, one can conclude that few iterations in this case give the best result. Observe that if we want to use the discrepancy principle for determining the stopping index, we can use the residual norm $\|y - Ax_j\|$ of the original equation in the stopping criterion. ◇

2.5 Notes and Comments

The literature on regularization of inverse problems is extensive. We refer to the textbooks [12], [35], [50], [130] and [139].

The truncated singular value decomposition has been treated widely in the literature. We refer to the book [56] and references therein concerning this topic.

The pseudoinverse $A^\dagger$ of a matrix $A \in \mathbb{R}^{n \times m}$ has several properties that sometimes are useful. We mention here the *Moore–Penrose* equations,

$$A^\dagger A A^\dagger = A^\dagger, \quad AA^\dagger A = A,$$
$$(A^\dagger A)^{\mathrm{T}} = A^\dagger A, \quad (AA^\dagger)^{\mathrm{T}} = AA^\dagger.$$

In fact, these equations characterize completely the pseudoinverse. The matrices $A^\dagger A$ and $AA^\dagger$ have a geometric interpretation as orthogonal projectors

$$A^\dagger A : \mathbb{R}^n \to \mathrm{Ker}(A)^\perp,$$
$$AA^\dagger : \mathbb{R}^m \to \mathrm{Ran}(A).$$

In view of the original works of Tikhonov concerning ill-posed problems (see, e.g., [129]), the term Tikhonov regularization is used somewhat loosely. Tikhonov considered the regularization of Fredholm equations of the first kind by minimizing the functional

$$F(x) = \|Ax - y\|^2 + \alpha^2 \Omega(x)$$

in a function space H. The penalty functional Ω was characterized by the property that the sets

$$\Omega_M = \{x \in H \mid \Omega(x) \leq M\}$$

are precompact in H. This condition quarantees the existence of the minimizer. In this sense, the Tikhonov regularization as defined here coincides with the original one only when H is finite-dimensional.

In large-scale inverse problems, the selection of the regularization parameter according to the discrepancy principle may be costly if one relies, e.g., on Newton's method. There are numerically effective methods to do the selection; see, e.g., [20].

In addition to Morozov's discrepancy principle, there are senveral other selection principles of the regularization parameters. We mention here the L-curve method (see [55]–[56]) and the generalized cross-validation (GCV) method ([39]).

The use of Kaczmarz iteration in tomographic problems has been discussed, e.g., in the book [96], which is a comprehensive representation of this topic in general.

At the end of Subsection 2.4.3, we considered the conjugate gradient iteration for nonsymmetric systems. Usually, when normal equations are considered, one avoids forming explicitly the matrix $A^{\mathrm{T}} A$. Since one works with the matrix A and its transpose, in comparison with the usual conjugate gradient, the algorithm requires one extra matrix-vector product per iteration. The algorithm has several acronyms, such as *conjugate gradient normal residual* (CGNR), *conjugate gradient normal equation* (CGNE) or *conjugate gradient least squares* (CGLS) methods. For references, see, e.g., the books [1] or [54].

In addition, for nonsymmetric problems other iterative solvers are available, e.g., *generalized minimal residual* (GMRES) method ([19], [109]). The various method differ from each other in the memory requirements, among other things.

3

Statistical Inversion Theory

This chapter is the central part of this book. We explain the statistical and, in particular, the Bayesian approach towards inverse problems and discuss the computational and interpretational issues that emerge from this approach.

The philosophy behind the statistical inversion methods is to recast the inverse problem in the form of statistical *quest for information*. We have directly observable quantities and others that cannot be observed. In inverse problems, some of the unobservable quantities are of primary interest. These quantities depend on each other through models. The objective of statistical inversion theory is to extract information and assess the uncertainty about the variables based on all available knowledge of the measurement process as well as information and models of the unknowns that are available prior to the measurement.

The statistical inversion approach is based on the following principles:

1. All variables included in the model are modelled as random variables.
2. The randomness describes our degree of information concerning their realizations.
3. The degree of information concerning these values is coded in the probability distributions.
4. The solution of the inverse problem is the posterior probability distribution.

The last item, in particular, makes the statistical approach quite different from the traditional approach discussed in the previous chapter. Regularization methods produce single estimates of the unknowns while the statistical method produces a distribution that can be used to obtain estimates that, loosely speaking, have different probabilities. Hence, the proper question to ask is not *what is the value of this variable?* but rather *what is our information about this variable?*

Classical regularization methods produce single estimates by a more or less ad hoc removal of the ill-posedness of the problem. The statistical method does not produce only single estimates. Rather, is an attempt to remove the ill-posedness by restating the inverse problem as a *well-posed extension* in a larger space of probability distributions. At the same time, it allows us to be explicit about the prior information that is often hidden in the regularization schemes. The similarities and differences between the statistical and the classical approaches are discussed later in Chapter 5.

In this chapter, the discussion is based on notions such as probability, random variables and probability densities. A brief review of these concepts is given in Appendix B.

3.1 Inverse Problems and Bayes' Formula

Although the results of this section hold in a more general setting, we assume here that all the random variables are absolutely continuous, that is, their probability distributions can be expressed in terms of probability densities.

As in the case of classical inverse problems, assume that we are measuring a quantity $y \in \mathbb{R}^m$ in order to get information about another quantity $x \in \mathbb{R}^n$. In order to relate these two quantities, we need a model for their dependence. This model may be inaccurate and it may contain parameters that are not well known to us. Furthermore, the measured quantity y always contains noise. In the traditional approach to inverse problems, we would typically write a model of the form

$$y = f(x, e), \tag{3.1}$$

where $f : \mathbb{R}^n \times \mathbb{R}^k \to \mathbb{R}^m$ is the model function and $e \in \mathbb{R}^k$ is vector containing all the poorly known parameters as well as the measurement noise.

One possible approach is to determine those components of e that do not change from measurement to measurement by appropriately chosen calibration measurements. The measurement noise, however, may be different from one instant to the other, and regularization methods are applied to cope with that part.

In statistical inverse problems, all parameters are viewed as random variables. As in Appendix B, we denote random variables by capital letters and their realizations by lowercase letters. Thus, the model (3.1) would lead to a relation

$$Y = f(X, E). \tag{3.2}$$

This is a relation tying together three random variables X, Y and E, and consequently their probability distributions depend on one another. However, as we shall later see in the examples, statistical inversion theory does not require even a model of the form (3.2) since the approach is based on relations between probability distributions. Before further elaborating this idea, let us introduce some nomenclature.

We call the directly observable random variable Y the *measurement*, and its realization $Y = y_\text{observed}$ in the actual measurement process the *data*. The nonobservable random variable X that is of primary interest is called the *unknown*. Those variables that are neither observable nor of primary interest are called *parameters* or *noise*, depending on the setting.

Assume that before performing the measurement of Y, we have some information about the variable X. In Bayesian theory, it is assumed that this information can be coded into a probability density $x \mapsto \pi_{\mathrm{pr}}(x)$ called the *prior density*. This term is self-explanatory: It expresses what we know about the unknown *prior* to the measurement.

Assume that, after analyzing the measurement setting as well as all additional information available about the variables, we have found the joint probability density of X and Y, which we denote by $\pi(x,y)$. Then, the marginal density of the unkown X must be

$$\int_{\mathbb{R}^m} \pi(x,y)dy = \pi_{\mathrm{pr}}(x).$$

If, on the other hand, we would know the value of the unknown, that is, $X = x$, the conditional probability density of Y given this information, would be

$$\pi(y \mid x) = \frac{\pi(x,y)}{\pi_{\mathrm{pr}}(x)}, \quad \text{if } \pi_{\mathrm{pr}}(x) \neq 0.$$

The conditional probability of Y is called the *likelihood function*, because it expresses the likelihood of different measurement outcomes with $X = x$ given.

Assume finally that the measurement data $Y = y_{\mathrm{observed}}$ is given. The conditional probability distribution

$$\pi(x \mid y_{\mathrm{observed}}) = \frac{\pi(x, y_{\mathrm{observed}})}{\pi(y_{\mathrm{observed}})}, \quad \text{if } \pi(y_{\mathrm{observed}}) = \int_{\mathbb{R}^n} \pi(x, y_{\mathrm{observed}}) dx \neq 0,$$

is called the *posterior distribution* of X. This distribution expresses what we know about X after the realized observation $Y = y_{\mathrm{observed}}$.

In the Bayesian framework, the inverse problem is expressed in the following way: *Given the data* $Y = y_{\mathrm{observed}}$, *find the conditional probability distribution* $\pi(x \mid y_{\mathrm{observed}})$ *of the variable* X.

We summarize the notations and results in the following theorem which can be referred to as the *Bayes' theorem of inverse problems*.

Theorem 3.1. *Assume that the random variable* $X \in \mathbb{R}^n$ *has a known prior probability density* $\pi_{\mathrm{pr}}(x)$ *and the data consist of the observed value* y_{observed} *of an observable random variable* $Y \in \mathbb{R}^k$ *such that* $\pi(y_{\mathrm{observed}}) > 0$. *Then the posterior probability distribution of* X, *given the data* y_{observed} *is*

$$\pi_{\mathrm{post}}(x) = \pi(x \mid y_{\mathrm{observed}}) = \frac{\pi_{\mathrm{pr}}(x)\pi(y_{\mathrm{observed}} \mid x)}{\pi(y_{\mathrm{observed}})}. \qquad (3.3)$$

In the sequel, we shall simply write $y = y_{\mathrm{observed}}$, and it is understood that when the posterior probability density is evaluated, we use the observed value of y.

In (3.3), the marginal density

$$\pi(y) = \int_{\mathbb{R}^n} \pi(x,y)dx = \int_{\mathbb{R}^n} \pi(y \mid x)\pi_{\mathrm{pr}}(x)dx$$

plays the role of a norming constant and is usually of little importance. Notice that it is in principle possible that $\pi(y) = 0$, that is, we get measurement data that have, vaguely speaking, zero probability. This is, in practice, technically seldom a problem. However, this would imply that the underlying models are not consistent with the reality.

In summary, looking at the Bayes' formula (3.3), we can say that solving an inverse problem may be broken into three subtasks:

1. Based on all the prior information of the unknown X, find a prior probability density π_{pr} that reflects judiciously this prior information.
2. Find the likelihood function $\pi(y \mid x)$ that describes the interrelation between the observation and the unknown.
3. Develop methods to explore the posterior probability density.

Each one of these steps may be a challenging problem on its own.

Before considering these problems in more detail, we discuss briefly how the statistical solution of an inverse problem can be used to produce single estimates as in the classical inversion methods.

3.1.1 Estimators

In the previous discussion, the solution of the inverse problem was defined to be the posterior distribution. As we shall see in the examples, if the unknown is a random variable with only few components, it is possible to visualize the posterior probability density as a nonnegative function of these variables. In most real-world inverse problems, the dimensionality of the inverse problem may be huge and, consequently, the posterior distribution lives in a very high-dimensional space, in which direct visualization is impossible. However, with a known posterior distribution, one can calculate different *point estimates* and *spread* or *interval estimates*. The point estimates answer to questions of the type *"Given the data y and the prior information, what is the most probable value of the unknown X?"* The interval estimates answer questions like *"In what interval are the values of the unknown with 90% probability, given the prior and the data?"*.

Before introducing the most common estimates, let us mention that classical inversion techniques can be seen as methodologies for producing single point estimates without a reference to any underlying statistical model. As we shall see later in Chapter 5, the statistical Bayesian model is useful also for analyzing these classical estimates. Thus, we emphasize here that although the statistical theory is sometimes viewed as a systematic means of producing point estimates, the theory is much more versatile.

One of the most popular statistical estimates is the *maximum a posteriori* estimate (MAP). Given the posterior probability density $\pi(x \mid y)$ of the unknown $X \in \mathbb{R}^n$, the MAP estimate x_{MAP} satisfies

$$x_{\text{MAP}} = \arg\max_{x \in \mathbb{R}^n} \pi(x \mid y),$$

provided that such maximizer exists. Observe that even when it exists, it may not be unique. The possible nonexistence and nonuniqueness indicate that the single-estimator based approaches to inverse problems may be unsatisfactory. The problem of finding a MAP estimate requires a solution of an *optimization problem*. Typically, the search for the maximizer is done by using iterative, often gradient-based, methods. As we shall see, in some cases this leads to the same computational problem as with the classical regularization methods. However, it is essential not to mix these two approaches since with the statistical approach the point estimates represent only part of the information on the unknowns.

Another common point estimate is the *conditional mean* (CM) of the unknown X conditioned on the data y, defined as

$$x_{\text{CM}} = \mathrm{E}\{x \mid y\} = \int_{\mathbb{R}^n} x \pi(x \mid y) dx,$$

provided that the integral converges. Finding the CM estimate generally amounts to solving an *integration problem*. A technical advantage of this estimate is in the fact that smoothness properties of the posterior distribution are not as crucial as in the MAP estimation problem, in particular when solved by gradient-based optimization. The main technical problem of CM estimation is that the integration is typically over a very high-dimensional space, in which common quadrature methods are not applicable. It is one of the main topics in this chapter to show alternative ways to perform the integration.

Before passing to the interval estimates, we mention the possibly most popular point estimate in statistics, the *maximum likelihood* (ML) estimate. This estimate x_{ML} which answers the question *"Which value of the unknown is most likely to produce the measured data y?"*, is defined as

$$x_{\text{ML}} = \arg\max_{x \in \mathbb{R}^n} \pi(y \mid x),$$

if such a maximizer exists. This is a non-Bayesian estimator, and from the point of view of ill-posed inverse problems, quite useless: As we shall see, it often corresponds to solving the classical inverse problem without regularization.

A typical spread estimator is the *conditional covariance*, defined as

$$\mathrm{cov}(x \mid y) = \int_{\mathbb{R}^n} (x - x_{\text{CM}})(x - x_{\text{CM}})^{\mathrm{T}} \pi(x \mid y) dx \in \mathbb{R}^{n \times n},$$

if the integral converges. Finding the conditional covariance is an integration problem.

As an example of interval estimates, consider the *Bayesian credibility set*. Given p, $0 < p < 100$, the credibility set D_p of $p\%$, is defined through the conditions

$$\mu(D_p \mid y) = \int_{D_p} \pi(x \mid y)dx = p/100, \quad \pi(x \mid y)\big|_{x \in \partial D_p} = \text{constant}.$$

Hence, the boundary of D_p is an equiprobability hypersurface enclosing $p\%$ of the mass of the posterior distribution.

Symmetric intervals of a given credibility with respect to single components of the unknown are also of interest. By defining the marginal density of the kth component X_k of X,

$$\pi(x_k \mid y) = \int_{\mathbb{R}^n} \pi(x_1, \cdots, x_n \mid y) dx_1 \cdots dx_{k-1} dx_{k+1} \ldots dx_n,$$

for a given p, $0 < p < 100$, we define $I_k(p) = [a, b] \subset \mathbb{R}$, where the endpoints a and b are determined from the conditions

$$\int_{-\infty}^{a} \pi(x_k) dx_k = \int_{b}^{\infty} \pi(x_k) dx_k = p/200.$$

Hence, $I_k(p)$ is the interval containing $p\%$ of the mass of the marginal density of X_k with equal probability mass in both tails of the density function.

The purpose of the following simple example is to show the problems related to single point estimators.

Example 1: Given the posterior distribution, which estimate gives a better idea of the posterior distribution? In general, given any estimate, it is possible to construct a distribution so that the estimate is misleading. To understand this claim, compare the estimates x_{MAP} and x_{CM} in the following simple one-dimensional case. Let $X \in \mathbb{R}$ and assume that the posterior distribution $\pi_{\text{post}}(x)$ of X is given by

$$\pi_{\text{post}}(x) = \pi(x) = \frac{\alpha}{\sigma_0} \phi\left(\frac{x}{\sigma_0}\right) + \frac{1-\alpha}{\sigma_1} \phi\left(\frac{x-1}{\sigma_1}\right),$$

where $0 < \alpha < 1$, $\sigma_0, \sigma_1 > 0$ and ϕ is the Gaussian distribution,

$$\phi(x) = \frac{1}{\sqrt{2\pi}} e^{-x^2/2}.$$

We observe that in this case,

$$x_{\text{CM}} = 1 - \alpha,$$

while

$$x_{\text{MAP}} = \begin{cases} 0, & \text{if } \alpha/\sigma_0 > (1-\alpha)/\sigma_1, \\ 1, & \text{if } \alpha/\sigma_0 < (1-\alpha)/\sigma_1. \end{cases}$$

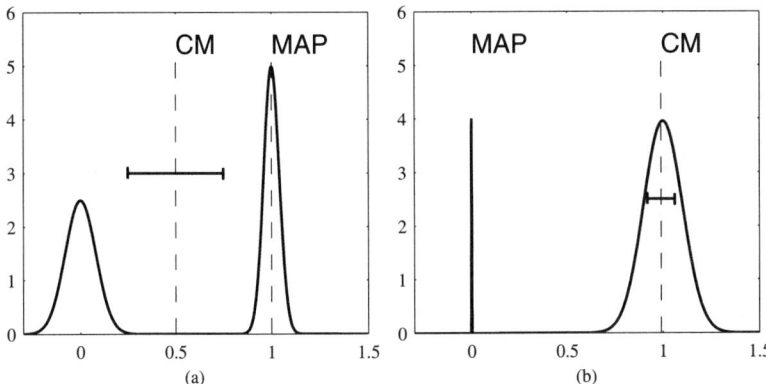

Figure 3.1. Examples of x_{CM} and x_{MAP}. The figures represent the probability densities $\pi(x)$ with parameter values $\alpha = 1/2$, $\sigma_0 = 0.08$ and $\sigma_1 = 0.04$ (a), and $\alpha = 0.01$, $\sigma_0 = 0.001 = \alpha\sigma_1$ and $\sigma_1 = 0.1$ (b).

If we have $\alpha = 1/2$ and σ_0 and σ_1 are small, it is highly unlikely that x takes any value near the estimate x_{CM}; see Figure 3.1 (a). On the other hand, if we take $\sigma_0 = \alpha\sigma_1$, then $\alpha/\sigma_0 = 1/\sigma_1 > (1-\alpha)/\sigma_1$ and therefore $x_{\mathrm{MAP}} = 0$. But this is a bad estimate for x if α is small, since the likelihood for x to take a value near $x = 0$ is less than α. Such a situation is depicted in Figure 3.1 (b). In the same figure, we have also shown the square root of the posterior variances (that is, the root mean squares), the posterior variances being

$$\sigma^2 = \int_{-\infty}^{\infty} (x - x_{\mathrm{CM}})^2 \pi(x) dx = \int_{-\infty}^{\infty} x^2 \pi(x) dx - x_{\mathrm{CM}}^2.$$

In this example, the variance can be computed analytically. We have

$$\sigma^2 = \alpha\sigma_0^2 + (1-\alpha)(\sigma_2^2 + 1) - (1-\alpha)^2.$$

Notice that when the conditional mean gives a poor estimate, this is reflected in wider variance, as seen in Figure 3.1 (a). ◇

3.2 Construction of the Likelihood Function

The construction of the likelihood function is often the most straightforward part in the statistical inversion. Therefore we discuss this part first before going on to the more subtle question of the construction of the priors. The likelihood function contains the forward model used in classical inversion techniques as well as information about the noise and other measurement and modelling uncertainties. We discuss some of the most frequent cases below. More complex models are discussed in more detail in Chapter 7.

3.2.1 Additive Noise

Most often - and this is true in particular in the classical inverse problems literature - the noise is modelled as additive and mutually independent of the unknown X. With the classical regularization methods, however, the mutual independence is usually implicit rather than a consciously modelled property. Hence, the stochastic model is

$$Y = f(X) + E,$$

where $X \in \mathbb{R}^n$, $Y, E \in \mathbb{R}^m$ and X and E are mutually independent. Assume that the probability distribution of the noise E is known, that is,

$$\mu_E(B) = \mathrm{P}\{E \in B\} = \int_B \pi_{\mathrm{noise}}(e) de.$$

If we fix $X = x$, the assumption of the mutual independence of X and E ensures that the probability density of E remains unaltered when conditioned on $X = x$. Therefore we deduce that Y conditioned on $X = x$ is distributed like E, the probability density being translated by $f(x)$, that is, the likelihood function is

$$\pi(y \mid x) = \pi_{\mathrm{noise}}(y - f(x)).$$

Hence, if the prior probability density of X is π_{pr}, from Bayes' formula (3.3) we obtain

$$\pi(x \mid y) \propto \pi_{\mathrm{pr}}(x) \pi_{\mathrm{noise}}(y - f(x)).$$

A slightly more complicated situation appears when the unknown X and the noise E are not mutually independent. In this case, we need to know the conditional density of the noise, that is,

$$\mu_E(B \mid x) = \int_B \pi_{\mathrm{noise}}(e \mid x) de.$$

In this case, we may write

$$\pi(y \mid x) = \int_{\mathbb{R}^m} \pi(y \mid x, e) \pi_{\mathrm{noise}}(e \mid x) de.$$

When both $X = x$ and $E = e$ are fixed, Y is completely specified, that is, $Y = y = f(x) + e$, so

$$\pi(y \mid x, e) = \delta(y - f(x) - e).$$

A substitution into the previous formula yields then

$$\pi(y \mid x) = \pi_{\mathrm{noise}}(y - f(x) \mid x),$$

and therefore

3.2 Construction of the Likelihood Function

$$\pi(x \mid y) \propto \pi_{\mathrm{pr}}(x)\pi_{\mathrm{noise}}(y - f(x) \mid x).$$

Before proceeding, we present a small numerical example to clarify the concepts. This example, as we shall see, does not have the typical properties of an ill-posed inverse problem.

Example 2: Consider the case in which we have additive independent measurement noise in our observation and this noise is mutually independent of the unknown. It is instructive to look at a simple example that can be easily visualized. Assume that $X \in \mathbb{R}^2$ and $Y, E \in \mathbb{R}^3$ and we have a linear model

$$Y = AX + E,$$

where $A \in \mathbb{R}^{3 \times 2}$ is

$$A = \begin{bmatrix} 1 & -1 \\ 1 & -2 \\ 2 & 1 \end{bmatrix}.$$

Let us assume that E has mutually independent components that are normally distributed with zero mean and variances equal to $\sigma^2 = 0.09$. Then,

$$\pi_{\mathrm{noise}}(e) \propto \exp\left(-\frac{1}{2\sigma^2}\|e\|^2\right).$$

Assume furthermore that we have the prior knowledge that the components of X satisfy

$$\mathrm{P}\{|X_j| > 2\} = 0, \quad j = 1, 2,$$

and no other information is available. Then, a natural choice of prior would be

$$\pi_{\mathrm{pr}}(x) \propto \chi_Q(x),$$

where χ_Q is the characteristic function of the cube $[-2, 2] \times [-2, 2]$. It follows that

$$\pi(x \mid y) \propto \chi_Q(x)\exp\left(-\frac{1}{2\sigma^2}\|y - Ax\|^2\right).$$

Assume that $x_0 = [1, 1]^{\mathrm{T}}$ is the true value of the unknown, and the data is $y = Ax_0 + e$, in which the realization e of the noise is drawn from π_{noise} above.

In Figure 3.2, we have plotted the posterior probability distribution of the variable $X \in \mathbb{R}^2$ as a grayscale image. The lines in the figure are the graphs of each linear equation of the system $Ax = y$, where y is the noisy data. Observe that the prior distribution has practically no effect here since the likelihood function, considered as a function of x, is already negligible when the condition $|x_j| < 2$ becomes active. This is an example of an inverse problem in which the maximum likelihood estimator would yield a feasible estimate of the unknown. The reason for this is that the problem is not ill-conditioned. Although the equation $y = Ax$ is overdetermined, and therefore no exact solution usually exists, the matrix A itself is well-conditioned. ◇

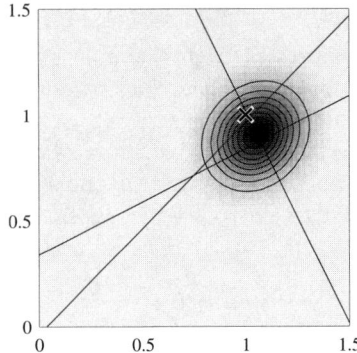

Figure 3.2. An example of the posterior density with noisy data. The true value x_0 corresponding to noiseless data is marked by a cross.

3.2.2 Other Explicit Noise Models

The additive noise model is not always an adequate one. Rather, we may have a more complicated relation tying the unknown, the observation and the noise together through a formula of the type (3.2). Formally, we may proceed as in the previous subsection and write

$$\pi(y \mid x, e) = \delta(y - f(x, e)),$$

and further

$$\pi(y \mid x) = \int_{\mathbb{R}^k} \delta(y - f(x, e)) \pi_{\text{noise}}(e \mid x) de. \tag{3.4}$$

This formula, however, may not be well defined or be of little practical value. Instead, let us consider some particular cases.

First, assume that we have a simple real-valued measurement, and the observation model contains multiplicative noise that is mutually independent with the unknown, that is, the stochastic model tying together $X \in \mathbb{R}$ and $Y \in \mathbb{R}$ is

$$Y = Ef(X),$$

where E is a real-valued noise and $f : \mathbb{R} \to \mathbb{R}$. Such noise could appear, for example, when a signal is amplified by a noisy amplifier. If π_{noise} is the probability density of E, the formula (3.4) gives

$$\pi(y \mid x) = \int_{\mathbb{R}} \delta(y - ef(x)) \pi_{\text{noise}}(e) de$$

$$= \frac{1}{f(x)} \int_{\mathbb{R}} \delta(y - \nu) \pi_{\text{noise}}\left(\frac{\nu}{f(x)}\right) d\nu = \frac{1}{f(x)} \pi_{\text{noise}}\left(\frac{y}{f(x)}\right),$$

and the posterior density is

3.2 Construction of the Likelihood Function

$$\pi(x \mid y) = \frac{\pi_{\text{pr}}(x)}{f(x)} \pi_{\text{noise}}\left(\frac{y}{f(x)}\right).$$

Consider now the case in which we have a noisy measurement with an incompletely known forward model. Let $A(v) \in \mathbb{R}^{m \times n}$ denote a matrix that depends on a parameter vector $v \in \mathbb{R}^k$, and assume that the deterministic model without measurement noise is $y = A(v)x$, $y \in \mathbb{R}^m$, $x \in \mathbb{R}^n$. Assume further that the actual measurement is corrupted by additive noise that is mutually independent with the unknown X and the parameter V. Thus, the statistical model in this case becomes

$$Y = A(V)X + E.$$

If π_{noise} is the probability density of E which is mutually independent with X and V, we have

$$\pi(y \mid x, v) = \pi_{\text{noise}}(y - A(v)x).$$

Furthermore, assuming that V and X are mutually independent and V has density π_{param}, we obtain the likelihood density

$$\pi(y \mid x) = \int_{\mathbb{R}^k} \pi(y \mid x, v)\pi_{\text{param}}(v)dv = \int_{\mathbb{R}^k} \pi_{\text{noise}}(y - A(v)x)\pi_{\text{param}}(v)dv. \tag{3.5}$$

An inverse problem that allows a formulation of this type is the *blind deconvolution problem* that we outline in the following example.

Example 3: Consider the following one-dimensional deconvolution problem. We want to estimate the function $f : \mathbb{R} \to \mathbb{R}$ from the noisy observations of the blurred image

$$g(t) = \phi(t) * f(t) = \int_{-\infty}^{\infty} \phi(t-s)f(s)ds.$$

We assume here that $f(t)$ vanishes outside a fixed interval I. This problem was already discussed in the context of the classical inversion methods in the previous chapter. Assume that the convolution kernel ϕ is symmetric but not known exactly. We seek to approximate the kernel in terms of Gaussian kernels, that is,

$$\phi(t) = \sum_{k=1}^{K} v_k \varphi_k(t),$$

where

$$\varphi_1(t) = \frac{1}{\sqrt{2\pi\sigma_1^2}} \exp\left(-\frac{1}{2\sigma_1^2}t^2\right),$$

$$\varphi_k(t) = \frac{1}{\sqrt{2\pi\sigma_k^2}} \exp\left(-\frac{1}{2\sigma_k^2}t^2\right) - \frac{1}{\sqrt{2\pi\sigma_{k-1}^2}} \exp\left(-\frac{1}{2\sigma_{k-1}^2}t^2\right), \; k \geq 2.$$

Here, we take the variances σ_k^2 as fixed and arranged in growing order. Furthermore, the coefficients v_k are poorly known.

Assume now that the observation is corrupted by additive Gaussian noise. Discretize the problem using a quadrature rule in I to approximate the convolution integral. If the discretization points are denoted by t_ℓ with corresponding weights w_ℓ, $1 \leq \ell \leq n$, the discrete observation model becomes

$$y = \sum_{k=1}^{K} v_k A_k x + e,$$

where $x = [f(t_1), \ldots, f(t_n)]^T \in \mathbb{R}^n$, $A_k \in \mathbb{R}^{m \times n}$, $A_k(i,j) = w_j \varphi_k(t_i - t_j)$. The vector $e \in \mathbb{R}^m$ represents the additive noise and $y \in \mathbb{R}^m$ is the noisy observation vector at points t_ℓ. We pass to the stochastic model. Assuming that the additive noise is zero mean white noise with variance γ^2, the likelihood is

$$\pi(y \mid x, v) \propto \exp\left(-\frac{1}{2\gamma^2} \left\| y - \sum_{k=1}^{K} v_k A_k x \right\|^2\right).$$

The exponent is quadratic in $v = [v_1, \ldots, v_K]^T$, that is, with respect to that parameter the above density is Gaussian. The possibility to find a closed form for the posterior density $\pi(x \mid y)$ by the formula (3.5) depends entirely on the density π_{param} of v. If we can choose this density to be Gaussian, an explicit integration is possible. We return to this example later in this section. ◇

3.2.3 Counting Process Data

In some cases, the likelihood function is not based on a model of the type $Y = f(X, E)$. Rather, we may know the probability density of the measurement itself, and the unknown defines the parameters of the density. We consider some typical examples below.

In many applications, the measurements are based on counting of events. As examples, we can think of an electron microscope in which the device counts the electrons arriving the detector over a given period of time. Similarly, in low-energy X-ray imaging, the data consists of photon counts. In other medical applications such as PET or SPECT imaging, the data is likewise a quantum count. In the latter two cases the number of counts in each observation might be very low, possibly zero.

In such cases, the signal can often be modelled as a *Poisson process*. Consider first a single channel data. We have a model for the expected or average observation, written in the deterministic form as

$$\bar{y} = f(x) \in \mathbb{R}.$$

In reality, the measurement Y is an integer-valued random variable with the expectation $\bar{y}$, that is,

3.2 Construction of the Likelihood Function

$$Y : \Omega \to \mathbb{N}, \quad Y \sim \text{Poisson}(\bar{y}),$$

so the likelihood function is

$$\pi(y \mid x) = \frac{f(x)^y}{y!} \exp\bigl(-f(x)\bigr).$$

In the case of multichannel data, the observation is a vector $y = [y_1, \ldots, y_M]^\mathrm{T}$. If we assume that each component has mutually independent fluctuation, we can write

$$\pi(y \mid x) = \prod_{i=1}^{m} \frac{f_i(x)^{y_i}}{y_i!} \exp\bigl(-f_i(x)\bigr)$$

$$\propto \exp\Bigl(y^\mathrm{T} \log f(x) - \|f(x)\|_1\Bigr),$$

where $\log f(x) = [\log f_1(x), \ldots, \log f_m(x)]^\mathrm{T}$, and $\|f(x)\|_1 = \sum |f_i(x)|$ denotes the ℓ^1-norm of $f(x)$. Since $f_i(x) \geq 0$, we can also write $\|f(x)\|_1 = \mathbf{1}^\mathrm{T} f(x)$ where $\mathbf{1} = [1, \ldots, 1]^\mathrm{T}$ Observe that in the above formula, the constant of proportionality depends on y but not on x. Since in the inverse problems we are interested in the dependence of the likelihood function on x, we have simply left the constant out. Remember that eventually y has a fixed value $y = y_{\text{observed}}$.

Often, a counting observation is corrupted by noise that is due to the measurement device. For example, assume that the measurement is a Poisson distributed counting process corrupted by mutually independent additive noise. In this case, we write the model as

$$Y = K + E, \quad K \sim \text{Poisson}(f(x)).$$

Write

$$\pi(y, k \mid x) = \pi(y \mid k, x) \pi(k \mid x).$$

If $K = k$ is given, the distribution of Y is determined by the distribution of E. Let us denote the probability density of E be $\pi_{\text{noise}}(e)$, and let the distribution of K with $X = x$ be as above. Then we can write

$$\pi(y, k \mid x) = \pi_{\text{noise}}(y - k) \frac{f(x)^k}{k!} \exp(-f(x)).$$

The intermediate variable k is not of interest to us here, so we calculate the marginal density of Y and arrive at the formula

$$\pi(y \mid x) = \sum_{k=0}^{\infty} \pi_{\text{noise}}(y - k) \frac{f(x)^k}{k!} \exp(-f(x)).$$

Often, such likelihood functions can be approximated by using Stirling's formula.

In addition to the Poisson distribution above, one can encounter other parametric observation models such as beta, gamma or log-normal likelihood models.

3.3 Prior Models

In the statistical theory of inverse problems, one can argue that the construction of the prior density is the most crucial step and often also the most challenging part of the solution. The major problem with finding an adequate prior density lies usually in the nature of the prior information. Indeed, it is often the case that our prior knowlege of the unknown is *qualitative* in nature. The problem then consists of transforming qualitative information into a *quantitative* form that can then be coded into the prior density. For example, a geophysicist who does subsurface electromagnetic sounding of the earth may expect to see layered structures with nonlayered inclusions and possibly some cracks. However, the use of a layered model can be too restrictive since it does not allow other structures that are of interest. Similarly, in medical imaging, one could be looking for a cancer that is known to be well located and having possibly a "characteristic" surface structure that the trained eye of a radiologist may recognize. These are qualitative descriptions of the belief of what one might see, but hard to translate into the language of densities.

The general goal in designing priors is to write down a density $\pi_{\rm pr}(x)$ with the following property. If E is a collection of *expectable* vectors x representing possible realizations of the unknown X and U is a collection of *unexpectable* ones, we should have

$$\pi_{\rm pr}(x) >> \pi_{\rm pr}(x') \text{ when } x \in E,\ x' \in U.$$

Thus, the prior probability distribution should be concentrated on those values of x that we expect to see and assign a clearly higher probability to them than to those that we do not expect to see.

In this section, we go through some explicit prior models that have been used successfully in inverse problems, with an emphasis on their qualitative properties. Also, we discuss how to construct priors based, for example, on sample data and on structural information, as well as how to explore and visualize priors to make their nature more tangible.

3.3.1 Gaussian Priors

The most commonly used probability densities in statistical inverse problems are undoubtedly Gaussian. This is due to the fact that they are easy to construct, yet they form a much more versatile class of densities than is usually believed. Moreover, they often lead to explicit estimators. For this reason, we dedicate a whole section (Section 3.4) to Gaussian densities.

3.3.2 Impulse Prior Densities

For the sake of definiteness, let us consider here an inverse problem in which the unknown is a two-dimensional pixel image that represents any discretized distributed physical parameter or coefficient in a body. Assume that the prior information is that the image contains small and well localized objects. This is the case, for example, when one tries to localize a tumor in X-ray imaging or the cortical activity by electric or magnetic measurements. In such cases, we can frequently use *impulse prior* models. The characteristic feature of these models is that they are concentrated on images with low average amplitude with few outliers standing out against the background. One such prior is the ℓ^1 *prior*. Let $x \in \mathbb{R}^n$ represent the pixel image, the component x_j being the intensity of the jth pixel. The ℓ^1 norm of x is defined as

$$\|x\|_1 = \sum_{j=1}^n |x_j|.$$

The ℓ^1 prior is defined as

$$\pi_{\mathrm{pr}}(x) = \left(\frac{\alpha}{2}\right)^n \exp\bigl(-\alpha\|x\|_1\bigr).$$

An even more enhanced impulse noise effect can be obtained by using a probability density of the form

$$\pi_{\mathrm{pr}}(x) \propto \exp\left(-\alpha \sum_{j=1}^n |x_j|^p\right), \quad 0 < p < 1.$$

Note that the lack of convexity is in principle not a problem when sampling is carried out.

Another density that produces few prominently different pixels with a low-amplitude background is the *Cauchy density*, defined as

$$\pi_{\mathrm{pr}}(x) = \left(\frac{\alpha}{\pi}\right)^n \prod_{j=1}^n \frac{1}{1+\alpha^2 x_j^2}.$$

To understand better the qualitative nature of these densities, let us make random draws of pixel images out of these densities. Consider first the ℓ^1 prior. Since we deal with images, it is natural to augment the density with a positivity constraint and write

$$\pi_{\mathrm{pr}}(x) = \alpha^n \pi_+(x) \exp\bigl(-\alpha\|x\|_1\bigr),$$

where $\pi_+(x) = 1$ if $x_j > 0$ for all j and $\pi_+(x) = 0$ otherwise. Since each component is independent of the others, random drawing can be performed componentwise by the following standard procedure: Define first the one-dimensional distribution function

$$\Phi(t) = \alpha \int_0^t e^{-\alpha s}\,ds = 1 - e^{-\alpha t}.$$

The mutually independent components x_j are now drawn by

$$x_j = \Phi^{-1}(t_j) = -\frac{1}{\alpha}\log(1-t_j),$$

where the t_j s are drawn randomly from the uniform distribution $\mathcal{U}([0,1])$.

Similarly, when we draw mutually independent components from the Cauchy distribution with the positivity constraint, we use the distribution function

$$\Phi(t) = \frac{2\alpha}{\pi}\int_0^t \frac{1}{1+\alpha^2 s^2}ds = \frac{2}{\pi}\arctan\alpha t.$$

In Figure 3.3, we have plotted 60×60 noise images randomly drawn from the ℓ^1 prior and from the Cauchy prior distribution. The parameter α appearing in the priors is chosen in both cases equal to unity. For comparison, Gaussian *white noise images* with a positivity constraint are also displayed. The Gaussian white noise density with positivity constraint is

$$\pi_{\mathrm{pr}}(x) \propto \pi_+(x)\exp\left(-\frac{1}{2\alpha^2}\|x\|^2\right),$$

the norm being the usual Euclidian norm. For the white noise prior with positivity constraint, the one-dimensional distribution function is

$$\Phi(t) = \frac{1}{\alpha}\sqrt{\frac{2}{\pi}}\int_0^t \exp\left(-\frac{1}{2\alpha^2}s^2\right)ds = \mathrm{erf}\left(\frac{t}{\alpha\sqrt{2}}\right),$$

where erf stands for the error function,

$$\mathrm{erf}(t) = \frac{2}{\sqrt{\pi}}\int_0^t \exp(-s^2)ds.$$

Here, too, we use $\alpha = 1$. The grayscale in each image is linear from the minimum to the maximum pixel value.

Let us introduce one more prior density that has been used successfully in image restoration problems in which one seeks few outstanding features in a low-amplitude background, namely the *entropy density*. Let $x \in \mathbb{R}^n$ denote a pixel image with positive entries, $x_j > 0$. The entropy of the image is defined as

$$\mathcal{E}(x) = -\sum_{j=1}^n x_j \log\left(\frac{x_j}{x_0}\right),$$

where $x_0 > 0$ is a given constant. The entropy density is of the form

$$\pi(x) \sim \exp(\alpha\mathcal{E}(x)).$$

The entropy density is closely related to a regularization technique called the *maximum entropy method* (MEM), in which the entropy is used as a regularizing functional in the generalized Tikhonov method.

3.3 Prior Models

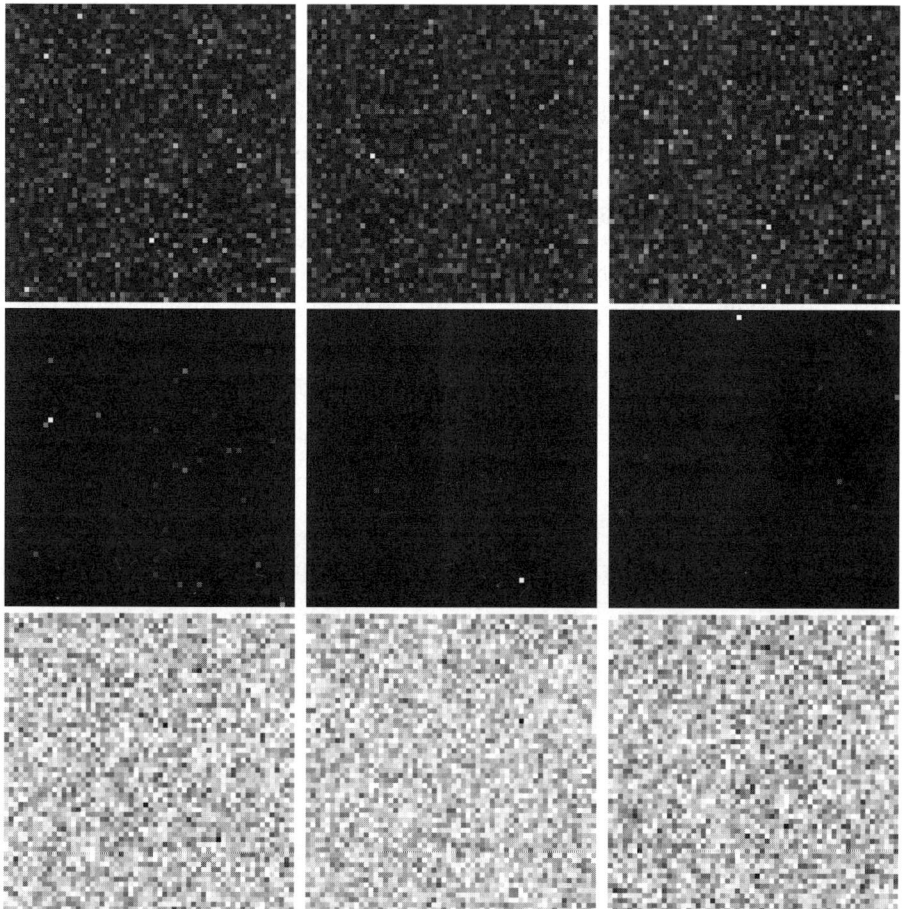

Figure 3.3. Random draws from ℓ^1 prior (top row), Cauchy prior (middle row) and white noise prior (bottom row). All priors include a positivity constraint.

The entropy density is qualitatively similar to the *lognormal density*, in which the logarithm of the unknown is normally distributed. For a single component, the lognormal density is given by

$$\pi(x) = \frac{1}{\sqrt{2\pi}\sigma x} \exp\left(-\frac{1}{2\sigma^2}(\log x - \log x_0)^2\right), \quad x > 0.$$

For vector-valued variables with mutually independent components, the density is naturally a product of the component densities.

3.3.3 Discontinuities

Often in inverse problems, one seeks an unknown that may contain discontinuities. The prior information may concern, for example, the size of the jumps of the unknown across the discontinuities or the probable occurrence rate of the discontinuities.

The impulse prior densities discussed above provide a tool for implementing this type of information in the prior. We consider here a one-dimensional example. Let us assume that the goal is to estimate a function $f : [0,1] \to \mathbb{R}$, $f(0) = 0$ from indirect observations. Our prior knowledge is that the function f may have a large jump at few places in the interval, but the locations of these places are unkown to us. One possible way to construct a prior in this case is to consider the finite difference approximation of the derivative of f and assume that it follows an impulse noise probability distribution. Let us discretize the interval $[0,1]$ by points $t_j = j/N$, $0 \leq j \leq N$ and write $x_j = f(t_j)$. Consider the density

$$\pi_{\mathrm{pr}}(x) = \left(\frac{\alpha}{\pi}\right)^N \prod_{j=1}^{N} \frac{1}{1+\alpha^2(x_j - x_{j-1})^2}.$$

It is again instructive to see what random draws from this distribution look like. To perform the drawing, let us define new random variables

$$\xi_j = x_j - x_{j-1}, \quad 1 \leq j \leq N. \tag{3.6}$$

The probability distribution of these variables is

$$\pi(\xi) = \left(\frac{\alpha}{\pi}\right)^N \prod_{j=1}^{N} \frac{1}{1+\alpha^2 \xi_j^2},$$

that is, they are independent from each other. Hence each ξ_j can be drawn from the one-dimensional Cauchy density. Having these mutually independent increments at hand, the vector x is determined from (3.6). Figure 3.4 shows four random draws from this distribution with discretization $N = 1200$.

The idea of this section is easy to generalize to higher dimensions. This generalization leads us naturally to Markov random fields.

3.3.4 Markov Random Fields

A rich class of prior distributions can be derived from the theory of *Markov random fields* (MRF), whose definition we give here only in the discrete case. Let X be a $\mathbb{R}^n$-valued random variable. A *neighborhood system* related to X is a collection of index sets,

$$\mathcal{N} = \{\mathcal{N}_i \mid 1 \leq i \leq n\}, \quad \mathcal{N}_i \subset \{1, 2, \ldots, n\},$$

with the following properties:

1. $i \notin \mathcal{N}_i$,
2. $i \in \mathcal{N}_j$ if and only if $j \in \mathcal{N}_i$.

3.3 Prior Models 67

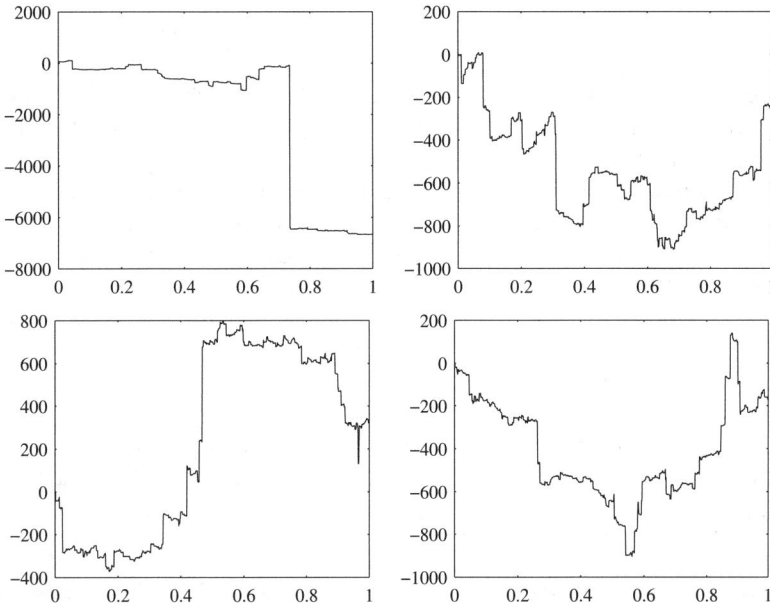

Figure 3.4. Four random draws from the density (3.6). The parameter is $\alpha = 1$.

The set $\mathcal{N}_i$ is the index set of the neighbors of the component X_i of X.

Let us consider the conditional distribution of the component X_i. We say that X is a discrete Markov random field with respect to the neighborhood system $\mathcal{N}$ if

$$\pi_{X_j}(B \mid x_1, \ldots, x_{j-1}, x_{j+1}, \ldots, x_n) = \pi_{X_j}(B \mid x_k, \ k \in \mathcal{N}_j).$$

In other words, the value of X_j depends on the values of the remaining components only through its neighbors. The probability densities of the MRFs are of a particular form. The *Hammersley–Clifford theorem* states that the probability density of an MRF is of the form

$$\pi(x) \propto \exp\left(-\sum_{j=1}^n V_j(x)\right),$$

where the functions V_j depend only on x_j and on components x_k with $k \in \mathcal{N}_j$. A useful MRF prior is the *total variation density*. We first define the concept of total variation of a function. Let $f : D \to \mathbb{R}$ be a function in $L^1(D)$, the space of integrable functions on $D \subset \mathbb{R}^n$. We define the total variation of f, denoted by $\mathrm{TV}(f)$ as

$$\mathrm{TV}(f) = \int_\Omega |\mathcal{D}f| = \sup\left\{\int_D f \nabla \cdot g \, dx \mid g = (g_1, \ldots, g_n)\right.$$

$$\in C_0^1(D, \mathbb{R}^n), \ \|g(x)\| \le 1 \bigg\}.$$

Here, the test function space $C_0^1(D, \mathbb{R}^n)$ consists of continuously differentiable vector-valued functions on D that vanish at the boundary. A function is said to have bounded variation if $\mathrm{TV}(f) < \infty$.

To understand the definition, consider a simple example. Assume that $D \subset \mathbb{R}^2$ is an open set and $B \subset D$ is a set bounded by a simple smooth curve not intersecting the boundary of D. Let $f : D \to \mathbb{R}$ be the characteristic function of B, that is, $f(x) = 1$ if $x \in B$ and $f(x) = 0$ otherwise. Let $g \in C_0^1(D, \mathbb{R}^2)$ be an arbitrary test function. By the divergence theorem we obtain

$$\int_D f \nabla \cdot g \, dx = \int_B \nabla \cdot g \, dx = \int_{\partial B} n \cdot g \, dS,$$

where n is the exterior unit normal vector of ∂B. Evidently, this integral attains its maximum, under the constraint $\|g(x)\| \le 1$, if we set $n \cdot g = 1$ identically. Hence,

$$\mathrm{TV}(f) = \mathrm{length}(\partial B).$$

Observe that the derivative of f is the Dirac delta of the boundary curve, which cannot be represented by an integrable function. Therefore, the space of functions with bounded variation differs from the corresponding Sobolev space.

To find a discrete analogue, let us consider the two-dimensional problem. Assume that $D \subset \mathbb{R}^2$ is bounded and divided into n pixels. A natural neighborhood system is obtained by defining two pixels as neighbors if they share a common edge. The total variation of the discrete image $x = [x_1, x_2, \ldots, x_n]^\mathrm{T}$ is now defined as

$$\mathrm{TV}(x) = \sum_{j=1}^n V_j(x), \quad V_j(x) = \frac{1}{2} \sum_{i \in \mathcal{N}_j} \ell_{ij} |x_i - x_j|,$$

where ℓ_{ij} is the length of the edge between the neighboring pixels. The discrete total variation density is given now as

$$\pi(x) \propto \exp(-\alpha \mathrm{TV}(x)).$$

It is well known that the total variation density is concentrated on images that are "blocky," that is, images consisting of blocks with short boundaries and small variation within each block. To understand this statement, consider Figure 3.5 depicting three different simple images with the same energy. The values of the pixels in these 7×7 images take three different values 0 (black), 1 (gray) or 2 (white). The total variations of these images are easy to calculate. We see that the blocky image on the left has the smallest total variation while the one on the right has the highest one. Hence, the left image has the highest probability to occur when measured by the total variation density.

3.3 Prior Models   69

Figure 3.5. Three 7×7 pixel images with equal energy but different total variation, from left to right 18, 28 and 40, respectively

To compare the total variation prior with the previously discussed ones, we draw randomly three noise images from the total variation prior equipped with the positivity constraint. These draws are displayed in Figure 3.6. Although not dramatically different from the white noise images in Figure 3.3, we can see here clear correlation between nearby pixels.

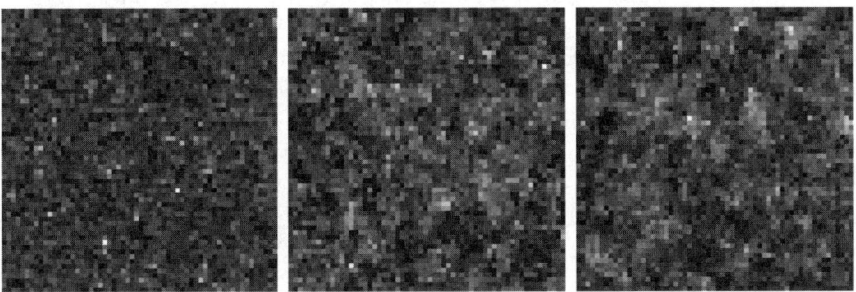

Figure 3.6. Three total variation noise images with positivity constraints.

The MRF priors serve as useful tools in designing *structural priors*. As an example, consider a medical imaging problem in which we know a priori the location of the organ boundaries. Such prior information may be based on anatomical data or on other imaging modalities such as X-ray or magnetic resonance data. In medical applications, this type of information is expressed by saying that we have a *segmented image* at our disposal. Assume that we have k different tissue types and a mapping $T : \{1, \ldots, n\} \to \{1, \ldots, k\}$ that classifies each pixel into one of the types. Now we can design a neighborhood system $\mathcal{N}$ in such a way that

$$T(i) \neq T(j) \Rightarrow j \notin \mathcal{N}_i,$$

that is, pixels of different type are never neighbors. Actually, the pixels are divided into *cliques* according to their type, each clique being independent of the others. Since there is no correlation between pixels of different cliques, even considerable jumps in the values across the organ boundaries become possible. Examples of these types will be discussed later.

3.3.5 Sample-based Densities

Often, the construction of the prior or likelihood densities is based on a large set of previously obtained samples of the random variable in question. For instance, measurement noise can be analyzed by performing an extensive series of calibration measurements with a known object. Similarly, the prior information, for example, in medical imaging applications, may be based on a large set of data obtained surgically or via another imaging modality. These data are sometimes called anatomical atlases.

Assume that $\pi = \pi(x)$ is the probability density of a random variable $X \in \mathbb{R}^n$, and that we have a large sample of realizations of X,

$$S = \{x^1, x^2, \ldots, x^N\},$$

where N is large. The objective is to approximate π based on S.

If X is a real-valued random variable and N is large, the estimation of π is relatively straightforward: One can use, for example, a spline approximation of the histogram of the samples. In higher dimensions, this approach is no longer applicable. In general, the problem can be seen as a *kernel estimation problem*. References on this widely studied topic can be found at the end of this chapter; see "Notes and Comments."

Assume that S consists of realizations of a random variable X. We can use the set S to estimate the moments of X. The second-order statistics of X is therefore

$$\mathrm{E}\{X\} \approx \overline{x} = \frac{1}{N} \sum_{j=1}^{N} x^j \quad \text{(mean)},$$

$$\mathrm{cov}(X) = \mathrm{E}\{XX^\mathrm{T}\} - \mathrm{E}\{X\}\mathrm{E}\{X\}^\mathrm{T}$$

$$\approx \Gamma = \frac{1}{N} \sum_{j=1}^{N} x^j (x^j)^\mathrm{T} - \overline{x}\,\overline{x}^\mathrm{T} \quad \text{(covariance)}.$$

Consider the eigenvalue decomposition of the matrix Γ,

$$\Gamma = UDU^\mathrm{T},$$

where U is an orthogonal matrix and D is diagonal with entries $d_1 \geq d_2 \geq \cdots \geq d_n \geq 0$. Typically, the vectors x^j are not very dissimilar from each other; thus the matrix Γ is singular or almost singular. Assume that $d_r > d_{r+1} \approx 0$, that is, only r first eigenvalues are significantly different from zero. This

suggests that $X - \mathrm{E}\{X\}$ could be a random variable with the property that its values lie with a high probability in the subspace spanned by the r first eigenvectors of Γ. Based on this analysis, we define the *subspace prior* given as
$$\pi(x) \propto \exp\bigl(-\alpha\|(1-P)(x-\bar{x})\|^2\bigr),$$
where P is the orthogonal projector $\mathbb{R}^n \to \mathrm{sp}\{u_1, \ldots, u_r\}$, in which the vectors u_j are the eigenvectors of Γ. If all the eigenvalues of Γ are significantly different from zero, the subspace prior gives no information. In such cases, we may still make use of the approximate low-order statistics. The most straightforward approximation for the probability density of X is now to use the Gaussian approximation, that is, we have
$$\pi_{\mathrm{pr}}(x) \propto \exp\left(-\frac{1}{2}(x-\bar{x})^{\mathrm{T}} \Gamma^{-1}(x-\bar{x})\right),$$

The success of this approximation can be evaluated, for example, by studying the higher-order statistics of X based on the sample. Also, Γ as a sample-based estimate for the covariance may well be numerically rank-deficient so that Γ^{-1} does not exist or is otherwise unreliable. In the following example we use a hybrid model of the subspace prior and the prior based on an estimated covariance matrix.

Example 4: Consider a two-dimensional image deblurring problem, in which the prior information of the image is that it consists of a round object on a constant background. Assuming that the image area is a unit square $D = [0,1] \times [0,1]$, the image is known to be of the type
$$f(p) = a\theta\bigl(r - |p - p_0|\bigr),$$
where θ is the Heaviside function, $a > 0$ is the amplitude of the image object, $p_0 \in D$ is the centerpoint and $r > 0$ is the radius. The parameters a, r and p_0 are unknown to us. However, we assume that we have a priori information of their distribution.

Naturally, one could formulate the inverse problem using these parameters directly. However, in this example we use the statistics of the parameters only to create a library of feasible images.

We divide the image area into $n = k \times k$ pixels and denote the pixelized images by a vector $x \in \mathbb{R}^k$. In the example below, $k = 40$. As the first step, we generate a sample of pixelized images based on the parameter distributions. In our example, we assume that the parameter distributions are
$$a \sim \mathcal{U}([0.8, 1.2]), \quad r \sim \mathcal{U}([0.2, 0.25]), \quad p_0 \sim \mathcal{N}((0.5, 0.5), (0.02)^2 I).$$

The sample size we use is $N = 500$. Having the sample $\{x^j\}$ of pixelized images, we calculate approximations for the mean $\bar{x}$ and the covariance Γ.

Figure 3.7 shows the 300 largest eigenvalues of the approximate covariance matrix. There is a significant gap between the eigenvalues d_r and d_{r+1}

where $r = 266$. The eigenvectors corresponding to these small eigenvalues can be described as noise along the annulus and should not be included in the prior model. We would then probably employ the singular covariance matrix obtained by truncating the eigenvalues at the cutoff level $\delta = 10^{-10}$ and approximate
$$D = \mathrm{diag}(d_1, d_2, \ldots, d_r, 0, \ldots, 0).$$
When the inverse covariance is needed, we would use the pseudoinverse instead. Figure 3.8 shows the computed mean and the eigenvectors corresponding to the five largest eigenvalues. ◇

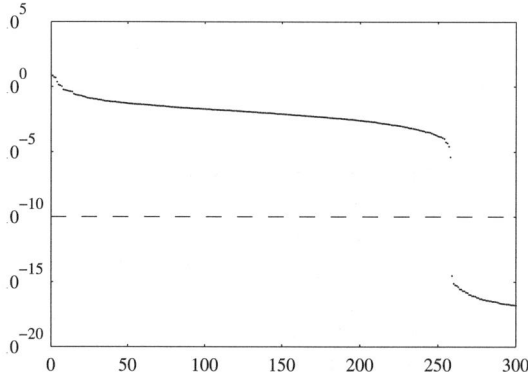

Figure 3.7. The 300 largest eigenvalues of the estimated covariance matrix.

3.4 Gaussian Densities

Gaussian probability densities have a special role in statistical inversion theory. First, they are relatively easy to handle and therefore they provide a rich source of tractable examples. But more importantly, due to the central limit theorem (see Appendix B), the Gaussian densities are often very good approximations to inherently non-Gaussian distributions when the observation is physically based on a large number of mutually independent random events.

We start with the definition of the Gaussian n-variate random variable.

Definition 3.2. *Let $x_0 \in \mathbb{R}^n$ and $\Gamma \in \mathbb{R}^{n \times n}$ be a symmetric positive definite matrix, denoted by $\Gamma > 0$ in the sequel. A Gaussian n-variate random variable X with mean x_0 and covariance Γ is a random variable with the probability density*
$$\pi(x) = \left(\frac{1}{2\pi|\Gamma|}\right)^{n/2} \exp\left(-\frac{1}{2}(x - x_0)^{\mathrm{T}} \Gamma^{-1}(x - x_0)\right),$$

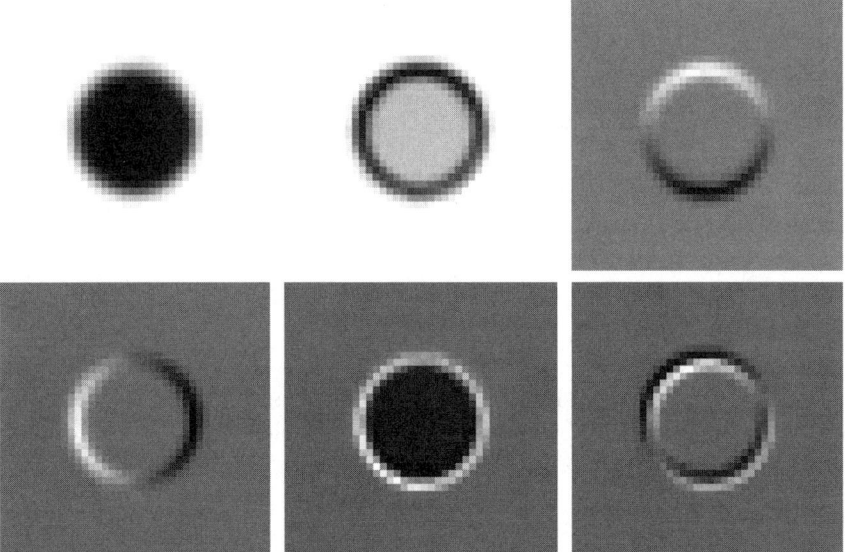

Figure 3.8. The mean (top left) and eigenvectors of Γ corresponding to five largest eigenvalues.

where $|\Gamma| = \det(\Gamma)$. *In such case, we use the notation*

$$X \sim \mathcal{N}(x_0, \Gamma).$$

Remark: In the literature, Gaussian random variables are often defined through the Fourier transform, or *characteristic function* as it is called in probability theory, as follows. A random variable X is Gaussian if

$$\mathrm{E}\{\exp(-\mathrm{i}\xi^\mathrm{T} X)\} = \exp\left(-\mathrm{i}\xi^\mathrm{T} x_0 - \frac{1}{2}\xi^\mathrm{T}\Gamma\xi\right), \quad (3.7)$$

where $x_0 \in \mathbb{R}^n$ and $\Gamma \in \mathbb{R}^{n \times n}$ is positive semidefinite, that is, $\Gamma \geq 0$. In this definition, it is not necessary to require that Γ is positive definite. Definition 3.2 is not more restrictive in the following sense. Let $d_1 \geq d_2 \geq \cdots \geq d_n$ denote the eigenvalues of the matrix Γ, and let $\{v_1, \ldots, v_n\}$ be the corresponding eigenbasis. Assume that the p first eigenvalues are positive and $d_{p+1} = \cdots = d_n = 0$. If X is a random variable with the property (3.7), one can prove that

$$X - x_0 \in \mathrm{sp}\{v_1, \ldots, v_p\}$$

almost certainly. To see this, observe that for any vector u perpendicular to the eigenvectors v_j, $1 \leq j \leq p$,

$$\mathrm{E}\{(u^\mathrm{T}(X - x_0))^2\} = u^\mathrm{T}\Gamma u = 0.$$

Therefore, by defining the orthogonal projection

$$P_p : \mathbb{R}^n \to \mathrm{sp}\{v_1, \ldots, v_p\} \cong \mathbb{R}^p,$$

the random variable $P_p X$ is a p-variate Gaussian random variable in the sense of Definition 3.2 with mean $x_0' = P_p x_0$ and covariance $\Gamma' = P_p^\mathrm{T} \Gamma P_p$. Using the notation $x = [P_p x; (1-P_p)x] = [x'; x'']$ and $x_0 = [P_p x_0; (1-P_p)x_0] = [x_0'; x_0'']$, we have in this case

$$\pi(x) = \left(\frac{1}{2\pi|\Gamma'|}\right)^{p/2} \exp\left(-\frac{1}{2}(x' - x_0')^\mathrm{T}(\Gamma')^{-1}(x' - x_0')\right) \delta(x'' - x_0''),$$

where δ denotes the Dirac delta in $\mathbb{R}^{n-p}$.

In this section, our aim is to derive closed formulas for the conditional means and covariances of Gaussian random variables. We start by recalling some elementary matrix properties.

Definition 3.3. *Let*

$$\Gamma = \begin{bmatrix} \Gamma_{11} & \Gamma_{12} \\ \Gamma_{21} & \Gamma_{22} \end{bmatrix} \in \mathbb{R}^{n \times n}$$

be a positive definite symmetric matrix, where $\Gamma_{11} \in \mathbb{R}^{k \times k}$, $\Gamma_{22} \in \mathbb{R}^{(n-k) \times (n-k)}$, $k < n$, and $\Gamma_{21} = \Gamma_{12}^\mathrm{T}$. We define the Schur complements $\tilde{\Gamma}_{jj}$ *of Γ_{jj}, $j = 1, 2$, by the formulas*

$$\tilde{\Gamma}_{22} = \Gamma_{11} - \Gamma_{12} \Gamma_{22}^{-1} \Gamma_{21}, \quad \tilde{\Gamma}_{11} = \Gamma_{22} - \Gamma_{21} \Gamma_{11}^{-1} \Gamma_{12}.$$

Observe that since the positive definiteness of Γ implies that Γ_{jj}, $j = 1, 2$, are also positive definite, the Schur complements are well defined.

Schur complements play an important role in calculating conditional covariances. We have the following matrix inversion lemma.

Lemma 3.4. *Let Γ be a matrix satisfying the assumptions of Definition 3.3. Then the Schur complements $\tilde{\Gamma}_{jj}$ are invertible matrices and*

$$\Gamma^{-1} = \begin{bmatrix} \tilde{\Gamma}_{22}^{-1} & -\tilde{\Gamma}_{22}^{-1} \Gamma_{12} \Gamma_{22}^{-1} \\ -\tilde{\Gamma}_{11}^{-1} \Gamma_{21} \Gamma_{11}^{-1} & \tilde{\Gamma}_{11}^{-1} \end{bmatrix}. \tag{3.8}$$

Proof: Consider the determinant of Γ,

$$|\Gamma| = \begin{vmatrix} \Gamma_{11} & \Gamma_{12} \\ \Gamma_{21} & \Gamma_{22} \end{vmatrix} \neq 0.$$

By substracting from the second row from the first one multiplied by $\Gamma_{21} \Gamma_{11}^{-1}$ from the left we find that

$$|\Gamma| = \begin{vmatrix} \Gamma_{11} & \Gamma_{12} \\ 0 & \Gamma_{22} - \Gamma_{21} \Gamma_{11}^{-1} \Gamma_{12} \end{vmatrix} = |\Gamma_{11}||\tilde{\Gamma}_{11}|, \tag{3.9}$$

implying that $|\tilde{\Gamma}_{11}| \neq 0$. Similarly we can prove that $\tilde{\Gamma}_{22}$ is invertible. Equation (3.9) is referred to as the *Schur identity*.

The proof of (3.8) follows from Gaussian elimination. Consider the linear system
$$\begin{bmatrix} \Gamma_{11} & \Gamma_{12} \\ \Gamma_{21} & \Gamma_{22} \end{bmatrix} \begin{bmatrix} x_1 \\ x_2 \end{bmatrix} = \begin{bmatrix} y_1 \\ y_2 \end{bmatrix}.$$

By eliminating x_2 from the second equation we get
$$x_2 = \Gamma_{22}^{-1}(y_2 - \Gamma_{21}x_1),$$

and substituting into the first equation we obtain
$$(\Gamma_{11} - \Gamma_{12}\Gamma_{22}^{-1}\Gamma_{21})x_1 = y_1 - \Gamma_{12}\Gamma_{22}^{-1}y_2$$

or, equivalently
$$x_1 = \tilde{\Gamma}_{22}^{-1}y_1 - \tilde{\Gamma}_{22}^{-1}\Gamma_{12}\Gamma_{22}^{-1}y_2.$$

Similarly we solve the second equation for x_2 and the claim follows. □

Observe that since Γ is a symmetric matrix, so is Γ^{-1}, and thus by setting the off-diagonal blocks of Γ^{-1} equal up to transpose, we obtain the identity
$$\Gamma_{22}^{-1}\Gamma_{21}\tilde{\Gamma}_{22}^{-1} = \tilde{\Gamma}_{11}^{-1}\Gamma_{21}\Gamma_{11}^{-1}. \tag{3.10}$$

We are now ready to prove the following result concerning the conditional probability densities of Gaussian random variables.

Theorem 3.5. *Let $X : \Omega \to \mathbb{R}^n$ and $Y : \Omega \to \mathbb{R}^k$ be two Gaussian random variables whose joint probability density $\pi : \mathbb{R}^n \times \mathbb{R}^k \to \mathbb{R}_+$ is of the form*
$$\pi(x,y) \propto \exp\left(-\frac{1}{2}\begin{bmatrix} x - x_0 \\ y - y_0 \end{bmatrix}^{\mathrm{T}} \begin{bmatrix} \Gamma_{11} & \Gamma_{12} \\ \Gamma_{21} & \Gamma_{22} \end{bmatrix}^{-1} \begin{bmatrix} x - x_0 \\ y - y_0 \end{bmatrix}\right). \tag{3.11}$$

Then the probability distribution of X conditioned on $Y = y$, $\pi(x \mid y) : \mathbb{R}^n \to \mathbb{R}_+$ is of the form
$$\pi(x \mid y) \propto \exp\left(-\frac{1}{2}(x - \overline{x})^{\mathrm{T}}\tilde{\Gamma}_{22}^{-1}(x - \overline{x})\right),$$

where
$$\overline{x} = x_0 + \Gamma_{12}\Gamma_{22}^{-1}(y - y_0).$$

Proof: By shifting the coordinate origin to $[x_0; y_0]$, we may assume that $x_0 = 0$ and $y_0 = 0$. By Bayes' formula, we have $\pi(x \mid y) \propto \pi(x,y)$, so we consider the joint probability density as a function of x. By Lemma 3.4 and the identity (3.10), we have
$$\pi(x,y) \propto \exp\left(-\frac{1}{2}(x^{\mathrm{T}}\tilde{\Gamma}_{22}^{-1}x - 2x^{\mathrm{T}}\tilde{\Gamma}_{22}^{-1}\Gamma_{12}\Gamma_{22}^{-1}y + y^{\mathrm{T}}\tilde{\Gamma}_{11}^{-1}y)\right).$$

Further, by completing the quadratic form in the exponential into squares, we can express the joint distribution as

$$\pi(x,y) \propto \exp\left(-\frac{1}{2}((x - \Gamma_{12}\Gamma_{22}^{-1}y)^{\mathrm{T}}\tilde{\Gamma}_{22}^{-1}(x - \Gamma_{12}\Gamma_{22}^{-1}y) + c)\right),$$

where

$$c = y^{\mathrm{T}}(\tilde{\Gamma}_{11}^{-1} - \Gamma_{22}^{-1}\Gamma_{21}\tilde{\Gamma}_{22}^{-1}\Gamma_{12}\Gamma_{22}^{-1})y$$

is mutually independent of x and can thus be factored out of the density. This completes the proof. □

Later, we need to calculate the marginal densities starting from joint probability densities. The following theorem produces the link.

Theorem 3.6. *Let X and Y be Gaussian random variables with joint probability density given by (3.11). Then the marginal density of X is*

$$\pi(x) = \int_{\mathbb{R}^k} \pi(x,y)dy \propto \exp\left(-\frac{1}{2}(x-x_0)^{\mathrm{T}}\Gamma_{11}^{-1}(x-x_0)\right).$$

Proof: Without loss of generality, we may assume that $x_0 = 0$, $y_0 = 0$. Letting $L = \Gamma^{-1}$ and using the obvious partition of the matrix L, we have

$$\pi(x,y) \sim \exp\left(-\frac{1}{2}(x^{\mathrm{T}}L_{11}x + 2x^{\mathrm{T}}L_{12}y + y^{\mathrm{T}}L_{22}y)\right).$$

Completing the term in the exponential to squares with respect to y yields

$$\pi(x,y) \propto \exp\bigg(-\frac{1}{2}((y + L_{22}^{-1}L_{21}x)^{\mathrm{T}}L_{22}(y + L_{22}^{-1}L_{21}x)$$

$$+ x^{\mathrm{T}}(L_{11} - L_{12}L_{22}^{-1}L_{21})x)\bigg).$$

By integrating with respect to y, we find that

$$\pi(x) \propto \exp\left(-\frac{1}{2}x^{\mathrm{T}}\tilde{L}_{22}x\right),$$

where the matrix $\tilde{L}_{22}$ is the Schur complement of L_{22}. Now the claim follows since, by Lemma 3.4, the inverse of the Schur complement $\tilde{L}_{22}$ of L_{22} is the first block of the inverse of L, that is,

$$\tilde{L}_{22}^{-1} = (L^{-1})_{11} = \Gamma_{11}.$$

□

Now we apply the above theorem to the following linear inverse problem. Assume that we have a linear model with additive noise,

3.4 Gaussian Densities

$$Y = AX + E, \tag{3.12}$$

where $A \in \mathbb{R}^{m \times n}$ is a known matrix, and $X : \Omega \to \mathbb{R}^n$, $Y, E : \Omega \to \mathbb{R}^m$, are random variables. Assume further that X and E are mutually independent Gaussian variables with probability densities

$$\pi_{\text{pr}}(x) \propto \exp\left(-\frac{1}{2}(x - x_0)^T \Gamma_{\text{pr}}^{-1}(x - x_0)\right),$$

and

$$\pi_{\text{noise}}(e) \propto \exp\left(-\frac{1}{2}(e - e_0)^T \Gamma_{\text{noise}}^{-1}(e - e_0)\right).$$

With this information, we get from Bayes' formula that the posterior distribution of x conditioned on y is

$$\pi(x \mid y) = \pi_{\text{pr}}(x)\pi_{\text{noise}}(y - Ax) \tag{3.13}$$

$$\propto \exp\left(-\frac{1}{2}(x - x_0)^T \Gamma_{\text{pr}}^{-1}(x - x_0) - \frac{1}{2}(y - Ax - e_0)^T \Gamma_{\text{noise}}^{-1}(y - Ax - e_0)\right).$$

It is straightforward to calculate the explicit form of the posterior distribution from this expression. However, the factorization approach derived in the previous theorem allows us to avoid the tedious matrix manipulations needed in this brute force approach. Indeed, since X and E are Gaussian, Y is also Gaussian, and we have

$$\mathrm{E}\left\{\begin{bmatrix} X \\ Y \end{bmatrix}\right\} = \begin{bmatrix} x_0 \\ y_0 \end{bmatrix}, \quad y_0 = Ax_0 + e_0.$$

Since

$$\mathrm{E}\left\{(X - x_0)(X - x_0)^T\right\} = \Gamma_{\text{pr}},$$

$$\mathrm{E}\left\{(Y - y_0)(Y - y_0)^T\right\} = \mathrm{E}\left\{(A(X - x_0) + (E - e_0))(A(X - x_0) + (E - e_0))^T\right\}$$
$$= A\Gamma_{\text{pr}}A^T + \Gamma_{\text{noise}},$$

and, furthermore,

$$\mathrm{E}\left\{(X - x_0)(Y - y_0)^T\right\} = \mathrm{E}\left\{(X - x_0)(A(X - x_0) + (E - e_0))^T\right\}$$
$$= \Gamma_{\text{pr}}A^T,$$

we find that

$$\mathrm{cov}\begin{bmatrix} X \\ Y \end{bmatrix} = \mathrm{E}\left\{\begin{bmatrix} X - x_0 \\ Y - y_0 \end{bmatrix}\begin{bmatrix} X - x_0 \\ Y - y_0 \end{bmatrix}^T\right\} = \begin{bmatrix} \Gamma_{\text{pr}} & \Gamma_{\text{pr}}A^T \\ A\Gamma_{\text{pr}} & A\Gamma_{\text{pr}}A^T + \Gamma_{\text{noise}} \end{bmatrix}.$$

Hence, the joint probability density of X and Y is of the form

$$\pi(x, y) \propto \exp\left(-\frac{1}{2}\begin{bmatrix} x - x_0 \\ y - y_0 \end{bmatrix}^T \begin{bmatrix} \Gamma_{\text{pr}} & \Gamma_{\text{pr}}A^T \\ A\Gamma_{\text{pr}} & A\Gamma_{\text{pr}}A^T + \Gamma_{\text{noise}} \end{bmatrix}^{-1} \begin{bmatrix} x - x_0 \\ y - y_0 \end{bmatrix}\right).$$

Thus, the following useful result is an immediate consequence of Theorem 3.5.

3 Statistical Inversion Theory

Theorem 3.7. *Assume that $X : \Omega \to \mathbb{R}^n$ and $E : \Omega \to \mathbb{R}^m$ are mutually independent Gaussian random variables,*

$$X \sim \mathcal{N}(x_0, \Gamma_{\mathrm{pr}}), \quad E \sim \mathcal{N}(e_0, \Gamma_{\mathrm{noise}}),$$

and $\Gamma_{\mathrm{pr}} \in \mathbb{R}^{n \times n}$ and $\Gamma_{\mathrm{noise}} \in \mathbb{R}^{k \times k}$ are positive definite. Assume further that we have a linear model (3.12) for a noisy measurement Y, where $A \in \mathbb{R}^{k \times n}$ is a known matrix. Then the posterior probability density of X given the measurement $Y = y$ is

$$\pi(x \mid y) \propto \exp\left(-\frac{1}{2}(x - \bar{x})^{\mathrm{T}} \Gamma_{\mathrm{post}}^{-1} (x - \bar{x})\right),$$

where

$$\bar{x} = x_0 + \Gamma_{\mathrm{pr}} A^{\mathrm{T}} (A \Gamma_{\mathrm{pr}} A^{\mathrm{T}} + \Gamma_{\mathrm{noise}})^{-1}(y - A x_0 - e_0), \quad (3.14)$$

and

$$\Gamma_{\mathrm{post}} = \Gamma_{\mathrm{pr}} - \Gamma_{\mathrm{pr}} A^{\mathrm{T}} (A \Gamma_{\mathrm{pr}} A^{\mathrm{T}} + \Gamma_{\mathrm{noise}})^{-1} A \Gamma_{\mathrm{pr}}. \quad (3.15)$$

Remark: As mentioned, the posterior density can be derived directly from formula (3.13) by arranging the quadratic form of the exponent according to the degree of x. Such a procedure gives an alternative representation for the posterior covariance matrix,

$$\Gamma_{\mathrm{post}} = \left(\Gamma_{\mathrm{pr}}^{-1} + A^{\mathrm{T}} \Gamma_{\mathrm{noise}}^{-1} A\right)^{-1}. \quad (3.16)$$

Furthermore, the posterior mean can be written as

$$\bar{x} = \left(\Gamma_{\mathrm{pr}}^{-1} + A^{\mathrm{T}} \Gamma_{\mathrm{noise}}^{-1} A\right)^{-1} \left(A^{\mathrm{T}} \Gamma_{\mathrm{noise}}^{-1} (y - e_0) + \Gamma_{\mathrm{pr}}^{-1} x_0\right). \quad (3.17)$$

The equivalence of formulas (3.15) and (3.16) can be shown with a tedious matrix manipulation, and is sometimes referred to as the *matrix inversion lemma*. Note that there are several different formulas which are referred to as matrix inversion lemmas.

The practical feasibility of the different forms depends on the dimensions n and m and also on the existence of the matrices Γ_{pr} and Γ_{noise} or their respective inverses. Especially, if the standard smoothness priors are employed, Γ_{pr} does not exist and the form (3.14) cannot be used.

Remark: In the purely Gaussian case, the centerpoint $\bar{x}$ is simultaneously the maximum a posteriori estimator and the conditional mean, that is,

$$\bar{x} = x_{\mathrm{CM}} = x_{\mathrm{MAP}}.$$

Similarly, Γ_{post} is the conditional covariance. Observe that in the sense of quadratic forms,

$$\Gamma_{\mathrm{post}} \leq \Gamma_{\mathrm{pr}},$$

that is, the matrix $\Gamma_{\mathrm{pr}} - \Gamma_{\mathrm{post}}$ is positive semidefinite. Since the covariance matrix of a Gaussian probability density expresses the width of the density, this inequality means that *a measurement can never increase the uncertainty.*

Example 5. Consider the simple case in which the prior has covariance proportional to the identity and mean zero, that is, $X \sim \mathcal{N}(0, \gamma^2 I)$. We refer to this prior as the *Gaussian white noise prior*. Similarly, assume that the noise is white noise, $E \sim \mathcal{N}(0, \sigma^2 I)$. In this particular case, we have

$$\overline{x} = \gamma^2 A^{\mathrm{T}}(\gamma^2 AA^{\mathrm{T}} + \sigma^2)^{-1} y = A^{\mathrm{T}}(AA^{\mathrm{T}} + \alpha I)^{-1} y,$$

where $\alpha = \sigma^2/\gamma^2$. This formula is known as the *Wiener filtered solution* of the ill-posed problem $y = Ax + e$. It can be expressed in terms of classical inverse problems that were discussed in Chapter 2. By using the singular value decomposition of the matrix A it is easy to see that,

$$\overline{x} = A^{\mathrm{T}}(AA^{\mathrm{T}} + \alpha I)^{-1} y = (A^{\mathrm{T}} A + \alpha I)^{-1} A^{\mathrm{T}} y.$$

Thus, the centerpoint of the posterior distribution is the Tikhonov regularized solution of the equation $Ax = y$ with regularization parameter α. This re-interpretation gives new insight into the choice of the parameter α: it is the ratio of the noise and prior variances. The statistical interpretation provides also information about the credibility of the result as well as cross-correlation information of the components of the estimator, of course given that the underlying model for X and E, and especially for their covariances, is a feasible one.

Observe that the above result is no surprise, since the maximum a posteriori estimator is, in the Gaussian case, also the minimizer of the *posterior potential* $V(x \mid y)$, defined by the relation

$$\pi(x \mid y) \propto \exp(-V(x \mid y)).$$

In this particular case, we have by formula (3.13)

$$V(x \mid y) = \frac{1}{2\gamma^2} \|x\|^2 + \frac{1}{2\sigma^2} \|y - Ax\|^2.$$

that is, the posterior potential is, up to a multiplicative constant, equal to the Tikhonov functional. ◇

The following is to be noted. In Tikhonov regularization the parameter α is usually selected without any special reference to σ^2. Thus, if we multiply $V(x \mid y)$ with any positive real number, we get the same minimizer, but the implied posterior covariance would be scaled according to this factor.

3.4.1 Gaussian Smoothness Priors

In this section we consider Gaussian smoothness priors and in particular those prior models that have *structural information* encoded in them. To make the discussion easier to follow, let us consider first an example in which the posterior potential and the Tikhonov functional are taken as a starting point.

Example 6: Assume that we seek to solve the equation $y = Ax + e$ by classical regularization methods. Suppose further that $x \in \mathbb{R}^n$ represents discretized values of some function $f : D \subset \mathbb{R}^n \mapsto \mathbb{R}$, which we know a priori to be twice continuously differentiable over D. To express this information as a constraint, we introduce the generalized Tikhonov functional

$$F_\alpha(x) = \|Ax - y\|^2 + \alpha \|Lx\|^2,$$

where $L : \mathbb{R}^n \to \mathbb{R}^k$ is a discrete approximation of the Laplacian in $\mathbb{R}^n$. As in Example 5, we might expect that there is a Gaussian prior probability density for x such that F_α is equal to the posterior potential $V(x \mid y)$. Indeed, if we assume that the data are corrupted by white noise with variance σ^2, and we set

$$V(x \mid y) = \frac{1}{2\sigma^2}\|y - Ax\|^2 + \frac{\alpha}{2\sigma^2}\|Lx\|^2 = \frac{1}{2\sigma^2}F_\alpha(x),$$

then minimizing F_α is tantamount to maximizing the conditional density $x \mapsto \exp(-V(x \mid y))$. Hence, a natural candidate for the prior distribution in this case would be

$$\pi_{\mathrm{pr}}(x) \propto \exp\left(-\frac{1}{2\gamma^2}\|Lx\|^2\right) \quad \gamma^2 = \frac{\sigma^2}{\alpha}. \tag{3.18}$$

We will refer to these types of prior as *smoothness priors* in the sequel. ◇

The aim is now to understand the nature of the smoothness from the statistical point of view. Hence, let $L \in \mathbb{R}^{k \times n}$ be a given matrix, and consider the density of the form

$$\pi_{\mathrm{pr}}(x) \propto \exp\left(-\frac{1}{2}\|L(x - x_0)\|^2\right) = \exp\left(-\frac{1}{2}(x - x_0)^\mathrm{T} L^\mathrm{T} L (x - x_0)\right).$$

The problem with the interpretation of this density as a Gaussian density according to Definition 3.2 is that, in general, the matrix L need not have full column rank, implying that the matrix $L^\mathrm{T} L$ is not necessarily invertible. Therefore, we seek the interpretation through a limiting process.

In the following, assume that the coordinate system is chosen so that $x_0 = 0$.

Definition 3.8. *(a) A random variable $W \in \mathbb{R}^k$ is called* pure or orthonormal white noise *if $W \sim \mathcal{N}(0, I)$, where $I \in \mathbb{R}^{k \times k}$ is the unit matrix. (b) Assume that $X \in \mathbb{R}^n$ is a Gaussian zero mean random variable. The matrix $L \in \mathbb{R}^{k \times n}$ is called a* whitening matrix *of X if*

$$LX = W \in \mathbb{R}^k$$

is pure white noise.

Assume that $X \in \mathbb{R}^n$ is Gaussian with positive definite covariance matrix $\Gamma \in \mathbb{R}^{n \times n}$, and let $\Gamma = CC^\mathrm{T}$ be its Cholesky factorization, where $C \in \mathbb{R}^{n \times n}$ is lower triangular matrix. Then the variable $Y = C^{-1}X$ is white noise. Indeed, we have

$$\mathrm{E}\{YY^\mathrm{T}\} = \mathrm{E}\{C^{-1}XX^\mathrm{T}(C^{-1})^\mathrm{T}\} = C^{-1}\Gamma(C^{-1})^\mathrm{T} = I.$$

Hence, the inverse of the Cholesky factor C of the covariance matrix provides a natural whitening matrix of the random variable X.

Conversely, assume that a matrix $L \in \mathbb{R}^{k \times n}$ is given. Our aim is to construct a Gaussian random variable $X \in \mathbb{R}^n$ such that the matrix L is *almost* a whitening matrix of X.

Assume that $L \in \mathbb{R}^{k \times n}$ has the singular value decomposition $L = UDV^\mathrm{T}$, where the diagonal elements of D are

$$d_1 \geq d_2 \geq \cdots d_p > d_{p+1} = \cdots = d_m = 0, \quad m = \min(k, n).$$

Partitioning $V = [v_1, \ldots, v_m]$ columnwise, we observe that

$$\mathrm{Ker}(L) = \mathrm{sp}\{v_{p+1}, \ldots, v_m\},$$

and let $Q = [v_{p+1}, \ldots, v_m] \in \mathbb{R}^{n \times (m-p)}$.

Lemma 3.9. *Let $W \in \mathbb{R}^k$ and $W' \in \mathbb{R}^{m-p}$ be two mutually independent white noise random variables, and let*

$$X = L^\dagger W + aQW', \tag{3.19}$$

where $L^\dagger$ is the pseudo-inverse of L and $a > 0$ is an arbitrary scalar. Then X has covariance

$$\Gamma = L^\dagger (L^\dagger)^\mathrm{T} + a^2 QQ^\mathrm{T},$$

and

$$\Gamma^{-1} = L^\mathrm{T} L + \frac{1}{a^2} QQ^\mathrm{T}. \tag{3.20}$$

Proof: By the mutual independence of W and W', we have

$$\mathrm{cov}(X) = \mathrm{E}\{XX^\mathrm{T}\} = L^\dagger \mathrm{E}\{WW^\mathrm{T}\}(L^\dagger)^\mathrm{T} + a^2 Q \mathrm{E}\{W'W'^\mathrm{T}\} Q^\mathrm{T}$$
$$= L^\dagger (L^\dagger)^\mathrm{T} + a^2 QQ^\mathrm{T} = \Gamma,$$

as claimed. This matrix is invertible. In fact, if $j \leq p$, we have

$$\Gamma v_j = L^\dagger (L^\dagger)^\mathrm{T} v_j = \frac{1}{d_j^2} v_j$$

from the definition of the pseudoinverse. On the other hand, for $j > p$, we obtain

$$\Gamma v_j = a^2 v_j.$$

It follows that the eigenvectors of Γ are v_j, $1 \leq j \leq n$ with positive eigenvalues and so

$$\Gamma^{-1} = \sum_{j=1}^{p} d_j^2 v_j v_j^{\mathrm{T}} + \frac{1}{a^2} \sum_{j=p+1}^{n} v_j v_j^{\mathrm{T}} = L^{\mathrm{T}} L + \frac{1}{a^2} Q Q^{\mathrm{T}},$$

completing the proof. □

For the random variable X defined by the formula (3.19) we have

$$LX = LL^{\dagger}W = UU^{\mathrm{T}}W = PW,$$

where P is the orthogonal projection operator onto $\mathrm{Ran}(L)$. The covariance matrix of PW is the identity operator on this subspace. Therefore, L is not a whitening operator in the sense of Definition 3.8, but as close an approximation as possible.

From formula (3.20) we see that, for large a, the random variable X has almost the desired probability density. However, in the direction of $\mathrm{Ker}(L)$, the variable X has huge variance, that is, the smoothness prior provides no information on the behavior of X in that direction. In fact, if $\mathrm{Ker}(L) \neq \{0\}$, the smoothness prior is not even a proper probability density. Indeed,

$$\pi_{\mathrm{pr}}(x) \propto \exp\left(-\frac{1}{2\gamma^2}\|Lx\|^2\right) = \exp\left(-\frac{1}{2\gamma^2} \sum_{j=1}^{p} d_j^2 (v_j^{\mathrm{T}} x)^2\right),$$

so by setting $H = \mathrm{sp}\{v_1, \ldots, v_p\}$, we have

$$\int_H \exp\left(-\frac{1}{2\gamma^2}\|Lx\|^2\right) dx = (2\pi)^{p/2} \frac{\gamma^p}{\prod_{j=1}^{p} d_j} < \infty.$$

However, if $n > p$,

$$\int_{\mathbb{R}^n} \pi_{\mathrm{pr}}(x) dx = \infty.$$

We call a prior density with the above nonintegrability property an *improper density*.

If the prior density is improper, we may encounter problems. Often, however, one can deal with inproper densities by compensating for the presence of a nontrival null space, as we shall see later in this chapter.

To get a more concrete idea of what a given smoothness prior means in practice, it is useful to consider random draws from the probability distribution. Lemma 3.9 provides us with a tool to do this drawing.

Example 7: Consider the smoothness prior obtained by discretizing the Laplacian. More specifically, let D denote the unit square $D = [0,1] \times [0,1] \subset \mathbb{R}^2$, and subdivide D into $N \times N$ equal square pixels. We approximate the Laplacian of a function f in D by using the five-point finite difference scheme, that is, if x_j is the value of f at an interior pixel p_j, we write

$$(Lx)_j = \frac{1}{N^2}\left(\sum_{k \sim j} x_k - 4x_j\right). \tag{3.21}$$

Here, the notation $k \sim j$ means that the pixels p_j and p_k share a common edge. For boundary pixels various conditions can be implemented depending on the prior information. Assume that we know that the function f is a restriction to D of a twice continuously differentiable function whose C^2-extension vanishes outside D. Then formula (3.21) can be used also at the boundary pixels. This leads to a matrix $L \in \mathbb{R}^{N \times N}$ that is a finite difference approximation of Δ with the Dirichlet boundary value, that is, the domain of definition of the Dirichlet Laplacian is

$$\mathcal{D}(\Delta_D) = \{f \in C^2(D) \mid f|_{\partial D} = 0\}.$$

It turns out that this matrix L has a trivial null space. Therefore, random draws from the smoothness prior can be done by the following procedure:

1. Draw a white noise vector $W \in \mathbb{R}^{N^2}$.
2. Solve $LX = \gamma W$ for X.

Observe that the whitening matrix L is sparse, and thus solving X is numerically not too expensive. Figure 3.9 shows three random draws from this Gaussian smoothness prior with the image size 60×60 pixels. ◇

Figure 3.9. Three random draws of 60×60 pixel images from the smoothness prior with the smoothness constraint defined by the Dirichlet Laplacian.

In the previous example, the smoothness prior was defined so that it became proper by implementing the Dirichlet boundary condition. This is not always justified, and in fact implementation of boundary conditions in the prior should be avoided unless the prior information dictates the presence of such condition. The lack of a boundary condition typically leads to improper priors. We shall consider such priors in later examples. First, we consider the effect of an improper prior on the posterior distribution.

Consider the posterior probability density of the linear observation model $y = Ax + e$ derived in Theorem 3.7. Clearly, when the prior density is improper, the formulas for the midpoint $\overline{x}$ and the posterior covariance Γ_{post} cannot be used as they require the knowledge of the prior covariance Γ_{pr} that does not exist. The difficulty can be overcome by the following result which provides an alternative formula for the posterior density.

Theorem 3.10. *Consider the linear observation model* $Y = AX + E$, $A \in \mathbb{R}^{m \times n}$, *where* $X \in \mathbb{R}^n$, $E \in \mathbb{R}^m$ *are mutually independent random variables and* E *is Gaussian,* $E \sim \mathcal{N}(0, \Gamma_{\text{noise}})$. *Let* $L \in \mathbb{R}^{n \times m}$ *be a matrix such that* $\text{Ker}(L) \cap \text{Ker}(A) = \{0\}$. *Then the function*

$$x \mapsto \pi_{\text{pr}}(x)\pi(y \mid x) \propto \exp\left(-\frac{1}{2}(\|Lx\|^2 + (y - Ax)^{\mathrm{T}}\Gamma_{\text{noise}}^{-1}(y - Ax))\right) \quad (3.22)$$

defines a Gaussian density over $\mathbb{R}^n$ *with center and covariance given by the formulas*

$$\overline{x} = (L^{\mathrm{T}}L + A^{\mathrm{T}}\Gamma_{\text{noise}}^{-1}A)^{-1}A^{\mathrm{T}}\Gamma_{\text{noise}}^{-1}y$$

and

$$\Gamma = (L^{\mathrm{T}}L + A^{\mathrm{T}}\Gamma_{\text{noise}}^{-1}A)^{-1}$$

Proof: We start by showing that the matrix $G = L^{\mathrm{T}}L + A^{\mathrm{T}}\Gamma_{\text{noise}}^{-1}A \in \mathbb{R}^{n \times n}$ is invertible. Indeed, if $x \in \text{Ker}(G)$, then

$$x^{\mathrm{T}}Gx = \|Lx\|^2 + \|\Gamma^{-1/2}Ax\|^2 = 0,$$

implying that $x \in \text{Ker}(L) \cap \text{Ker}(A) = \{0\}$, that is, $x = 0$. Hence, $\text{Ker}(G) = \{0\}$ and G is invertible.

Consider the quadratic functional in the exponent of (3.22). Refomulating it in terms of a positive definite quadratic form, we obtain

$$\|Lx\|^2 + (y - Ax)^{\mathrm{T}}\Gamma_{\text{noise}}^{-1}(y - Ax) = x^{\mathrm{T}}Gx - 2x^{\mathrm{T}}A^{\mathrm{T}}\Gamma_{\text{noise}}^{-1}y + y^{\mathrm{T}}\Gamma_{\text{noise}}^{-1}y$$
$$= (x - \overline{x})^{\mathrm{T}}G(x - \overline{x}) + c,$$

where $c = c(y)$ is mutually independent of x. Since G is a positive definite symmetric matrix, the above expression is a potential defining a Gaussian density with the claimed center and covariance. □

It follows immediately from the proof of the previous theorem that if the matrices A and L have a nontrivial common kernel, the posterior density does not define a probability density, and consequently the inverse problem becomes *underdetermined*, that is, the prior and the measurement together do not provide enough information to produce a proper posterior density, and no solution to the problem exists.

In the following examples, we show how the smoothness priors can be used to implement structural prior information to inverse problems. In addition, we demonstrate how the improper priors can be explored.

Example 8: Consider a geophysical inverse problem in which one seeks to estimate the underground structure, for example, from electromagnetic or elastoacoustic sounding. Let us consider here a well-to-well logging problem, in which two parallel bore holes are drilled in the earth. A high-frequency radio transmitter is placed into one of the bore holes and a receiver into the other. Using different transmitter–receiver positions, the attenuation between the holes is measured. We are interested in the two-dimensional section between the holes.

Often, structural prior information is available. For instance, the geophysical history of the area may indicate the presence of layered structures such as sediments and layered deposits. Combined, for example, with the drilling data, an overall idea of the direction of the layering may be available. Assume that we have the prior information that in the plane defined by the bore holes, the layers are perpendicular to a known vector u. Let v be a vector in the borehole plane such that v is perpendicular to u. We assume that the material parameter that we are estimating, denoted by $\varrho = \varrho(p)$, does not oscillate much along the layer structures, while in the direction u perpendicular to the layers, large and rapid changes are possible. Hence, we may assume that

$$\mathcal{L}_v \varrho(p) = v \cdot \nabla \varrho(p)$$

should be restricted.

To discretize the problem, we subdivide the area between the boreholes in pixels and denote by x_j the approximate value of ϱ in the jth pixel. Let L_v denote a finite difference approximation of the operator $\mathcal{L}_v$. The prior density that we consider here is now of the form

$$\pi_{\text{pr}}(x) \sim \exp\left(-\alpha \|L_v x\|^2\right).$$

The parameter $\alpha > 0$ is at this point is not specified. We return to the selection of the parameters later.

The discretization scheme used here is as follows: Consider first derivatives in the horizontal and vertical directions, that is, $v = e_1$ and $v = e_2$, respectively. We use a symmetric finite difference approximation for derivatives in the interior pixels and one-sided finite difference approximation at boundary pixels. For a unit vector $v = \cos\theta e_1 + \sin\theta e_2$, we write simply

$$L_v = \cos\theta L_{e_1} + \sin\theta L_{e_2}.$$

It is easy to see that in general, the prior thus defined becomes improper. For instance, $L_{e_1} x = 0$ for every pixel image x that is constant along the rows. Therefore, the visualization of the prior is not as simple as in the previous example. What one can do is to fix the values of the image in a number of pixels and consider the prior density of the remaining ones conditioned on the fixed ones. By possibly rearranging the ordering of the pixels, we may assume that the vector $x \in \mathbb{R}^n$ is written as $x = [x'; x'']$, where $x' \in \mathbb{R}^k$ contains the fixed pixel values while $x'' \in \mathbb{R}^{n-k}$ contains the unspecified pixel values. By partitioning accordingly the matrix

$$L_v^{\text{T}} L_v = B = \begin{bmatrix} B_{11} & B_{12} \\ B_{21} & B_{22} \end{bmatrix}, \qquad (3.23)$$

we find that the conditional prior of X'' conditioned on $X' = x'$ has the density

86   3 Statistical Inversion Theory

$$\pi_{\mathrm{pr}}(x'' \mid x') \propto \exp\left(-\alpha\left(x'^{\mathrm{T}}B_{11}x' + x'^{\mathrm{T}}B_{12}x'' + x''^{\mathrm{T}}B_{21}x' + x''^{\mathrm{T}}B_{22}x''\right)\right)$$

$$\propto \exp\left(-\alpha\left(x'' + B_{22}^{-1}B_{21}x'\right)^{\mathrm{T}} B_{22}\left(x'' + B_{22}^{-1}B_{21}x'\right)\right)$$

In the expression above, the block B_{22} needs to be invertible. This can be achieved by fixing sufficiently many pixels to guarantee that the null space of the prior of the remaining ones becomes trivial.

To demonstrate this idea, assume that the image area is a square divided in $N \times N$ equally sized pixels. We consider the conditional prior by fixing the pixel values at two boundary edges, one vertical and one horizontal. Figure 3.10 represents three random draws with $N = 80$ from the conditional prior density with different layering directions. Pixel values at the left and bottom edges are fixed to zero. The direction vector u perpendicular to the sediment layers is indicated in each figure. ◇

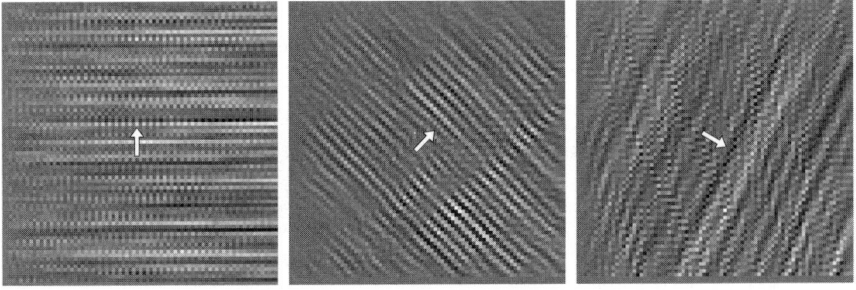

Figure 3.10. Random draws from the conditional smoothness prior density with different sedimentation directions. The pixel values on the left and bottom edges have been fixed to value $x_i = 0$.

Remark: Using the idea of conditioning, it is possible to create Gaussian smoothness priors that are proper. In the previous example, assume that we define a proper Gaussian probability density for the variables $X' \in \mathbb{R}^k$,

$$X' \sim \mathcal{N}(x_0', \Gamma').$$

Then, we obtain a new density for X by writing

$$\widetilde{\pi}_{\mathrm{pr}}(x) = \pi_{\mathrm{pr}}(x'' \mid x')\pi_0(x')$$

$$\propto \exp\left(-\alpha\left(x'' + B_{22}^{-1}B_{21}x'\right)^{\mathrm{T}} B_{22}\left(x'' + B_{22}^{-1}B_{21}x'\right)\right.$$

$$\left. -\frac{1}{2}(x' - x_0')^{\mathrm{T}}(\Gamma')^{-1}(x' - x_0')\right)$$

$$= \exp\left(-\frac{1}{2}(x - x_0)^{\mathrm{T}}\Gamma^{-1}(x - x_0)\right),$$

where the midpoint and covariance matrices are

$$x_0 = \begin{bmatrix} x_0' \\ -B_{22}^{-1}B_{21}]x_0' \end{bmatrix}, \quad \Gamma = \begin{bmatrix} 2\alpha B_{12}B_{22}^{-1}B_{21} + (\Gamma')^{-1} & 2\alpha B_{21} \\ 2\alpha B_{12} & 2\alpha B_{22} \end{bmatrix}^{-1}.$$

This construction will be used in later chapters of the book.

Example 9: In this example, we implement structural information by using direction-sensitive prior distributions. Assume that our prior information is that the image we are seeking to estimate has rapid changes or even discontinuities at *known* locations. Such a situation could occur, for example, in medical imaging applications: A different imaging modality or anatomical information of the organ locations could give us at least an approximate idea of in which rapid changes in material parameters might occur. Another application is image enhancing in remote sensing, in which geographical information of, for example, shorelines or urban area boundaries is available.

Let us denote by $D \subset \mathbb{R}^2$ the image area and let $f : D \to \mathbb{R}$ denote the unkown function to be estimated. We define a matrix field over D,

$$A : D \mapsto \mathbb{R}^{2 \times 2}, \quad A(p)^\mathrm{T} = A(p), \quad A(p) \geq 0.$$

Let $A = V\Lambda V^\mathrm{T}$ be the spectral decomposition of A, where $\Lambda = \Lambda(p) = \mathrm{diag}(\lambda_1(p), \lambda_2(p))$ with $\lambda_j(p) \geq 0$ and $V = V(p) = [v_1(p), v_2(p)]$ is the matrix whose columns are the orthonormal eigenvectors of A. The vectors v_j define a *frame field* in D.

Consider the functional

$$W : f \mapsto \int_D \|A(p)\nabla f(p)\|^2 dp.$$

In terms of the spectral factorization of A, we have

$$W(f) = \sum_{j=1}^{2} \int_D |\lambda_j(p) v_j(p)^\mathrm{T} \nabla f(p)|^2 dp = \sum_{j=1}^{2} \int_D |\mathcal{L}_{\lambda_j v_j} f(p)|^2 dp,$$

where the operator $\mathcal{L}_{\lambda_j v_j}$ is the Lie derivative along the flow generated by the vector field $\lambda_j v_j$. We observe that if the flow line of this vector field crosses a curve where f changes significantly, the contribution to the integral is large unless λ_j becomes small. This simple observation is the key in designing the prior density.

Assume that the structural information of the domain D is given in the form of a function $\gamma : D \to \mathbb{R}$, $\gamma \in C^1(D)$. This function could be, for example, a smooth approximation of a level set function describing the structure of D. The aim is to construct a matrix-valued function A with the following properties: If $u \in \mathbb{R}^2$ is perpendicular to $\nabla \gamma \neq 0$, then $Au = u$. On the other hand, if u is collinear with $\nabla \gamma$, then $Au = \delta(\|\nabla \gamma\|)u$, where the function $\delta(t)$ should be small as t is large. By choosing, for example, $\delta(t) = 1/(1+t^2)$, a matrix function with the desired properties is

$$A(p) = I - \frac{1}{(1+\|\nabla\gamma\|^2)} \nabla\gamma(p)(\nabla\gamma(p))^{\mathrm{T}}.$$

Indeed, the eigenvalues of A are

$$\lambda_1(p) = 1, \quad \lambda_2(p) = \frac{1}{1+\|\nabla\gamma(p)\|^2}$$

corresponding to eigenvectors

$$v_1(p) \perp \nabla\gamma(p), \quad v_2(p) = \frac{1}{\|\nabla\gamma(p)\|} \nabla\gamma(p),$$

when $\nabla\gamma(p) \neq 0$.

To consider the discrete problem, assume that the image area D is discretized into n pixels and denote by x_k the pixel value of f in the kth pixel. Figure 3.11 shows the level set function that corresponds to the prior information. We assume it to be known a priori that the pixels inside and outside the shape are correlated, while the correlation across the boundary should be negligible.

Figure 3.11. The level set function defining the structure.

Let $L_j \in \mathbb{R}^{k \times n}$ denote the discrete matrix approximation of the operator $\mathcal{L}_{\lambda_j v_j}$ such that

$$\sum_{j=1}^{2} \int_D |\mathcal{L}_{\lambda_j v_j} f(p)|^2 dp \approx \sum_{j=1}^{2} x^{\mathrm{T}} L_j^{\mathrm{T}} L_j x = x^{\mathrm{T}} B x.$$

The matrix B is symmetric and positive semidefinite. In this example, choose the function δ such that when $\nabla\gamma \neq 0$, then $\delta(\|\nabla\gamma\|) = 0$. In the discrete approximation, this means that for pixels x_k and x_ℓ whose center points are on different sides of the boundary curve, we set

$$(L_j)_{k,\ell} = (L_j)_{\ell,k} = 0, \quad j = 1, 2,$$

that is, the discrete directional derivative across the boundary curve vanishes. In the sense of Markov random fields, the pixels are divided into two cliques according to which side of the boundary their center point is. The structural prior density is defined as

$$\pi_{\mathrm{pr}}(x) \sim \exp\left(-\alpha\frac{1}{2}x^{\mathrm{T}}Bx\right),$$

where $\alpha > 0$ is a parameter whose value in this example is not important. This prior now has the important property that pixel values on different sides of a boundary line where the original function f is allowed to jump should not be correlated.

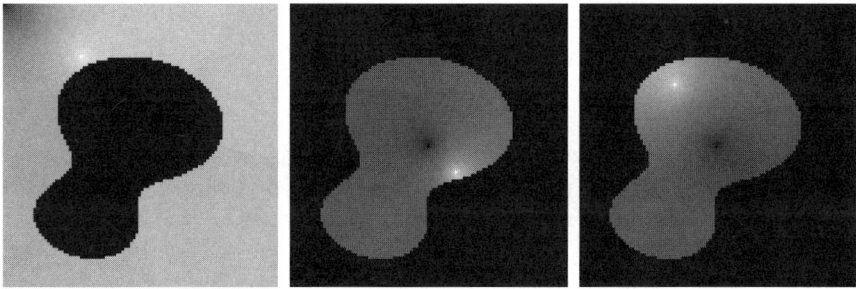

Figure 3.12. Three examples of the conditional correlations between pixels. The two fixed pixels (upper left corner, a pixel at the center of the image) are marked with black, and the highest correlation (that is, autocorrelation) corresponds to white.

To give a statistical interpertation of this prior, we consider the conditional covariances of various pixels. First, observe that the dimension of the kernel of the operator B is 2. Indeed, $Bx = 0$ if and only if x represents a constant function inside and outside the set plotted in Figure 3.11. Therefore, it suffices to fix the value of x at two pixels, one inside and one outside. We pick two such pixels, and consider the conditional prior of the remaining pixels. The fixed pixels are at the upper left corner of the image and right at the center of it. Using the partitioning in formula (3.23), the conditional covariance of the remaining pixels is B_{22}^{-1}. Each column of this matrix tells how strongly the corresponding pixel is correlated with other pixels. We calculate four columns of this matrix corresponding to pixels in different places of the image and plot the correlation in Figure 3.12. The fixed pixels are marked in these picures with black. These pictures reveal that the resulting prior density works as it should: Pixels at different sides of the discontinuities are not correlated at all.
◇

3.5 Interpreting the Posterior Distribution

The third task in the statistical inversion technique is to develop methods for exploring the posterior probability densities. Before discussing such tools in the next section, let us make some general remarks about how the posterior density should be viewed in the context of inverse problems.

The interpretation of the posterior distribution as the solution of an inverse problem is a subtle issue. To properly grasp its meaning, it is important to understand thoroughly the meaning of the credibility sets and intervals. For example, while it would seem natural to jump into conclusions like *"The probability that the true value of the unknown is in the set D_p is p%.,"* this interpretation is *incorrect* as the following example shows.

Example 10: Consider the following trivial inverse problem: Determine the value of $x \in \mathbb{R}$ by measuring directly x with some additive mutually independent noise. Here, the statistical model is

$$Y = X + E.$$

Let us assume that the prior probability density of X is zero mean Gaussian with unit variance while the density of the noise E is zero mean Gaussian with variance σ^2. Thus, the posterior probability density of X is

$$\pi(x \mid y) \propto \exp\left(-\frac{1}{2}x^2 - \frac{1}{2\sigma^2}(y-x)^2\right).$$

This density is Gaussian with respect to x. Indeed, by completing in squares and ignoring factors that depend on y only, we find that, for given data y,

$$\pi(x \mid y) \propto \exp\left(-\frac{1+\sigma^2}{2\sigma^2}\left(x - \frac{y}{1+\sigma^2}\right)^2\right).$$

From this expression we can immediately see the conditional mean x_{CM} and variance γ^2 of x to be

$$x_{\text{CM}} = \frac{y}{1+\sigma^2}, \quad \gamma^2 = \frac{\sigma^2}{1+\sigma^2}.$$

In this case it is easy to calculate credibility intervals. For instance, the 90% credibility interval $I(90)$ is

$$I(90) \approx [x_{\text{CM}} - \alpha\gamma, x_{\text{CM}} + \alpha\gamma], \quad \alpha \approx 1.64.$$

Assume now that the data corresponds to a "true" value of x, say $x = x_0 > 0$ and moreover, the additive noise realization in the data acquisition happens to be negligible. In this case, we have approximately $y \approx x_0 > 0$. Now it may happen that in this case $x_0 \notin I(90)$. In fact, if $y\gamma > \alpha$, we have

$$x_{\text{CM}} + \alpha\gamma < x_{\text{CM}} + \gamma^2 y = y \approx x_0.$$

To understand this result, notice that the condition $x_0 > \alpha/\gamma$ is rather improbable. Indeed, we have

$$\int_{\alpha/\gamma}^{\infty} \pi_{\mathrm{pr}}(x) dx < \int_{\alpha}^{\infty} \pi_{\mathrm{pr}}(x) dx < 0.05.$$

This example shows that one needs to be careful with interpretations and not to forget the role of the prior density. It also underlines the importance of modelling the prior distribution correctly, so that it truly respects the prior information about the unknown. This is particularly important in cases when the prior is informative, that is, when it has a strong biasing effect on the posterior distribution. ◇

3.6 Markov Chain Monte Carlo Methods

The abstract definition of the solution of an inverse problem as the posterior probability distribution is not very helpful in practice, if we have no means of exploring it. Various random sampling methods have been proposed, and we discuss here an effective class of these, known as the Markov chain Monte Carlo techniques (MCMC for short).

In Section 3.1.1, in which we discussed briefly single estimates based on the posterior distribution, it was shown that the maximum a posteriori estimate leads to an optimization problem, while the conditional mean and conditional covariances require integration over the space $\mathbb{R}^n$ where the posterior density is defined. It is clear that if the dimension n of the parameter space $\mathbb{R}^n$ is large, the use of numerical quadrature rules is out of the question: An m-point rule for each direction would require m^n integration points, exceeding the computational capacity of most computers. Another problem with quadrature rules is that they require a relatively good knowledge of the support of the probability distribution, which is usually part of the information that we are actually looking for.

An alternative way to look at the problem is the following. Instead of evaluating the probability density at given points, let the density itself determine a set of points, a *sample*, that supports well the distribution. These sample points can then be used for approximate integration. The MCMC methods are at least on the conceptual level relatively simple algorithms to generate sample ensembles for Monte Carlo integration.

3.6.1 The Basic Idea

Before going into detailed analysis, we discuss first the basic idea behind Monte Carlo integration.

Let μ denote a probability measure over $\mathbb{R}^n$. Further, let f be a scalar or vector-valued measurable function, integrable over $\mathbb{R}^n$ with respect to the measure μ, that is, $f \in L^1(\mu(dx))$. Assume that the objective is to estimate the integral of f with respect to the measure μ. In numerical quadrature methods, one defines a set of support points $x_j \in \mathbb{R}^n$, $1 \leq j \leq N$ and the corresponding weights w_j to get an approximation

$$\int_{\mathbb{R}^n} f(x)\mu(dx) \approx \sum_{j=1}^{N} w_j f(x_j).$$

Typically, the quadrature methods are designed so that they are accurate for given functions spanning a finite-dimensional space. Typically, these functions are polynomials of restricted degree.

In Monte Carlo integration, the support points x_j are generated randomly by drawing from some probability density and the weights w_j are then determined from the distribution μ. Ideally, the support points are drawn from the probability distribution determined by the measure μ itself. Indeed, let $X \in \mathbb{R}^n$ denote a random variable such that μ is its probability distribution. If we had a random generator such that repeated realizations of X could be produced, we could generate a set of points distributed according to μ. Assume that $\{x_1, x_2, \ldots, x_N\} \subset \mathbb{R}^n$ is such a representative ensemble of samples distributed according to the distribution μ. We could then seek to approximate the integral of f by the so-called ergodic average,

$$\int_{\mathbb{R}^n} f(x)\mu(dx) = \mathrm{E}\{f(X)\} \approx \frac{1}{N}\sum_{j=1}^{N} f(x_j). \qquad (3.24)$$

The MCMC methods are systematic ways of generating a sample ensemble such that (3.24) holds. We need some basic tools from probability theory to do this.

Let $\mathfrak{B} = \mathfrak{B}(\mathbb{R}^n)$ denote the Borel sets over $\mathbb{R}^n$. A mapping $P: \mathbb{R}^n \times \mathfrak{B} \to [0,1]$ is called a *probability transition kernel*, if

1. for each $B \in \mathfrak{B}$, the mapping $\mathbb{R}^n \to [0,1]$, $x \mapsto P(x,B)$ is a measurable function;
2. for each $x \in \mathbb{R}^n$, the mapping $\mathfrak{B} \to [0,1]$, $B \mapsto P(x,B)$ is a probability distribution.

A *discrete time stochastic process* is an ordered set $\{X_j\}_{j=1}^{\infty}$ of random variables $X_j \in \mathbb{R}^n$. A *time-homogenous Markov chain* with the transition kernel P is a stochastic process $\{X_j\}_{j=1}^{\infty}$ with the properties

$$\mu_{X_{j+1}}(B_{j+1} \mid x_1, \ldots, x_j) = \mu_{X_{j+1}}(B_{j+1} \mid x_j) = P(x_j, B_{j+1}). \qquad (3.25)$$

In words, the first equality states that the probability for $X_{j+1} \in B_{j+1}$ conditioned on observations $X_1 = x_1, \ldots, X_j = x_j$ equals the probability conditioned on $X_j = x_j$ alone. This property is stated often by saying that "the future depends on the past only through the present." The second equality says that time is homogenous in the sense that the dependence of adjacent moments does not vary in time. Let us emphasize that in (3.25) the kernel P does not depend on time j.

More generally, we define the transition kernel that propagates k steps forward in time, setting

$$P^{(k)}(x_j, B_{j+k}) = \mu_{X_{j+k}}(B_{j+k} \mid x_j)$$
$$= \int_{\mathbb{R}^n} P(x_{j+k-1}, B_{j+k}) P^{(k-1)}(x_j, dx_{j+k-1}), \quad k \geq 2,$$

where it is understood that $P^{(1)}(x_j, B_{j+1}) = P(x_j, B_{j+1})$. In particular, if μ_{X_j} denotes the probability distribution of X_j, the distribution of X_{j+1} is given by

$$\mu_{X_{j+1}}(B_{j+1}) = \mu_{X_j} P(B_{j+1}) = \int_{\mathbb{R}^n} P(x_j, B_{j+1}) \mu_{X_j}(dx_j). \tag{3.26}$$

The measure μ is an *invariant measure* of $P(x_j, B_{j+1})$ if

$$\mu P = \mu, \tag{3.27}$$

that is, the distribution of the random variable X_j before the time step $j \to j+1$ is the same as the variable X_{j+1} after the step.

We still need to introduce few concepts concerning the transition kernels. Given a probability measure μ, the transition kernel P is *irreducibile* (with respect to μ) if for each $x \in \mathbb{R}^n$ and $B \in \mathfrak{B}$ with $\mu(B) > 0$ there exists an integer k such that $P^{(k)}(x, B) > 0$. Thus, regardless of the starting point, the Markov chain generated by the transition kernel P visits with a positive probability any set of positive measure.

Let P be an irreducible kernel. We say that P is *periodic* if, for some integer $m \geq 2$, there is a set of disjoint nonempty sets $\{E_1, \ldots, E_m\} \subset \mathbb{R}^n$ such that for all $j = 1, \ldots, m$ and all $x \in E_j$, $P(x, E_{j+1(\bmod m)}) = 1$. In other words, a periodic transition kernel generates a Markov chain that remains in a periodic loop for ever. A kernel P is an *aperiodic* kernel if it is not periodic.

The following result is of crucial importance for MCMC methods. The proof of this theorem will be omitted. For references, see "Notes and Comments" at the end of the chapter.

Proposition 3.11. *Let μ be a probability measure in $\mathbb{R}^n$ and $\{X_j\}$ a time-homogenous Markov chain with a transition kernel P. Assume further that μ is an invariant measure of the transition kernel P, and that P is irreducible and aperiodic. Then for all $x \in \mathbb{R}^n$,*

$$\lim_{N \to \infty} P^{(N)}(x, B) = \mu(B) \text{ for all } B \in \mathfrak{B}, \tag{3.28}$$

and for $f \in L^1(\mu(dx))$,

$$\lim_{N\to\infty}\frac{1}{N}\sum_{j=1}^{N}f(X_j) = \int_{\mathbb{R}^n} f(x)\mu(dx) \qquad (3.29)$$

almost certainly.

The above theorem gives a clear indication how to explore a given probability distribution: One needs to construct an invariant, aperiodic and irreducible transition kernel P and draw a sequence of sample points $x_1, x_2, \ldots$ using this kernel, that is, one needs to *calculate a realization of the Markov chain.*

The property (3.29) is the important ergodicity property used in Monte Carlo integration. The convergence in (3.28), stating that μ is a limit distribution for the transition kernel P, can be stated also in a slightly stronger form; see "Notes and Comments."

In the following sections, we discuss how to construct transition kernels with the desired properties. The two most common procedures are the *Metropolis–Hastings algorithm* and the *Gibbs sampler.* A large number of variants of these basic methods have been proposed.

3.6.2 Metropolis–Hastings Construction of the Kernel

Let μ denote the target probability distribution in $\mathbb{R}^n$ that we want to explore by the sampling algorithm. To avoid measure-theoretic notions, we assume that μ is absolutely continuous with respect to the Lebesgue measure, $\mu(dx) = \pi(x)dx$. We wish to determine a transition kernel $P(x, B)$ such that μ is its invariant measure.

Let P denote any transition kernel. When a point $x \in \mathbb{R}^n$ is given, we can postulate that the kernel either proposes a move to another point $y \in \mathbb{R}^n$ or it proposes no move away from x. This allows us to split the kernel into two parts,

$$P(x, B) = \int_B K(x,y)dy + r(x)\chi_B(x),$$

where χ_B is the characteristic function of the set $B \in \mathfrak{B}$. Although $K(x,y) \geq 0$ is actually a density, we may think of $K(x,y)dy$ as the probability of the move from x to the infinitesimal set dy at y while $r(x) \geq 0$ is the probability of x remaining inert. The characteristic function χ_B of B appears since if $x \notin B$, the only way for x of reaching B is through a move.

The condition $P(x, \mathbb{R}^n) = 1$ implies that

$$r(x) = 1 - \int_{\mathbb{R}^n} K(x,y)dy. \qquad (3.30)$$

In order for $\pi(x)dx$ to be an invariant measure of P, we must have the identity

$$\mu P(B) = \int_{\mathbb{R}^n}\left(\int_B K(x,y)dy + r(x)\chi_B(x)\right)\pi(x)dx$$

$$= \int_B \left(\int_{\mathbb{R}^n} \pi(x) K(x,y) dx + r(y)\pi(y) \right) dy$$

$$= \int_B \pi(y) dy$$

for all $B \in \mathfrak{B}$, implying that

$$\pi(y)(1 - r(y)) = \int_{\mathbb{R}^n} \pi(x) K(x,y) dx. \tag{3.31}$$

By formula (3.30), this is tantamount to

$$\int_{\mathbb{R}^n} \pi(y) K(y,x) dx = \int_{\mathbb{R}^n} \pi(x) K(x,y) dx. \tag{3.32}$$

This condition is called the *balance equation*. In particular, if K satisfies the *detailed balance equation*, which is given by

$$\pi(y) K(y,x) = \pi(x) K(x,y) \tag{3.33}$$

for all pairs $x, y \in \mathbb{R}^n$, then the balance equation holds a fortiori. Conditions (3.32) and (3.33) constitute the starting point in constructing the Markov chain transition kernels used for stochastic sampling.

In the Metropolis–Hastings algorithm, the aim is to construct a transition kernel K that satisfies the detailed balance equation (3.33).

Let $q: \mathbb{R}^n \times \mathbb{R}^n \to \mathbb{R}_+$ be a given function with the property $\int q(x,y) dy = 1$. The kernel q is called the *proposal distribution* or *candidate-generating kernel* for reasons explained later. Such a function q defines a transition kernel,

$$Q(x, A) = \int_A q(x,y) dy.$$

If q happens to satisfy the detailed balance equation, we set simply $K(x,y) = q(x,y)$, $r(x) = 0$ and we are done. Otherwise, we correct the kernel by a multiplicative factor and define

$$K(x,y) = \alpha(x,y) q(x,y), \tag{3.34}$$

where α is a correction term to be determined. Assume that for some x, $y \in \mathbb{R}^n$, instead of the detailed balance we have

$$\pi(y) q(y,x) < \pi(x) q(x,y). \tag{3.35}$$

Our aim is to choose α so that

$$\pi(y)\alpha(y,x) q(y,x) = \pi(x)\alpha(x,y) q(x,y). \tag{3.36}$$

This is achieved if we set

$$\alpha(y,x) = 1, \quad \alpha(x,y) = \frac{\pi(y)q(y,x)}{\pi(x)q(x,y)} < 1. \tag{3.37}$$

By reversing x and y, we see that the kernel K defined through (3.34) satisfies (3.33) if we define

$$\alpha(x,y) = \min\left(1, \frac{\pi(y)q(y,x)}{\pi(x)q(x,y)}\right). \tag{3.38}$$

This transition kernel is called the Metropolis–Hastings kernel.

The above derivation does not shed much light on how to implement the method. Fortunately, the algorithm turns out to be relatively simple. Usually, it is carried out in practice through the following steps.

1. Pick the initial value $x_1 \in \mathbb{R}^n$ and set $k = 1$.
2. Draw $y \in \mathbb{R}^n$ from the proposal distribution $q(x_k, y)$ and calculate the acceptance ratio

$$\alpha(x_k, y) = \min\left(1, \frac{\pi(y)q(y, x_k)}{\pi(x_k)q(x_k, y)}\right).$$

3. Draw $t \in [0,1]$ from uniform probability density.
4. If $\alpha(x_k, y) \geq t$, set $x_{k+1} = y$, else $x_{k+1} = x_k$. When $k = K$, the desired sample size, stop, else increase $k \to k+1$ and go to step 2.

Before presenting some examples, a couple of remarks are in order. First, if the candidate-generating kernel is symmetric, that is,

$$q(x,y) = q(y,x)$$

for all $x, y \in \mathbb{R}^n$, then the acceptance ratio α simplifies to

$$\alpha(x,y) = \min\left(1, \frac{\pi(y)}{\pi(x)}\right).$$

Hence, we accept immediately moves that go towards higher probability and sometimes also moves that take us to lower probabilities. In particular, the symmetry condition is satisfied if the proposal distribution corresponds to a *random walk model*, that is,

$$q(x,y) = g(x-y),$$

for some nonnegative even function $g : \mathbb{R}^n \to \mathbb{R}_+$, that is, $g(x) = g(-x)$.

An important but difficult issue is the stopping criterion, that is, how to decide when the sample is large enough. This question, as well as the convergence issues in general, will be discussed later in the light of examples.

Example 11: In this example, we test the Metropolis–Hastings algorithm with a low-dimensional density. Consider a density π in $\mathbb{R}^2$ defined as

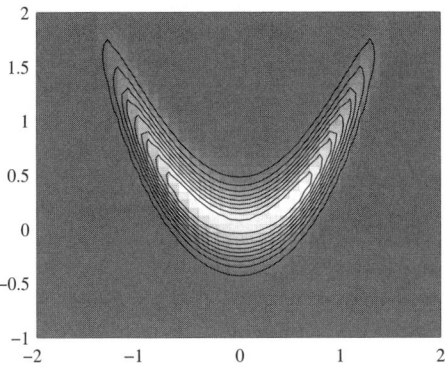

Figure 3.13. Contour plot of the density given in formula (3.39).

$$\pi(x) \propto \exp\left(-10(x_1^2 - x_2)^2 - \left(x_2 - \frac{1}{4}\right)^4\right). \tag{3.39}$$

The contour plot of this density is shown in Figure 3.13.

We construct a Metropolis–Hastings sequence using the random walk proposal distribution to explore this density. Let us define

$$q(x,y) \propto \exp\left(-\frac{1}{2\gamma^2}\|x-y\|^2\right).$$

In other words, we assume that the random step from x to y is distributed as white noise,

$$W = Y - X \sim \mathcal{N}(0, \gamma^2 I).$$

This choice of the proposal distribution gives rise to the following updating scheme:

Pick initial value x_1. Set $x = x_1$
for $k = 2 : K$ do
    Calculate $\pi(x)$
    Draw $w \sim \mathcal{N}(0, \gamma^2 I)$, set $y = x + w$
    Calculate $\pi(y)$
    Calculate $\alpha(x, y) = \min(1, \pi(y)/\pi(x))$
    Draw $u \sim \mathcal{U}([0, 1])$
    if $u < \alpha(x, y)$
        Accept: Set $x = y$, $x_k = x$,
    else
        Reject: Set $x_k = x$,
    end
end

Figure 3.14 shows the outcomes of four runs of the algorithm with different values of the constant $\gamma > 0$ controlling the random step length. In the figures, we have marked the sample points with dots. The iteration was started at the lower right corner from where the algorithm walks by random steps towards the ridge of the density.

From these scatter plots, one can observe the following features of the algorithm. When the step length is small, the process explores the density very slowly. In particular, we see in the top left figure that the chain has not yet visited large parts of the support of the density. The density of the points indicate that new proposals are accepted frequently, and the sampler moves slowly but steadily. When the step length is increased, the coverage increases. On the other hand, the increasing step length has a prize. When the step length becomes larger, the ratio of accepted proposals gets smaller. In the runs shown in this example, the proportion of accepted proposals are 96% for the smallest step length corresponding to $\gamma = 0.01$, 87% for $\gamma = 0.05$, 78% for $\gamma = 0.1$ and only 16% for $\gamma = 1$.

Another feature that can be observed in the scatter plots is that when the chain is started (in our case from the lower right corner at $x = [1, 5, -0.8]$), it takes a while before the chain starts to properly sample the distribution. In particular, when the step length is small, the chain has a long tail from the starting value to the proper support of the density. Usually, these warming-up draws represent poorly the distribution to be explored and one should try to remove them from the sample. The beginning of the chain is often called the *burn-in*.

The questions that arise are, which step size is the best, and also, how long is the burn-in phase. Some insight can be obtained by considering, for example, the time evolution of single components. Figure 3.15 shows the evolution of the first component in our example. The curve in 3.15(a) clearly shows a long burn-in period: Up to roughly 500 draws, there is a clear drift towards a value around which the curve starts to oscillate. In the three other curves (b)–(d), one could conclude that the burn-in is faster. Another feature worth mentioning is the correlation between different draws. Clearly, in curves (a)–(c), the correlation length is quite significant, that is, there are low-frequency components in these curves. This means that the draws are not mutually independent, a feature that slows down the convergence of the Markov chain integration. This issue will be discussed later in more detail. As a rule of thumb, one can say that if the diagnostic curves look like "fuzzy worms", such as in Figure 3.15(d), the step size is large enough. ◊

3.6.3 Gibbs Sampler

A slightly different algorithm is obtained if the candidate-generating kernel is defined by using the density π directly and a block partitioning of the vectors in $\mathbb{R}^n$.

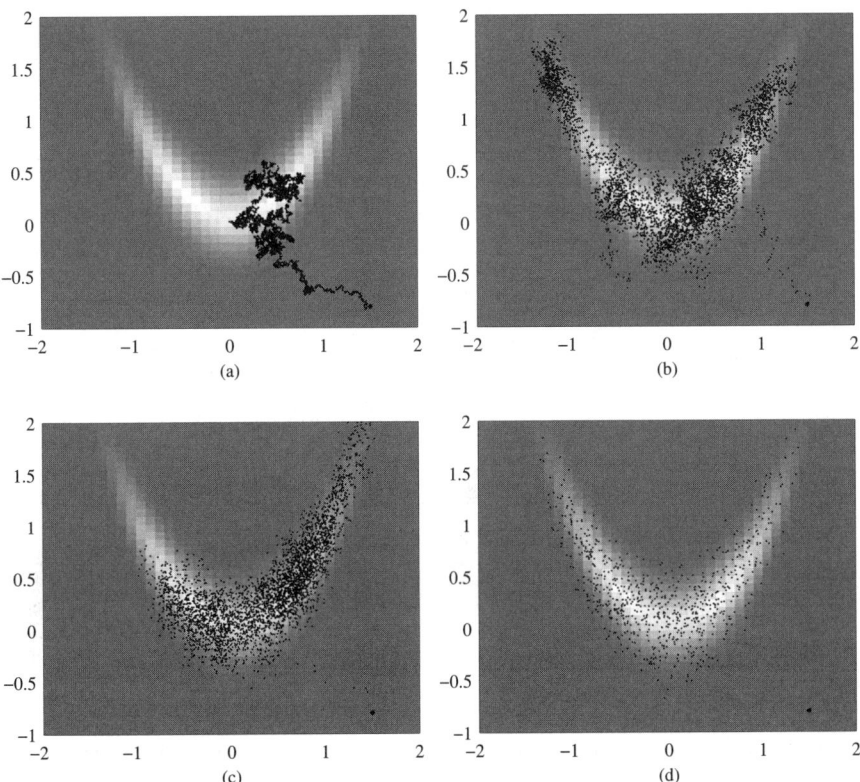

Figure 3.14. Outcomes of the random walk Metropolis–Hastings runs with different step sizes, (a) $\gamma = 0.01$, (b) $\gamma = 0.05$, (c) $\gamma = 0.1$ and and (d) $\gamma = 1$. The sample size in each figure is 5000.

We introduce first some notations. Let $I = \{1, 2, \ldots, n\}$ be the index set of $\mathbb{R}^n$ and $I = \cup_{j=1}^m I_m$ a partitioning of the index set into disjoint nonempty subsets. We denote by k_j the number of the elements in I_j, that is, $k_j = \#I_j$. By definition, we have $k_1 + \cdots + k_m = n$. Accordingly, we may partition $\mathbb{R}^n$ as $\mathbb{R}^n = \mathbb{R}^{k_1} \times \cdots \times \mathbb{R}^{k_m}$ and correspondingly

$$x = [x_{I_1}; \ldots; x_{I_m}] \in \mathbb{R}^n, \ x_{I_j} \in \mathbb{R}^{k_j}, \tag{3.40}$$

where the components of x are rearranged so that $x_i \in \mathbb{R}$ is a component of the vector x_{I_j} if and only if $i \in I_j$.

The following notation is used: We write

$$x_{-I_j} = [x_{I_1}; \ldots; \widehat{x_{I_j}}; \ldots; x_{I_m}]$$
$$= [x_{I_1}, \ldots; x_{I_{j-1}}, x_{I_{j+1}}; \ldots; x_{I_m}];$$

in other words, the negative subindex or the hat denotes that the corresponding elements are deleted from the vector.

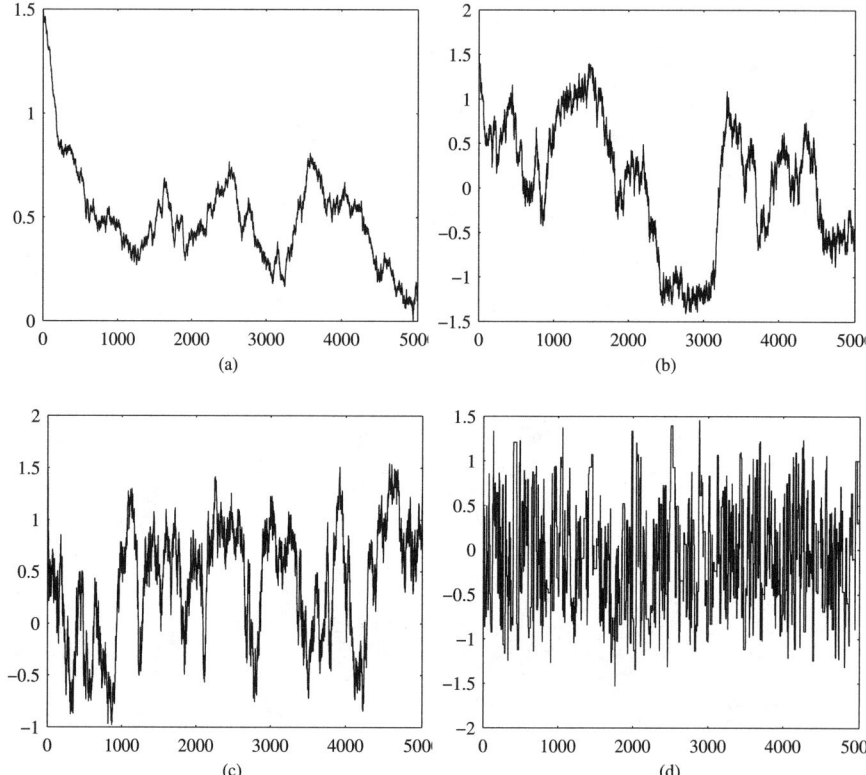

Figure 3.15. Convergence diagnostics for the component x_1 with the values of γ as in Figure 4.12.

Note that if we choose $k_j = 1$ for all j, then $m = n$ and the vector x_{I_j} is simply the jth component of the vector x. This is often the case in practice, and the notation x_{-j} is used for the $n-1$-vector with the jth element deleted.

If X is an n-variate random variable with the probability density π, the probability density of the ith block X_{I_i} conditioned on all $X_{I_j} = x_{I_j}$ for which $i \neq j$, is given by

$$\pi(x_{I_i} \mid x_{-I_i}) = C_i \pi(x_{I_1}, \ldots, x_{I_{i-1}}, x_{I_i}, x_{I_{i+1}}, \ldots, x_{I_m}),$$

where C_i is a normalization constant. With these notations, we can define a transition kernel K by the formula

$$K(x,y) = \prod_{i=1}^{m} \pi(y_{I_i} \mid y_{I_1}, \ldots, y_{I_{i-1}}, x_{I_{i+1}}, \ldots, x_{I_m}), \qquad (3.41)$$

and we set

$$r(x) = 0.$$

Although this transition kernel does not in general satisfy the detailed balance equation (3.33), it satisfies the weaker but sufficient balance condition (3.32). To prove this claim, consider first the left side of (3.32). We observe that since by definition the conditional probability densities are normed to be proper probability densities, we have

$$\int_{\mathbb{R}^{k_i}} \pi(x_{I_i} \mid x_{I_1}, \ldots, x_{I_{i-1}}, y_{I_{i+1}}, \ldots, y_{I_m}) dx_{I_i} = 1$$

for all $i = 1, \ldots, m$. It follows that

$$\int_{\mathbb{R}^{k_m}} K(y,x) dx_{I_m} = \int_{\mathbb{R}^{k_m}} \prod_{i=1}^{m} \pi(x_{I_i} \mid x_{I_1}, \ldots, x_{I_{i-1}}, y_{I_{i+1}}, \ldots, y_{I_m}) dx_{I_m}$$

$$= \prod_{i=1}^{m-1} \pi(x_{I_i} \mid x_{I_1}, \ldots, x_{I_{i-1}}, y_{I_{i+1}}, \ldots, y_{I_m}) \int_{\mathbb{R}^{k_m}} \pi(x_{I_m} \mid y_{I_1}, \ldots, y_{I_{m-1}}) dx_{I_m}$$

$$= \prod_{i=1}^{m-1} \pi(x_{I_i} \mid x_{I_1}, \ldots, x_{I_{i-1}}, y_{I_{i+1}}, \ldots, y_{I_m}).$$

Inductively, by integrating always with respect to the last block of x, we obtain

$$\int_{\mathbb{R}^n} K(y,x) dx = 1.$$

This observation implies that

$$\int_{\mathbb{R}^n} \pi(y) K(y,x) dx = \pi(y) \int_{\mathbb{R}^n} K(y,x) dx = \pi(y).$$

Consider now the right side of (3.32). Observing that $K(x,y)$ is independent of x_{I_1}, by the definition of marginal probability density, we have

$$\int_{\mathbb{R}^{k_1}} \pi(x) K(x,y) dx_{I_1} = K(x,y) \int_{\mathbb{R}^{k_1}} \pi(x) dx_{I_1} = K(x,y) \pi(x_{I_2}, \ldots, x_{I_m}).$$

By substituting K in the above formula we see that

$$\int_{\mathbb{R}^{k_1}} \pi(x) K(x,y) dx_{I_1}$$

$$= \left(\prod_{i=2}^{m} \pi(y_{I_i} \mid y_{I_1}, \ldots, y_{I_{i-1}}, x_{I_{i+1}}, \ldots, x_{I_m}) \right)$$

$$\times \pi(y_{I_1} \mid x_{I_2}, \ldots, x_{I_m}) \pi(x_{I_2}, \ldots, x_{I_m})$$

$$= \left(\prod_{i=2}^{m} \pi(y_{I_i} \mid y_{I_1}, \ldots, y_{I_{i-1}}, x_{I_{i+1}}, \ldots, x_{I_m}) \right) \pi(y_{I_1}, x_{I_2}, \ldots, x_{I_m}).$$

We continue inductively, integrating next with respect to x_{I_2}. The general step in this process is

$$\int_{\mathbb{R}^{k_i}} \pi(y_{I_1},\ldots,y_{I_{i-1}},x_{I_i},\ldots x_{I_m})\pi(y_{I_i} \mid y_{I_1},\ldots,y_{I_{i-1}},x_{I_{i+1}},\ldots,x_{I_m})dx_{I_i}$$

$$= \pi(y_{I_1},\ldots,y_{I_{i-1}},x_{I_{i+1}},\ldots x_{I_m})\pi(y_{I_i} \mid y_{I_1},\ldots,y_{I_{i-1}},x_{I_{i+1}},\ldots,x_{I_m})$$

$$= \pi(y_{I_1},\ldots,y_{I_{i-1}},y_{I_i},x_{I_{i+1}},\ldots,x_{I_m}).$$

By integrating over all blocks of x, we arrive at the desired formula,

$$\int_{\mathbb{R}^n} \pi(x)K(x,y)dx = \pi(y),$$

showing that (3.32) holds. The algorithm just obtained is known as the *Gibbs sampler*. The special case when $m = n$ and $k_1 = \cdots = k_m = 1$, the algorithm is called the *single component Gibbs sampler*.

Although the construction of the Gibbs sampler, as opposed to the Metropolis–Hastings kernel, is not based on the requirement that the transition kernel satisfies the detailed balance equation, it is often seen as a special case of the Metropolis–Hastings algorithm. While algorithmically they yield essentially the same method, the main difference is that the proposal is always accepted in the Gibbs sampler.

Before presenting examples illustrating the convergence of the method, we summarize the practical steps needed for actual implementation of the algorithm.

1. Pick the initial value $x_1 \in \mathbb{R}^n$ and set $k = 1$.
2. Set $x = x_k$. For $1 \leq j \leq m$, draw $y_{I_j} \in \mathbb{R}^{k_j}$ from the k_j-dimensional distribution $\pi(y_{I_j} \mid y_{I_1},\ldots,y_{I_{j-1}},x_{I_{j+1}},\ldots,x_{I_m})$.
3. Set $x_{k+1} = y$. When $k = K$, the desired sample size, stop. Else, increase $k \to k+1$ and repeat from step 2.

As mentioned above, the crucial difference between the Metropolis–Hastings and Gibbs sampler algorithms is that the proposal in Gibbs sampler is always accepted. On the other hand, the drawing process in the latter may be more complicated and time consuming than in the Metropolis–Hastings algorithm, where the proposal distribution can be simpler to handle.

Example 12: We consider the same two-dimensional distribution as in Example 11, this time using the single component Gibbs sampler. In this example, the algorithm can be written along the following lines:

Pick initial value x_1. Set $x = x_1$
for $j = 2 : N$ do
    Calculate $\Phi_1(t) = \int_{-\infty}^t \pi(x^1,x^2)dx^1$
    Draw $u \sim \mathcal{U}([0,1])$, set $x^1 = \Phi_1^{-1}(u)$
    Calculate $\Phi_2(t) = \int_{-\infty}^t \pi(x^1,x^2)dx^2$

Draw $u \sim \mathcal{U}([0,1])$, set $x^2 = \Phi_2^{-1}(u)$
Set $x_k = x$
end

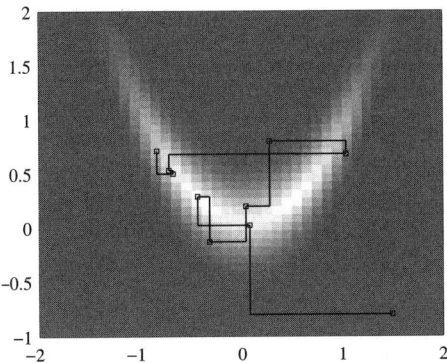

Figure 3.16. The first 10 draws produced by the Gibbs sampler. The samples are marked by a square, and the componentwise updating paths from sample to sample are drawn.

In Figure 3.16, the first 10 draws are shown. To emphasize the drawing process, the segments joining consecutive sample points are also plotted. ◇

Example 13: As a less trivial example than the one considered above, we consider the blind deconvolution problem introduced in Example 3 of this chapter. The problem is to estimate the pair $(x, v) \in \mathbb{R}^n \times \mathbb{R}^K$, where x represents the image and v consists of the weights of the Gaussian kernels that constitute the unknown convolution kernel. The observation $y \in \mathbb{R}^N$ with additive Gaussian noise of variance γ^2 is linked to the unknowns via the likelihood function

$$\pi(y \mid x, v) \propto \left(-\frac{1}{2\gamma^2} \left\| y - \sum_{k=1}^K v_k A_k x \right\|^2 \right).$$

Let us assume that the random variables X and V are mutually independent. The prior probability density of X is assumed to be a white noise prior, $X \sim \mathcal{N}(0, \alpha^2 I)$, where $\alpha > 0$ is fixed.

The prior density of the parameter vector is likewise assumed to be Gaussian. We set

$$\pi_{\text{param}}(v) \propto \exp\left(-\frac{1}{2}\beta^2 \|L(v - v_0)\|^2\right),$$

with $\beta > 0$ known. Here, $v_0 \in \mathbb{R}^K$ corresponds to an approximate kernel and L is a first-order difference matrix. This means that we assume that the first basis function is a good guess for the convolution kernel. Using the differences of Gaussian kernels as basis functions and the difference matrix as L are due to numerical stability.

Hence, the joint probability density is

$$\pi(x,v,y) \propto \exp\left(-\frac{1}{2\gamma^2}\left\|y - \sum_{k=1}^{K} v_k A_k x\right\|^2 - \frac{1}{2\alpha^2}\|x\|^2 - \frac{1}{2}\beta^2\|L(v-v_0)\|^2\right).$$

We observe that while the density is not Gaussian with respect to the vector $[x;v]$, the conditional probabilities are. We have

$$\pi(x \mid v,y) \propto \exp\left(-\frac{1}{2}\left\|\begin{bmatrix}\gamma^{-1}\sum_{k=1}^{K} v_k A_k \\ \alpha^{-1} I\end{bmatrix} x - \begin{bmatrix}\gamma^{-1} y \\ 0\end{bmatrix}\right\|^2\right).$$

Similarly, by defining

$$S = \begin{bmatrix} A_1 x, A_2 x, \ldots, A_K x \end{bmatrix} \in \mathbb{R}^{n \times K},$$

we can write

$$\pi(v \mid x,y) \propto \exp\left(-\frac{1}{2}\left\|\begin{bmatrix}\gamma^{-1} S \\ \beta L\end{bmatrix} v - \begin{bmatrix}\gamma^{-1} y \\ \beta L v_0\end{bmatrix}\right\|^2\right).$$

Now we may design a block form Gibbs sampling algorithm with the following simple steps.

1. Choose initial values for v and x, for example, by

$$v = v_0, \quad x = x_0 = \begin{bmatrix}\gamma^{-1}\sum_{k=1}^{K} v_{0,k} A_k \\ \sqrt{2\alpha} L\end{bmatrix}^{\dagger} \begin{bmatrix}\gamma^{-1} y \\ 0\end{bmatrix},$$

and set $j = 0$.

2. Update v: Draw $\eta \in \mathbb{R}^{n+K}$, $\eta \sim \mathcal{N}(0, I)$ and set

$$v_{j+1} = \begin{bmatrix}\gamma^{-1} S \\ \beta L\end{bmatrix}^{\dagger}\left(\begin{bmatrix}\gamma^{-1} y \\ \beta L v_0\end{bmatrix} + \eta\right).$$

3. Update x: Draw $\xi \in \mathbb{R}^{2n}$, $\xi \sim \mathcal{N}(0, I)$ and set

$$x_{j+1} = \begin{bmatrix}\gamma^{-1}\sum_{k=1}^{K} v_{j+1,k} A_k \\ \sqrt{2\alpha} L\end{bmatrix}^{\dagger}\left(\begin{bmatrix}\gamma^{-1} y \\ 0\end{bmatrix} + \xi\right).$$

4. Set $j \leftarrow j + 1$ and repeat from step 2 until the desired sample size is reached.

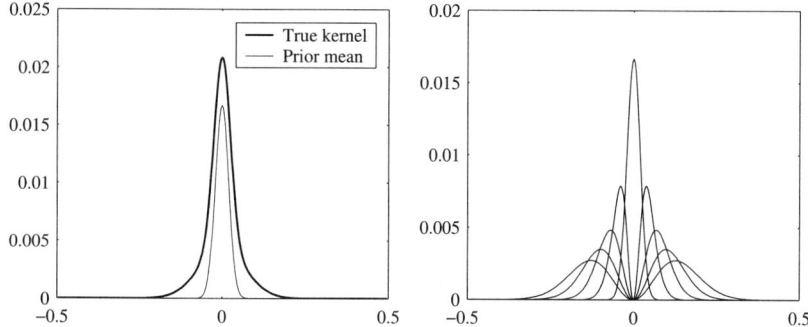

Figure 3.17. The true convolution kernel and the representation kernels ϕ_k, $k = 1, \ldots, 5$.

We apply the above algorithm to the one-dimensional blind deconvolution problem, in which the true convolution kernel is a Gaussian with STD $\sigma = 0.025$. The basis functions ϕ_k, $1 \leq k \leq 5$ are chosen as in Example 3 with $(\sigma_1, \ldots, \sigma_5) = (0.02, 0.04, 0.06, 0.08, 0.1)$. The basis functions are plotted in Figure 3.17. The parameter values defining the prior densities are $\alpha = 0.08$ and $\beta = 100$, and the vector v_0 defining the center of the prior is $v_0 = [1; 0; 0; 0; 0]$. The standard deviation γ of the additive Gaussian noise is 1% of the maximum value of the noiseless data vector.

Figure 3.18 shows the true image and true convolution kernels, the initial guesses (v_0, x_0) as well as the computed CM estimates using a sample sixe of 5000.

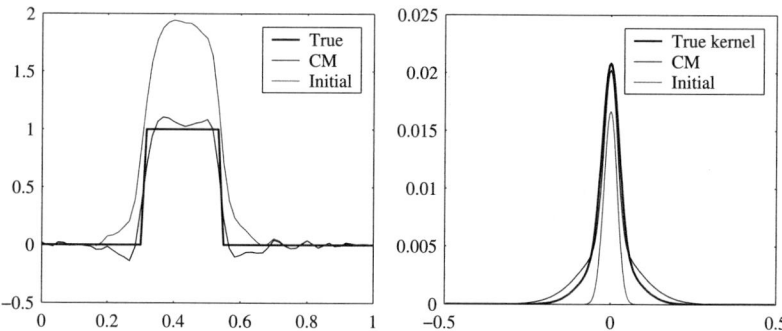

Figure 3.18. The conditional mean estimates of the image and the kernel.

For comparison, let us seek the MAP estimate using the following sequential optimization scheme:

1. Choose initial values (v_0, x_0) and set $j = 0$.

2. Update $v_j \leftarrow v_{j+1}$ by maximizing $\pi(v \mid x^j, y)$.
3. Update $x_j \leftarrow x_{j+1}$ by maximizing $\pi(x \mid v^{j+1}, y)$.
4. Set $j \leftarrow j + 1$ and repeat from step 2 until convergence.

The updating formulas are particularly simple in the present case. We have

$$v_{j+1} = \begin{bmatrix} \gamma^{-1}S \\ \beta L \end{bmatrix}^\dagger \begin{bmatrix} \gamma^{-1}y \\ \beta L v_0 \end{bmatrix}, \quad x_{j+1} = \begin{bmatrix} \gamma^{-1}\sum_{k=1}^{K} v_{j+1,k} A_k \\ \sqrt{2\alpha}L \end{bmatrix}^\dagger \begin{bmatrix} \gamma^{-1}y \\ 0 \end{bmatrix}.$$

Comments concerning the convergence of this scheme can be found in "Notes and Comments" at the end of this chapter. Here we display the results obtained by merely 20 iterations; see Figure 3.19. ◇

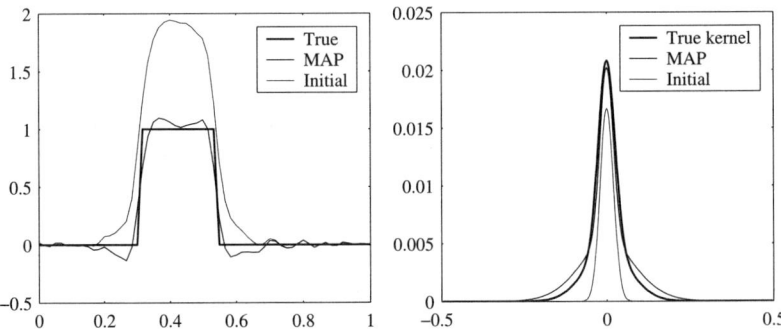

Figure 3.19. The MAP estimates of the image and the kernel.

Further examples of both the Metropolis–Hastings and Gibbs sampler are given later in Chapters 5 and 7.

3.6.4 Convergence

As indicated in the previous sections, it is not simple to decide when the MCMC sample is large enough to cover the probability distribution in question. Proposition 3.11 states that

$$\lim_{N \to \infty} \frac{1}{N} \sum_{n=1}^{N} f(X_n) = \int f(x)\pi(x)dx$$

provided that the transition kernel P satisfies certain conditions. Hence, the success of the Metropolis–Hasting and Gibbs sampler algorithms depends largely on whether they satisfy the assumptions of Proposition 3.11. There are known sufficient conditions concerning the density π that guarantee the ergodicity of these methods. In the following proposition, we give some relatively general conditions. The proof is omitted here. For the literature, we refer to "Notes and Comments."

Proposition 3.12. *(a) Let $\pi : \mathbb{R}^n \to \mathbb{R}_+$ be a probability density, and $E_+ = \{x \in \mathbb{R}^n \mid \pi(x) > 0\}$, and let $q : \mathbb{R}^n \times \mathbb{R}^n \to \mathbb{R}_+$ be a candidate-generating kernel and $Q(x, A)$ be the corresponding transition probability function. If Q is aperiodic, then the Metropolis–Hastings chain is also aperiodic. Further, if Q is irreducible and $\alpha(x, y) > 0$ for all $(x, y) \in E_+ \times E_+$, then the Metropolis–Hastings chain is irreducible.*

(b) Let π be a lower semicontinuous density and $E_+ = \{x \in \mathbb{R}^n \mid \pi(x) > 0\}$. The Gibbs sampler defines an irreducible and aperiodic transition kernel if E_+ is connected and each $(n-1)$-dimensional marginal $\pi(x_{-j}) = \int_\mathbb{R} \pi(x) dx_j$ is locally bounded.

An example of a probability density in which the Gibbs sampler does not converge is given by

$$\pi(x) = \pi(x_1, x_2) \propto \chi_{[0,1]\times[0,1]} + \chi_{[2,3]\times[2,3]},$$

where $\chi_{[a,b]}$ denotes the characteristic function of the interval $[a, b]$. This density is bimodal and the jump from one of the squares to the other happens with probability zero.

In practical applications, the crucial question is to decide whether a sample is large enough. A partial answer is obtained by the central limit theorem; see Appendix B. Assume that the variables $Y_n = f(X_n)$ are mutually independent and equally distributed with $\mathrm{E}(Y_j) = y$ and $\mathrm{cov}(Y_j) = \sigma^2$, and define

$$\tilde{Y}_N = \frac{1}{N} \sum_{n=1}^N Y_n$$

and further,

$$Z_N = \frac{\sqrt{N}(\tilde{Y}_N - y)}{\sigma}.$$

Then, $\tilde{Y}_N \to y$ almost surely and, asymptotically, Z_N is normally distributed, that is,

$$\lim_{N \to \infty} \mathrm{P}\{Z_n \leq z\} = \frac{1}{\sqrt{2\pi\sigma^2}} \int_{-\infty}^z \exp\left(-\frac{1}{2\sigma^2} s^2\right) ds.$$

The above result says that the approximation error behaves as

$$\frac{1}{N} \sum_{n=1}^N f(X_n) - \int f(x)\pi(x)dx \sim \frac{\sigma}{\sqrt{N}}$$

provided that the samples are independent. In practice, however, it often happens that consecutive elements show correlation. The shorter the correlation length, the faster the approximation can be assumed to converge. Another problem with the asymptotic formula above is that the variance σ is hard to estimate.

The independence of the consecutive draws can be estimated from the sample itself. Assume that we are interested in the convergence of the integral of $f(x) \in \mathbb{R}$ with respect to the probability density $\pi(x)$. Let us denote $z_j = f(x_j)$, where $\{x_1, \ldots, x_N\} \subset \mathbb{R}^n$ is a MCMC sample and let $\bar{z} = N^{-1}\sum z_j$. The sample-based estimate of the scaled *time series autocovariance* of z_j can be computed as

$$\gamma'_k = \frac{1}{(N-k)\gamma_0} \sum_{j=1}^{N-k} (z_j - \bar{z})(z_{j+k} - \bar{z}) \qquad (3.42)$$

where $\gamma_0 = N^{-1}\|z\|^2$. Later in this book, we illustrate by examples the use of the ideas in this section.

3.7 Hierarcical Models

We have pointed out that many of the the classical regularization methods for solving ill-posed problems can be viewed as constructing estimators based on the posterior distribution. This issue will be discussed further in Chapter 5. A large part of the literature discussing regularization techniques is devoted to the problem of selecting the regularization parameters, the Morozov discrepancy principle being the most commonly used one. A question frequently asked is, whether the Bayesian approach provides any similar tools. In particular, in the examples presented so far, the prior densities typically depend on parameters such as variance and midpoint that are always assumed to be known. From the point of view of classical methods, this corresponds to knowing ahead of time the regularization parameters.

In the Bayesian framework the answer to the question how the parameters should be chosen is: *If a parameter is not known, it is a part of the inference problem*. This approach leads to *hierarchical models*, also known as *hyperprior models*. We shall discuss this topic by considering an example.

Example 14: Consider a linear inverse problem with additive Gaussian noise,

$$Y = AX + E, \quad E \sim \mathcal{N}(0, \Gamma_{\text{noise}}).$$

Assume that the prior model for X is also a Gaussian,

$$X \sim \mathcal{N}\left(x_0, \frac{1}{\alpha}\Gamma_{\text{pr}}\right),$$

where Γ_{pr} is a known symmetric positive definite matrix, but $\alpha > 0$ is poorly known. Hence, we write a conditional prior for $X \in \mathbb{R}^n$, assuming that α was known, as

$$\pi_{\text{pr}}(x \mid \alpha) = \frac{\alpha^{n/2}}{\sqrt{(2\pi)^n |\Gamma_{\text{pr}}|}} \exp\left(-\frac{1}{2}\alpha x^{\mathrm{T}} \Gamma_{\text{pr}}^{-1} x\right),$$

3.7 Hierarcical Models

that is, the prior density is conditioned on the knowledge of α. Assume further that we have a hyperprior density $\pi_{\mathrm{h}}(\alpha)$ for the parameter α. We assume here that this density is a Rayleigh distribution,

$$\pi_{\mathrm{h}}(\alpha) = \frac{\alpha}{\alpha_0^2} \exp\left(-\frac{1}{2}\left(\frac{\alpha}{\alpha_0}\right)^2\right), \quad \alpha > 0, \qquad (3.43)$$

where $\alpha_0 > 0$ is the centerpoint of the hyperprior which we take to be fixed. This issue, however, is elaborated below. The joint probability distribution of α and X is therefore

$$\pi(\alpha, x) = \pi_{\mathrm{pr}}(x \mid \alpha)\pi_{\mathrm{h}}(\alpha)$$

$$\propto \alpha^{(n+2)/2} \exp\left(-\frac{1}{2}\alpha x^{\mathrm{T}}\Gamma_{\mathrm{pr}}^{-1}x - \frac{1}{2}\left(\frac{\alpha}{\alpha_0}\right)^2\right).$$

Considering now both α and X as unknowns, we write Bayes' formula conditioned on the observation $Y = y$ as

$$\pi(x, \alpha \mid y) \propto \alpha^{(n+2)/2} \exp\bigg(-\frac{1}{2}\alpha x^{\mathrm{T}}\Gamma_{\mathrm{pr}}^{-1}x$$

$$-\frac{1}{2}\left(\frac{\alpha}{\alpha_0}\right)^2 - \frac{1}{2}(y - Ax)^{\mathrm{T}}\Gamma_{\mathrm{noise}}^{-1}(y - Ax)\bigg).$$

This formula allows us to estimate α and X simultaneously.

To demonstrate the use of this formula, let us consider a deconvolution problem in two dimensions. More precisely, we consider the problem introduced in Chapter 2, Example 5. Thus, the matrix A is a $32^2 \times 32^2 = 1024 \times 1024$ matrix approximation of the convolution with the kernel $K(t) = \exp(-a|t|)$, where $a = 20$, and the true image in the unit square $[0,1] \times [0,1]$ is the same used in the aforementioned example.

First, we study the effect of the hyperparameter α_0. To this end, we calculate the MAP estimates for the pair (X, α) for various values of α_0. In particular, we are interested in the effect of this parameter on the estimated value of α.

The MAP estimate is found by seeking the minimizer of the functional $F : \mathbb{R} \times \mathbb{R}^n \to \mathbb{R}$, given by

$$F(\alpha, x) = \frac{1}{2}\alpha x^{\mathrm{T}}\Gamma_{\mathrm{pr}}^{-1}x + \frac{1}{2}\left(\frac{\alpha}{\alpha_0}\right)^2 + \frac{1}{2}(y - Ax)^{\mathrm{T}}\Gamma_{\mathrm{noise}}^{-1}(y - Ax) - \frac{n+2}{2}\log\alpha. \qquad (3.44)$$

In our example, we use the same noise and prior structure as in the Example 5 of Chapter 2, that is,

$$\Gamma_{\mathrm{noise}} = \sigma^2 I, \quad \Gamma_{\mathrm{pr}} = I,$$

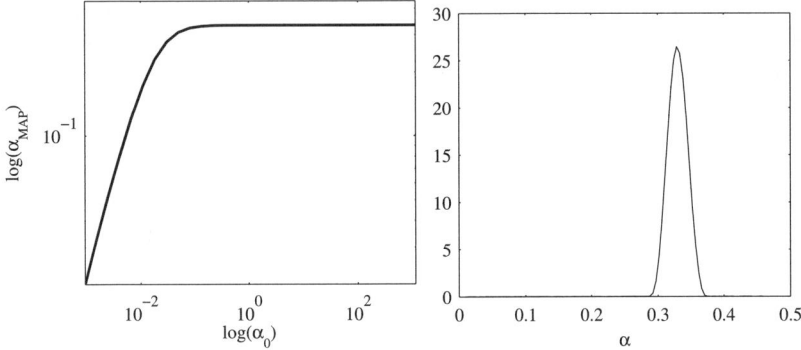

Figure 3.20. The MAP estimate of the prior parameter α as a function of the hyperparameter α_0 (left). The histogram of α based on 5000 MCMC draws with $\alpha_0 = 2$ (right).

where the noise level σ is 0.5% of the maximum entry of the noiseless signal.

Figure 3.20 shows the MAP estimate of α as a function of the hyperparameter α_0. The result demonstrates that the MAP estimate is rather insensitive to the hyperparameter provided that it is not very small. Large values correspond to wide hyperpriors. Observe that if instead of the Rayleigh distribution (3.43) we had chosen merely a flat positivity constraint prior,

$$\pi_{\mathrm{h}}(\alpha) = \pi_{+}(\alpha), \qquad (3.45)$$

the function F_+ corresponding to (3.45) would be

$$F_+(\alpha, x) = \frac{1}{2}\alpha\, x^{\mathrm{T}} \Gamma_{\mathrm{pr}}^{-1} x + \frac{1}{2}(y - Ax)^{\mathrm{T}} \Gamma_{\mathrm{noise}}^{-1}(y - Ax) - \frac{n}{2}\log\alpha,$$

that is, essentially similar to F with $\alpha_0 \to \infty$.

To assess the reliability of the MAP estimate of the hyperparameter, we study the probability distribution of α by using MCMC techniques. The MCMC is based on a block version of the Gibbs sampler. First, we fix the hyperparameter $\alpha_0 = 2$. As an initial point for the calculation, we choose

$$\alpha = \alpha_0, \quad x = \begin{bmatrix} \sigma^{-1}A \\ \sqrt{\alpha}I \end{bmatrix}^{\dagger} \begin{bmatrix} \sigma^{-1}y \\ 0 \end{bmatrix},$$

that is, the initial value of x is the MAP estimate conditioned on $\alpha = \alpha_0$. Having the initial values, we start the iteration: A new value for α is drawn from the one-dimensional density

$$\pi(\alpha \mid x, y) \propto \exp\left(-\frac{1}{2}\alpha\|x\|^2 - \frac{1}{2}\left(\frac{\alpha}{\alpha_0}\right)^2 + \frac{n+2}{2}\log\alpha\right).$$

Having updated the parameter α, we then update the value of x by drawing its value from the density

$$\pi(x \mid \alpha, y) \propto \exp\left(-\frac{1}{2}\alpha\|x\|^2 - \frac{1}{2\sigma^2}\|y - Ax\|^2\right).$$

Observe that this density is Gaussian, so the drawing can be performed in a single batch operation. Indeed, we perform the drawing by the formula

$$x = \begin{bmatrix} \sigma^{-1}A \\ \sqrt{\alpha}I \end{bmatrix}^\dagger \left(\xi + \begin{bmatrix} \sigma^{-1}y \\ 0 \end{bmatrix}\right),$$

where ξ is a white noise vector in $\mathbb{R}^{2n}$.

We calculate a sample of 5000 draws. Figure 3.20 shows the distribution of the parameters α. The distribution is rather narrow, so the MAP estimate is rather informative.

It is interesting and instructive to compare the results to the standard Morozov discrepancy principle of choosing the Tikhonov regularization parameter. Recall that by writing the Tikhonov regularized solution as

$$x_\delta = \arg\min\left(\|Ax - y\|^2 + \delta\|x\|^2\right),$$

the parameter δ is chosen by the condition

$$\|Ax_\delta - y\|^2 = \operatorname{Tr}(\Gamma_n) = n\sigma^2,$$

and the corresponding parameter α is then computed by

$$\alpha_{\text{Morozov}} = \frac{\delta}{\sigma^2}.$$

This procedure produces a value $\alpha_{\text{Morozov}} = 19.2$, a value that is clearly larger than the CM or MAP estimate. Figure 3.21 shows the Tikhonov regularized estimate with this choice of the parameter α, and for comparison, the CM estimate based on the previously explained MCMC run.

When comparing these results, in the terminology of the classical regularization theory one is tempted to say that the CM estimate is "underregularized" when compared to the Tikhonov regularization. However, from the statistical point of view, the CM estimate is *consistent with the prior*. If one uses a white noise prior with unknown variance, the CM estimate is expected to display the characteristics of a noise image. In our example, the true image is not a typical white noise image. It is possible to demonstrate that if we draw a large number of images from the white noise distribution with a given variance, as an estimate for the true variance the CM estimate is superior to that given by the Morozov discrepancy principle.

The estimation theory is discussed in more detail in Chapter 5 of this book. We also show results with hyperprior models where the true images are more consistent with the prior. ◇

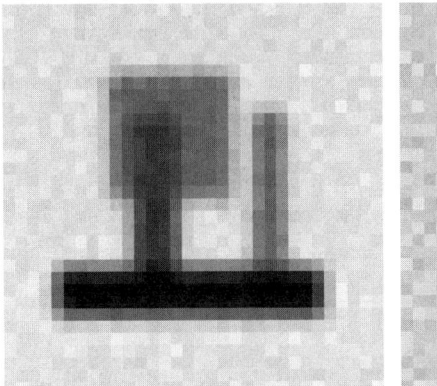 

Figure 3.21. The Tikhonov regularized solution with regulariation parameter selected by the discrepancy principle (left) and the CM estimate based on 5000 MCMC draws with $\alpha_0 = 2$ (right).

3.8 Notes and Comments

Traditionally, the field of inverse problems has been developed from a basis of deterministic models. The notion of statistical inference in inverse problems has been advocated by some outhors, particularly in geophysical inverse problems in which these ideas have been cultivated for some time; see, e.g., [126], [125], [82]. More recent reviews on statistical inversion include [90, 91, 92, 124].

A classic reference on statistical inference is Laplace's Memoir on inverse probability from 1774; see [122]). It is interesting to note that in Laplace's work, the idea of statistical inverse problems is not very far.

The literature concerning Bayesian statistics is enormous. For the central concepts and methodology, we refer to [15] which is a classical textbook on Bayesian statistics. We also note that the statistics community is somewhat divided between the Bayesian school and the so-called frequentist school. The frequentists' criticism concerns mostly the use of prior models. A classical text on the frequentist methodology and interpretation is [25], while the philosophical differences between the two schools are discussed for example in [84]. Within the frequentist paradigm, we also refer to so-called robust methods [61], in which the error model related to the likelihood is iteratively adjusted based on the actual data.

In contrast to the deterministic inversion approaches and the underlying philosophy in which a solution is always forced, the statistical approach is somewhat different in the following sense. Given the prior information and models as well as the measurements, we may still end up in a situation in which the posterior distribution is improper. Formally we would have infinite variances with respect to certain subspaces or manifolds. This would imply that we if we are fair, we acknowledge that the prior information and the measurements are not enough to produce any reliable estimate.

As for the everyday notion of the word *prior*, we note that this does not necessarily refer to "temporally prior" although this would usually be the case. Rather, the word refers to the prior modelling being carried out *without the data* that is to be used in the likelihood.

The posterior mean formulas (3.14) and (3.17) are often not used computationally if the covariances are not to be computed. For example, for (3.17) we can write

$$\overline{x} = \arg\min_x \left\| \begin{bmatrix} L_{\text{noise}} A \\ L_{\text{prior}} \end{bmatrix} x - \begin{bmatrix} L_{\text{noise}}(y - e_0) \\ L_{\text{prior}} x_0 \end{bmatrix} \right\|,$$

where L_{noise} and L_{prior} are the Cholesky factors of Γ_{noise} and Γ_{pr}, respectively. This least squares problem can be solved by using the QR decomposition.

When applying the statistical approach, it is important that the unknown is the one of primary interest. For instance, if we want to estimate a variable $Z = G(X)$ and the observation model is $Y = F(X) + E$, E being noise, we should not try to estimate first X and then compute Z: If G is nonlinear, we have usually $G(x_{\text{CM}}) \neq (G(x))_{\text{CM}}$, and the estimation error variance is often larger for $G(x_{\text{CM}})$.

It is possible, at least in the Gaussian case, to develop a statistical inversion theory in Hilbert spaces of infinite dimensions or, more generally, in distribution spaces. We refer to the articles [83] and [86].

The total variation prior has been used in classical approach to inverse problems as a Tikhonov-type penalty term; see, e.g., [30] and [139]. For the properties of functions with bounded total variation, we refer to the book [45].

The impulse noise priors such as the ℓ^1-prior and the maximum entropy prior have been analyzed in the article [31], where it was demonstrated that in image denoising, they favor estimates that are almost black.

A simplified proof of the Hammersley–Clifford theorem can be found in the article [13]. As a reference concerning continuous Markov random fields, see the textbook [107].

A reference on kernel estimation methods that are useful in constructing sample-based prior densities is [127]. For the construction of sample- or simulation-based priors and their application to electrical impedance tomography, see [132, 135].

Structural priors, as presented in this book, were discussed in the article [66].

For Markov chain Monte Carlo methods, we refer to the monograph [85] and the book [44]. The former contains a wealth of alternative algorithms for the basic Metropolis–Hastings and Gibbs sampler algorithms presented in this book. The original references concerning these methods are [40, 41, 57, 89].

The convergence issues and mixing properties of the MCMC methods were not discussed in detail. These topics are relatively involved, and we refer to the articles [117] and [128] for further discussion. The central ideas in those articles go back to [97].

In the blind deconvolution problem of Example 13, we used a sequential optimization method for computing the MAP estimate. The convergence of this iteration scheme is discussed in the article [14]. The blind deconvolution problem is widely studied because of its importance in many applied fields. Often, it is treated as a total least squares problem. For a recent reference, see [102].

There are numerous cases in which the construction of an explicit model for the prior can be exceedingly cumbersome. One possibility is to use Bayesian neural networks; see for example [78, 79].

4
Nonstationary Inverse Problems

In several applications, one encounters a situation in which measurements that constitute the data of an inverse problem are done in a nonstationary environment. More precisely, it may happen that the physical quantities that are the focus of our primary interest are time dependent and the measured data depends on these quantities at different times. For example, consider the continuous monitoring of the human heart by measuring the magnetic fields outside the body due to bioelectromagnetism. In these measurements, the observed signal is weak and the noise level high. If the target was a static source, one could average a sequence of measurements to reduce the noise level. However, measurements at different time instances give information of the state of the heart at that particular moment and an averaging of the measurements at different times gives a more or less useless averaged signal.

Inverse problems of this type are called here *nonstationary inverse problems*. In this section, we derive a Bayesian model for a class of nonstationary problems and discuss the corresponding statistical inversion methods. These methods are referred to as *Bayesian filtering methods* for historic reasons. Undoubtedly, the most famous and also the most widely used of these methods is the Kalman filter that we derive as a special case.

4.1 Bayesian Filtering

In this section, as in the previous chapter, we limit the discussion to finite-dimensional models. Much of the material in this section could immediately be extended to infinite-dimensional Hilbert spaces. Also, for the sake of clarity, we use discrete time evolution models. In some applications, the discrete evolution model is derived from a continuous evolution model, for example, from a stochastic differential equation. We shall consider such cases in the preliminary example below and especially in Chapter 7.

4.1.1 A Nonstationary Inverse Problem

Let us introduce an example that clarifies the concepts discussed in this chapter.

Example 1: Consider a waveform $\psi : \mathbb{R} \to \mathbb{R}$ that moves with a constant velocity c along the real axis, that is, we have a function

$$u(s,t) = \psi(s - ct), \quad t > 0, \; s \in \mathbb{R}. \tag{4.1}$$

We assume that at fixed locations $s = z_\ell$, $0 \leq z_1 < z_2 < \cdots < z_L \leq 1$, the amplitude of u through a blurring kernel is observed. More precisely, we assume that the observed data at time t is

$$g(z_\ell, t) = \int_{-\infty}^{\infty} K(z_\ell - s) u(s,t) dt + v_\ell(t), \quad 1 \leq \ell \leq L,$$

where the convolution kernel K is given as

$$K(s) = 1 - \left(\frac{s}{d}\right)^2, \text{ if } |s| \leq d, \quad K(s) = 0, \text{ if } |s| > d,$$

and $d > 0$ is the width parameter of the blurring kernel. The observations are corrupted by additive noise $v_\ell(t)$ with relatively high amplitude. Furthermore, assume that the time instants of the observations are t_k, $0 \leq t_0 < t_1 < \cdots$. The problem is to estimate the function u in the interval $[0,1]$ from this data.

The degree of difficulty of this problem depends heavily on what our prior information of the underlying situation is. We shall consider two cases:

(i) We assume that $u(s, t_k)$ and $u(s, t_{k+1})$ are not very different functions.
(ii) We know that $u(s,t)$ is a wave moving to the right, but the wave speed and the waveform are not exactly known. Hence, we assume that u satisfies the equation

$$\frac{\partial u}{\partial s} + \frac{1}{\bar{c}} \frac{\partial u}{\partial t} \approx 0,$$

where $\bar{c} > 0$ is the guessed wave speed. The fact that the above equation may not be exact is due to the uncertainty about the wave speed and whether the linear wave propagation model is correct.

For later use, we discretize the problem. Let $s_0 = 0 < s_1 < \ldots < s_N = 1$ be a discretization of the interval $[0,1]$ into equal intervals of length $1/N$. The discretized observation model is then a linear one. By using, for example, the trapezoid rule, we write the discrete model as

$$Y_k = G X_k + V_k, \quad k \geq 0,$$

where

$$X_k = [u(s_1, t_k), u(s_2, t_k), \ldots, u(s_N, t_k)]^{\mathrm{T}} \in \mathbb{R}^N,$$

and $G \in \mathbb{R}^{L \times N}$ is the matrix with elements

$$G_{\ell,n} = w_j K(z_\ell - s_n),$$

where w_ℓ is the weight of the quadrature rule, and $V_k \in \mathbb{R}^L$ is the noise vector,

$$V_k = [v_1(t_k), \ldots, v_L(t_k)]^\mathrm{T}.$$

Despite of the simplicity of the observation model, the high noise level makes the problem difficult when no prior information is present. In fact, each time slice constitutes a separate inverse problem and the sequential nature of the measurement provides no help.

However, if the sampling frequency is high compared to the propagation velocity of the wave, one could try to average few consecutive observations and thus reduce the noise level, hoping that the underlying function u does not change much during the averaging window. Of course, if the propagation velocity c is high, averaging causes severe blurring in the reconstruction.

To understand how the prior information helps here, consider the case (i) above. We assume that the underlying dynamics of the vectors X_k is unknown to us. A possible "black box" model to account for the unknown dynamics could be a random walk model, that is, we write

$$X_{k+1} = X_k + W_{k+1},$$

where W_{k+1} is a Gaussian random vector. In this model, the variance of W_{k+1} is a parameter that controls the difference between the state vectors at consecutive time instants. The modelling of the statistics of W_k might be based on available prior infomation or might be chosen ad hoc.

Finally, let us consider the case (ii) in which we have approximate information about the true underlying dynamics of the state. We use the approximate model

$$\frac{\partial u}{\partial t} = -\bar{c}\frac{\partial u}{\partial s} + e(s,t),$$

where the term e is a small but unknown function representing the modelling error. We discretize this equation first with respect to s by a finite difference scheme. By writing

$$\frac{\partial u}{\partial s}(s_n, t) \approx N(u(s_n, t) - u(s_{n-1}, t)),$$

we end up with a semidiscrete model,

$$\frac{dX(t)}{dt} = LX(t) + Z(t) + E(t), \tag{4.2}$$

where $L \in \mathbb{R}^{N \times N}$ is the finite difference matrix,

$$L = -\bar{c}N \begin{bmatrix} 1 & 0 & \ldots & 0 \\ -1 & 1 & & 0 \\ \vdots & & \ddots & \vdots \\ 0 & \ldots & -1 & 1 \end{bmatrix},$$

the vector $Z(t) \in \mathbb{R}^N$ represents the unknown value of u at the left end of the interval,
$$Z(t) = \bar{c}N \left[u(s_0, t), 0, \ldots, 0\right]^\mathrm{T},$$
and $E(t) \in \mathbb{R}^N$ corresponds to the modelling error $e(s,t)$,
$$E(t) = [e(s_1, t), \ldots, e(s_N, t)]^\mathrm{T}.$$

By assuming that the time step between consecutive observation instants is $\tau > 0$, we discretize equation (4.2) by the implicit Euler scheme, yielding the evolution equation
$$X_{k+1} = FX_k + W_{k+1},$$
where
$$F = (I - \tau L)^{-1} \in \mathbb{R}^{N \times N},$$
and the vector $W_{k+1} \in \mathbb{R}^N$ is
$$W_{k+1} = F\big(Z(t_{k+1}) + E(t_{k+1})\big).$$

The numerical discussion of this example is postponed until we have developed the statistical machinery a bit further. ◇

4.1.2 Evolution and Observation Models

Motivated by the previous example, we shall now define the Bayesian filtering problem for discrete time stochastic processes.

Let $\{X_k\}_{k=0}^\infty$ and $\{Y_k\}_{k=1}^\infty$ be two stochastic processes. The random vector $X_k \in \mathbb{R}^{n_k}$ represents the quantities that we are primarily interested in, and it is called the *state vector*. The vector $Y_k \in \mathbb{R}^{m_k}$ represents the measurement. We refer to it as the *observation* at the kth time instant. For the sake of definiteness, we assume that the probability distributions are absolutely continuous with respect to the Lebesgue measure so that we can talk about probability densities rather than distributions. We postulate the following three properties for these processes:

1. The process $\{X_k\}_{k=0}^\infty$ is a Markov process, that is,
$$\pi(x_{k+1} \mid x_0, x_1, \ldots, x_k) = \pi(x_{k+1} \mid x_k).$$

2. The process $\{Y_k\}_{k=1}^\infty$ is a Markov process with respect to the history of $\{X_k\}$, that is,
$$\pi(y_k \mid x_0, x_1, \ldots, x_k) = \pi(y_k \mid x_k).$$

3. The process $\{X_k\}_{k=0}^\infty$ depends of the past observations only through its own history, that is,
$$\pi(x_{k+1} \mid x_k, y_1, y_2, \ldots, y_k) = \pi(x_{k+1} \mid x_k).$$

These postulates can be illustrated by the following dependency scheme (or neighborhood system):

$$\begin{array}{ccccccc} X_0 & \to & X_1 & \to & X_2 & \to \cdots \to & X_n & \to \cdots \\ & & \downarrow & & \downarrow & & \downarrow & \\ & & Y_1 & & Y_2 & & Y_n & \end{array}$$

If the stochastic processes $\{X_n\}_{n=0}^{\infty}$ and $\{Y_n\}_{n=1}^{\infty}$ satisfy conditions 1–3 above, we call this pair an *evolution–observation model*. Evidently, for the evolution–observation model to be completely specified, we need to specify the following:

1. The probability density of the initial state X_0.
2. The Markov transition kernels $\pi(x_{k+1} \mid x_k)$, $k = 0, 1, 2, \ldots$.
3. The likelihood functions $\pi(y_k \mid x_k)$, $k = 1, 2, \ldots$.

To avoid possible confusion due to our shorthand notation, let us emphasize here that the Markov chain $\{X_k\}_{k=0}^{\infty}$ needs not be time-homogenous, that is, the transition kernels $\pi(x_{k+1} \mid x_k)$ can vary in time. Similarly, it is important that the likelihood functions $\pi(y_k \mid x_k)$ are allowed change in time.

To better understand the assumptions above, consider the case that is often the starting point in practice: Assume that we have a Markov model describing the evolution of the states X_k and an observation model for vectors Y_k depending on the current state X_k,

$$X_{k+1} = F_{k+1}(X_k, W_{k+1}), \quad k = 0, 1, 2, \ldots, \tag{4.3}$$

$$Y_k = G_k(X_k, V_k). \quad k = 1, 2, \ldots. \tag{4.4}$$

Here the functions F_{k+1} and G_k are assumed to be known functions, and the random vectors $W_{k+1} \in \mathbb{R}^{p_{k+1}}$ and $V_k \in \mathbb{R}^{q_k}$ are called the *state noise* and *observation noise*, respectively. The equation (4.3) is called the *state evolution equation* while (4.4) is called the *observation equation*.

In order that the processes $\{X_k\}_{k=0}^{\infty}$ and $\{Y_k\}_{k=0}^{\infty}$ are an evolution–observation model, we make the following assumptions concerning the state noise and observation noise processes:

1. For $k \neq \ell$, the noise vectors W_k and W_ℓ as well as V_k and V_ℓ are mutually independent and also mutually independent of the initial state X_0.
2. The noise vectors W_k and V_ℓ are mutually independent for all k, ℓ.

We emphasize that having a system of the form (4.3)–(4.4) is not necessary for the Bayesian approach.

The inverse problem considered in this chapter is to extract information of the state vectors X_k based on the measurements Y_k. Similarly to the discussion in the previous chapter, in the Bayesian approach we do not *only* try to find a single estimate of the state but rather get the posterior distribution of the state vector conditioned on the observations. To this end, let us denote

$$D_k = \{y_1, y_2, \ldots, y_k\},$$

The conditional probability of the state vector x_k conditioned on all the measurements $y_1, \ldots, y_n$ is denoted as

$$\pi(x_k \mid y_1, \ldots, y_n) = \pi(x_k \mid D_n).$$

It is understood that $\pi(x_k \mid D_0) = \pi(x_k)$.

Several different problems can be considered here. To set the terminology straight, we give the following classification that is in agreement with the classical literature.

Definition 4.1. *Assume that the stochastic processes $\{X_k\}_{k=0}^{\infty}$ and $\{Y_k\}_{k=1}^{\infty}$ form an evolution–observation model. The problem of determining the conditional probability*

1. $\pi(x_{k+1} \mid D_k)$ *is called a* prediction problem;
2. $\pi(x_k \mid D_k)$ *is called a* filtering problem;
3. $\pi(x_k \mid D_{k+p})$, $p \geq 1$ *is called a* p-lag (fixed-lag) smoothing problem;
4. *when the complete measurement sequence $D_K = \{y_1, \ldots, y_K\}$ is finite, $\pi(x_k \mid D_K)$, $1 \leq k \leq K$ is called a* (fixed interval) smoothing problem.

Clearly, the problems listed above serve different purposes. In problems in which real-time information of the underlying state x_k is vital, the filtering approach is important. This is the case, for example, in control problems in which one has to act based on the current state. On the other hand, if one has a possibility to perform a longer series of observations before estimating the state x_k, the smoothing approaches might be preferred since they yield smaller estimation errors than filtering. The prediction problem may be interesting per se, for example, when future decisions need to be made based on current knowledge as in financial problems. Often, however, the prediction problem is just an intermediate step for the filtering step. It is also possible to predict over longer horizons than a single time step.

We begin our the discussion by considering the Bayesian filtering problems and by deriving the basic updating formulas for conditional probability distributions. More precisely, it is our goal to derive formulas that allow us to construct the following recursive updating scheme:

$$\pi(x_0)$$
$$\downarrow$$
$$\pi(x_1|x_0) \longrightarrow \text{evolution updating}$$
$$\downarrow$$
$$\pi(x_1)$$
$$\downarrow$$
$$\text{observation updating} \longleftarrow \pi(y_1|x_1)$$
$$\downarrow$$
$$\pi(x_1 \mid D_1)$$
$$\downarrow$$
$$\pi(x_2|x_1) \longrightarrow \text{evolution updating}$$
$$\downarrow$$
$$\pi(x_2 \mid D_1)$$
$$\downarrow$$
$$\text{observation updating} \longleftarrow \pi(y_2 \mid x_2)$$
$$\downarrow$$
$$\pi(x_2 \mid D_2)$$
$$\downarrow$$
$$\vdots$$

Other common terms for the evolution and observation updates especially in the engineering literature are time and measurements updates, respectively. In this type of recursive scheme, the state evolution equation is used for solving the prediction problem from the filtering problem of the previous time level, while the new observations are used to update the predicted probability distribution. Therefore, we need to find formulas for the following updating steps:

1. Time evolution updating: Given $\pi(x_k \mid D_k)$, find $\pi(x_{k+1} \mid D_k)$ based on the Markov transition kernel $\pi(x_{k+1} \mid x_k)$.
2. Observation updating: Given $\pi(x_{k+1} \mid D_k)$, find $\pi(x_{k+1} \mid D_{k+1})$ based on the new observation y_{k+1} and the likelihood function $\pi(y_{k+1} \mid x_{k+1})$.

The updating equations are given in the following theorem.

Theorem 4.2. *Assume that the pair $\{X_k\}_{k=0}^{\infty}$, $\{Y_k\}_{k=0}^{\infty}$ of stochastic processes is an evolution–observation model. Then the following updating formulas apply:*

1. *Time evolution updating: We have*

$$\pi(x_{k+1} \mid D_k) = \int \pi(x_{k+1} \mid x_k)\pi(x_k \mid D_k)dx_k. \quad (4.5)$$

2. *Observation updating: We have*

$$\pi(x_{k+1} \mid D_{k+1}) = \frac{\pi(y_{k+1} \mid x_{k+1})\pi(x_{k+1} \mid D_k)}{\pi(y_{k+1} \mid D_k)}, \quad (4.6)$$

4 Nonstationary Inverse Problems

where

$$\pi(y_{k+1} \mid D_k) = \int \pi(y_{k+1} \mid x_{k+1})\pi(x_{k+1} \mid D_k)dx_{k+1}.$$

Proof: To prove the identity (4.5), consider the probability density

$$\pi(x_{k+1}, x_k, D_k) = \pi(x_{k+1} \mid x_k, D_k)\pi(x_k, D_k)$$
$$= \pi(x_{k+1} \mid x_k)\pi(x_k \mid D_k)\pi(D_k),$$

where we used the fact that based on the evolution equation, if x_k is known, x_{k+1} is conditionally independent of the previous observations. Further,

$$\pi(x_{k+1}, D_k) = \int \pi(x_{k+1}, x_k, D_k)dx_k,$$

so by substituting the previous identity we find that

$$\pi(x_{k+1} \mid D_k) = \frac{\pi(x_{k+1}, D_k)}{\pi(D_k)} = \int \pi(x_{k+1} \mid x_k)\pi(x_k \mid D_k)dx_k.$$

This proves (4.5).

To derive the observation updating formula (4.6), we write first

$$\pi(x_{k+1} \mid D_{k+1}) = \frac{\pi(x_{k+1}, D_{k+1})}{\pi(D_{k+1})}. \tag{4.7}$$

Consider the joint probability distribution $\pi(x_{k+1}, D_{k+1})$. Then

$$\pi(x_{k+1}, D_{k+1}) = \pi(y_{k+1}, x_{k+1}, D_k) = \pi(y_{k+1} \mid x_{k+1}, D_k)\pi(x_{k+1}, D_k).$$

From the equation (4.4) and the mutual independence of V_k, we deduce that y_{k+1} is conditionally independent of the previous observation if x_{k+1} is known, that is,

$$\pi(y_{k+1} \mid x_{k+1}, D_k) = \pi(y_{k+1} \mid x_{k+1}).$$

Substituting these expressions in formula (4.7) we have

$$\pi(x_{k+1} \mid D_{k+1}) = \frac{\pi(y_{k+1} \mid x_{k+1})\pi(x_{k+1}, D_k)}{\pi(D_{k+1})}.$$

Furthermore, since

$$\pi(x_{k+1}, D_k) = \pi(x_{k+1} \mid D_k)\pi(D_k),$$

and

$$\pi(D_{k+1}) = \pi(D_k, y_{k+1}) = \pi(y_{k+1} \mid D_k)\pi(D_k),$$

it follows now that

$$\pi(x_{k+1} \mid D_{k+1}) = \frac{\pi(x_{k+1}, D_{k+1})}{\pi(D_{k+1})} = \frac{\pi(y_{k+1} \mid x_{k+1})\pi(x_{k+1} \mid D_k)}{\pi(y_{k+1} \mid D_k)},$$

proving equation (4.6). □

Before discussing the computational issues, it is useful to give an interpretation for the derived equations. Consider first equation (4.5). The integrand in this formula is simply the joint probability distribution of the variables x_k and x_{k+1} conditioned on the observations D_k. Hence, formula (4.5) is simply the marginal probability distribution of x_{k+1}, that is, the probability distribution of x_{k+1} regardless of what the value of x_k was.

Similarly, the equation (4.6) is easy to interpret in terms of the statistical inversion. Assume that we have computed the distribution $\pi(x_{k+1} \mid D_k)$, and we consider this distribution as the prior distribution for x_{k+1} when the new observation y_{k+1} arrives. Then equation (4.6) is nothing other than the Bayes formula.

With these interpretations in mind, we consider next the simplest of all models, the Gaussian linear case.

4.2 Kalman Filters

4.2.1 Linear Gaussian Problems

Theorem 4.2 of the previous section leads to the well known formulas of the Kalman predictor and Kalman filter, if the following special conditions hold: First, the state equations are linear with additive noise processes, that is,

$$X_{k+1} = F_{k+1} X_k + W_{k+1}, \quad k = 0, 1, 2, \ldots, \tag{4.8}$$

$$Y_k = G_k X_k + V_k \quad k = 1, 2, \ldots. \tag{4.9}$$

Here, we assume that F_{k+1} and G_k are known matrices. Further, the noise vectors W_{k+1} and V_k are Gaussian with known means and covariances. Without a loss of generality, we may assume that they are zero mean vectors. Third, the noise vectors are mutually independent, that is,

$$W_k \perp W_\ell, \quad V_k \perp V_\ell, \quad k \neq \ell,$$

and

$$W_k \perp V_\ell.$$

Finally, the probability distribution of X_0 is known and Gaussian. Again, without a loss of generality, we can assume that X_0 is zero mean.

Under these rather restrictive conditions, the following theorem holds.

Theorem 4.3. *Assume that the above assumptions are valid. Denote $x_{k|\ell} = \mathrm{E}\,(x_k \mid D_\ell)$ and $\Gamma_{k|\ell} = \mathrm{cov}\,(x_k \mid D_\ell)$ and define $\pi(x_0) = \pi(x_0 \mid D_0)$. Then the time evolution and observation updating formulas take the following forms:*

1. *Time evolution updating: Assume that we know the Gaussian distribution*

$$\pi(x_k \mid D_k) \sim \mathcal{N}(x_{k|k}, \Gamma_{k|k}).$$

Then,

$$\pi(x_{k+1} \mid D_k) \sim \mathcal{N}(x_{k+1|k}, \Gamma_{k+1|k}),$$

where

$$x_{k+1|k} = F_{k+1} x_{k|k}, \qquad (4.10)$$

$$\Gamma_{k+1|k} = F_{k+1} \Gamma_{k|k} F_{k+1}^{\mathrm{T}} + \Gamma_{w_{k+1}}, \qquad (4.11)$$

2. *Observation updating: Assume that we know the Gaussian distribution*

$$\pi(x_{k+1} \mid D_k) \sim \mathcal{N}(x_{k+1|k}, \Gamma_{k+1|k}).$$

Then,

$$\pi(x_{k+1} \mid D_{k+1}) \sim \mathcal{N}(x_{k+1|k+1}, \Gamma_{k+1|k+1}),$$

where

$$x_{k+1|k+1} = x_{k+1|k} + K_{k+1}(y_{k+1} - G_{k+1} x_{k+1|k}), \qquad (4.12)$$

$$\Gamma_{k+1|k+1} = (1 - K_{k+1} G_{k+1}) \Gamma_{k+1|k}, \qquad (4.13)$$

and the matrix K_{k+1}, known as the Kalman gain matrix, is given by the formula

$$K_{k+1} = \Gamma_{k+1|k} G_{k+1}^{\mathrm{T}} (G_{k+1} \Gamma_{k+1|k} G_{k+1}^{\mathrm{T}} + \Gamma_{v_{k+1}})^{-1},$$

Proof: As explained at the end of Section 4.1, the product $\pi(x_{k+1} \mid x_k)\pi(x_k \mid D_k)$ is the joint probability distribution of x_k and x_{k+1}. The claimed formula for the evolution update follows now from the Theorem 3.6 in Chapter 3. The claim concerning the observation updating is a direct consequence of Theorem 3.7 in Chapter 3 Indeed, by interpreting $\pi(x_{k+1} \mid D_k) = \pi_{\mathrm{pr}}(x_{k+1})$ and $\pi(y_{k+1} \mid x_{k+1}) = \pi_{\mathrm{noise}}(y_{k+1} - G_{k+1} x_{k+1})$ where π_{noise} is the probability distribution of the observation noise v_{k+1}, the result follows by direct substitution. □

Usually the initial state is not very well known. This is handled typically by setting the initial covariance large. This uncertainty is reflected in the variances of the first estimates, but this transition effect usually soon wears out.

As an application of the Kalman filters, we return to the introductory Example 1 of the previous subsection.

Example 1 (continued): Having developed the necessary tools for discussing linear Gaussian nonstationary inverse problems, we continue the Example 1 described in the beginning of this chapter.

Assume that the waveform ψ consists of two Gaussian humps,

$$\psi(s) = a_1 \exp\left(-\frac{1}{s_1^2}(s-c_1)^2\right) + a_2 \exp\left(-\frac{1}{s_2^2}(s-c_2)^2\right),$$

where $a_1 = 1.5$, $a_2 = 1.0$, $s_1 = 0.08$, $s_2 = 0.04$ and $c_1 = 0.1$, $c_2 = 0.25$. The propagation velocity is $c = 0.04$, and the observations are done at unit time intervals. We assume that there are 11 evenly distributed observation points, and the wave is followed for 20 time units. The additive observation noise is Gaussian white noise with standard deviation of 5% of the maximum noiseless signal. Figure 4.1 shows the true wave during the observation sequence.

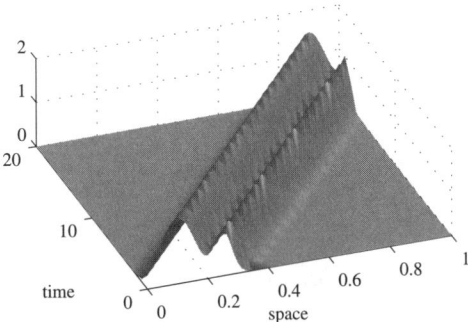

Figure 4.1. The true wave $u(s - ct)$.

We estimate the wave based on the dynamic observation using the random walk model and the wave propagation model. In the random walk model, the state noise is $W_{k+1} \sim \mathcal{N}(0, \gamma^2 I)$, where $\gamma = 0.1$.

In the wave propagation model, we consider two cases. First, we assume that the propagation speed is known and second, that it is known only approximately so that the employed propagation speed is $\bar{c} = 1.5c$, that is, we grossly overestimate the speed. Note also that we estimate $u(s,t)$ rather than $\psi(s)$ since we are not necessarily sure that the overall linear wave propagation model is otherwise completely correct.

The state noise covariance in this case is

$$\mathrm{E}\{W_{k+1} W_{k+1}^\mathrm{T}\} = F\left(\mathrm{E}\{Z(t_{k+1}) Z(t_{k+1})^\mathrm{T}\} + \mathrm{E}\{E(t_{k+1}) E(t_{k+1})^\mathrm{T}\}\right) F^\mathrm{T}$$
$$= F(\alpha^2 e_1 e_1^\mathrm{T} + \gamma^2 I) F^\mathrm{T},$$

where $e_1 = [1, 0, \ldots, 0]^\mathrm{T}$, α^2 is the variance of the boundary value at the left end of the interval and γ^2 is the variance of the modelling error. We use tha value $\alpha = 0.1$ in both cases. For the correct wave speed, the modelling error is assumed to be smaller than with incorrect wave speed. In the calculations, we use $\gamma = 10^{-4}$ for the correct wave speed and $\gamma = 0.1$ for the incorrect one. It turns out that the method is quite insensitive to these parameters.

Figure 4.2 shows the results of the Kalman filter-based estimates. Not surprisingly, the wave propagation model gives a smoother estimate, since it

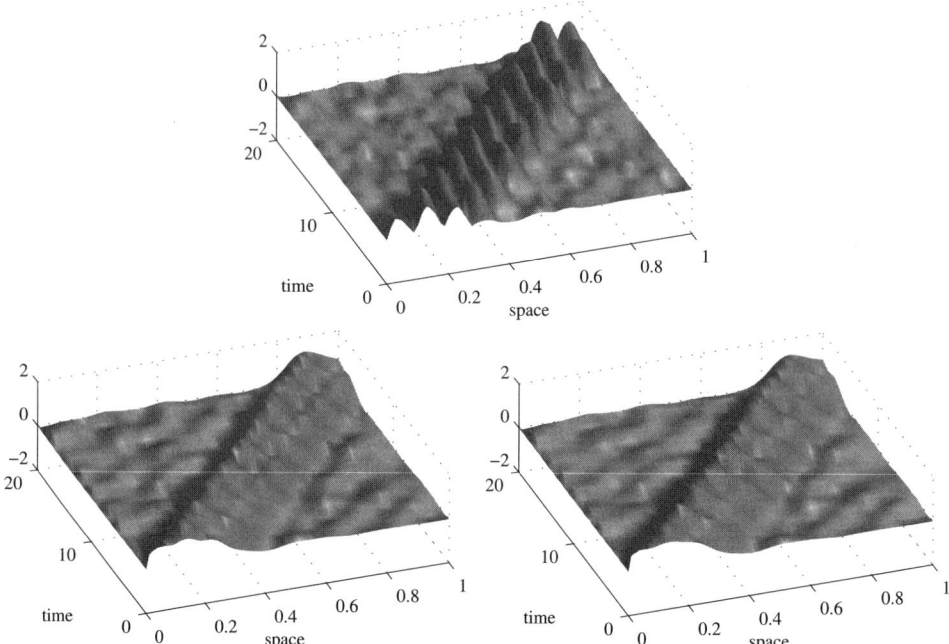

Figure 4.2. Kalman filtered estimates. The top one is based on the random walk model, the right one on wave propagation model with correct speed and the left one with overestimated speed.

has a built-in smoothing prior. Notice also that the method is quite robust with respect to propagation speed estimate. The estimates with correct and incorrect wave speed estimates are not equal, but the difference is too small to be visible. In practice, this is an important feature in many applications, see especially the the similar but much more complex example in Chapter 7.
◊

In the following sections, we consider Bayes filtering when the linearity and/or normality assumptions are not satisfied.

4.2.2 Extended Kalman Filters

The effectiveness of Kalman filtering is based on the fact that Gaussian densities remain Gaussian in linear transformations. Hence, the density updating is achieved by updating only the mean and the covariance. For nonlinear models, this is no longer true. However, one can try to approximate the densities by Gaussian densities and follow the evolution of these approximations. The success depends strongly on the properties of the nonlinear evolution models. Such approximation is called *extended Kalman filtering* (EKF).

Consider an evolution–observation model

$$X_{k+1} = F_{k+1}(X_k) + W_{k+1}, \quad k = 0, 1, 2, \ldots, \tag{4.14}$$

$$Y_k = G_k(X_k) + V_k, \quad k = 1, 2, \ldots, \tag{4.15}$$

where the mappings F_{k+1} and G_k are assumed to be differentiable. The state noise W_k and observation noise V_k are as in the previous section, that is, Gaussian, zero mean and mutually independent. We assume that the initial distribution $\pi(X_0)$ is known. As usual, we interpret $\pi(X_0 \mid D_0) = \pi(X_0)$ for the empty observation D_0.

In the following discussion, we use the following notations. We denote by $\pi_G(x_j \mid D_k)$ a Gaussian approximation of the conditional density $\pi(x_j \mid D_k)$, $j = k, k+1$. The quality of the approximation is not assessed in any way here. Instead, we give the updating steps for computing the Gaussian approximations based on the evolution–observation model (4.14)–(4.15).

Assume that $\pi_G(x_k \mid D_k)$ is known,

$$\pi_G(x_k \mid D_k) \sim \mathcal{N}(x_{k|k}, \Gamma_{k|k}).$$

The first step is to approximate $\pi(x_{k+1} \mid D_k)$. We write the time evolution updating formula (4.5) as

$$\pi(x_{k+1} \mid D_k) = \int \pi(x_{k+1} \mid x_k) \pi(x_k \mid D_k) dx_k$$

$$\approx \int \pi(x_{k+1} \mid x_k) \pi_G(x_k \mid D_k) dx_k.$$

To obtain the Gaussian approximation for the transition kernel $\pi(x_{k+1} \mid x_k)$, we use the linearized approximation of F_{k+1} around the center $x_{k|k}$,

$$X_{k+1} = F_{k+1}(X_k) + W_{k+1}$$
$$\approx F_{k+1}(x_{k|k}) + \mathcal{D}F_{k+1}(x_{k|k})(X_k - x_{k|k}) + W_{k+1},$$

where $\mathcal{D}F$ denotes the Jacobian of F. The propagation step of the standard Kalman filter gives immediately the updated Gaussian approximation. We write

$$\pi_G(x_{k+1} \mid D_k) \sim \mathcal{N}(x_{k+1|k}, \Gamma_{k+1|k}),$$

where

$$x_{k+1|k} = F_{k+1}(x_{k|k}),$$
$$\Gamma_{k+1|k} = (\mathcal{D}F_{k+1}) \Gamma_{k|k} (\mathcal{D}F_{k+1})^{\mathrm{T}} + \Gamma_{w_k+1},$$

where we denoted for brevity $\mathcal{D}F_{k+1} = \mathcal{D}F_{k+1}(x_{k|k})$.

To derive the observation updating, we use the Bayes formula (4.6) and write

$$\pi(x_{k+1} \mid D_{k+1}) \propto \pi(y_{k+1} \mid x_{k+1})\pi(x_{k+1} \mid D_k) \tag{4.16}$$

$$\approx \pi(y_{k+1} \mid x_{k+1})\pi_G(x_{k+1} \mid D_k)$$

$$\propto \exp\Big(-\frac{1}{2}(x_{k+1} - x_{k+1|k})^T \Gamma_{k+1|k}^{-1}(x_{k+1} - x_{k+1|k})$$

$$-\frac{1}{2}\big(y_{k+1} - G_{k+1}(x_{k+1})\big)^T \Gamma_{v_{k+1}}^{-1}\big(y_{k+1} - G_{k+1}(x_{k+1})\big)\Big).$$

To find a Gaussian approximation for this density, the most straightforward method is to approximate G_{k+1} by its linearization about $x_{k+1|k}$, that is,

$$G_{k+1}(x_{k+1}) \approx G_{k+1}(x_{k+1|k}) + \mathcal{D}G_{k+1}(x_{k+1|k})(x_{k+1} - x_{k+1|k}).$$

This approximation yields immediately a Gaussian approximation,

$$\pi_G(x_{k+1} \mid D_{k+1}) \sim \mathcal{N}(x_{k+1|k+1}, \Gamma_{k+1|k+1}),$$

where

$$x_{k+1|k+1} = x_{k+1|k} + K_{k+1}\big(y_{k+1} - G_{k+1}(x_{k+1|k})\big), \tag{4.17}$$

$$\Gamma_{k+1|k+1} = \big(1 - K_{k+1}\mathcal{D}G_{k+1}\big)\Gamma_{k+1|k}, \tag{4.18}$$

and the gain matrix is

$$K_{k+1} = \Gamma_{k+1|k}\mathcal{D}G_{k+1}^T\big(\mathcal{D}G_{k+1}\Gamma_{k+1|k}\mathcal{D}G_{k+1}^T + \Gamma_{v_{k+1}}\big)^{-1}.$$

Above, we have denoted $\mathcal{D}G_{k+1} = \mathcal{D}G_{k+1}(x_{k+1|k+1})$.

This is the simplest and most straightforward version of EKF. When the computation time is not a crucial issue, one can refine the observation updating step by including an inner iteration loop as follows.

Instead of accepting $x_{k+1|k}$ as the linearization point in (4.16), one can first seek iteratively a minimizer of the exponent, that is,

$$x_* = \arg\min\big(f(x)\big),$$

where

$$f(x) = (x - x_{k+1|k})^T \Gamma_{k+1|k}^{-1}(x - x_{k+1|k})$$

$$+ \big(y_{k+1} - G_{k+1}(x)\big)^T \Gamma_{v_{k+1}}^{-1}\big(y_{k+1} - G_{k+1}(x)\big).$$

The minimization can be done with desired accuracy.

An iterative Gauss–Newton minimization starts with the initial value $x^0 = x_{k+1|k}$, and the updating step $x^j \to x^{j+1}$ is the following. We linearize G_{k+1} about the current point x^j,

$$G_{k+1}(x) \approx G_{k+1}(x^j) + \mathcal{D}G_{k+1}(x^j)(x - x^j).$$

By defining

$$\delta x = x - x_{k+1|k}, \quad y^j = y_{k+1} - G_{k+1}(x^j), \quad \mathcal{D}G^j_{k+1} = \mathcal{D}G_{k+1}(x^j),$$

the approximation of f becomes

$$f(x) \approx \delta x^T \Gamma_{k+1|k} \delta x$$
$$+ \left(y^j - \mathcal{D}G^j_{k+1}(x_{k+1|k} - x^j - \delta x)\right)^T \Gamma^{-1}_{v_{k+1}} \left(y^j - \mathcal{D}G^j_{k+1}(x_{k+1|k} - x^j - \delta x)\right).$$

The minimizer of this quadratic expression is

$$\delta x = \left(\Gamma^{-1}_{k+1|k} + (\mathcal{D}G^j_{k+1})^T \Gamma^{-1}_{v_{k+1}} \mathcal{D}G^j_{k+1}\right)^{-1} \left(y^j - \mathcal{D}G_{k+1}(x_{k+1|k} - x^j)\right).$$

The updated value is defined as $x^{j+1} = x_{k+1|k} + \delta x$. This can be expressed in terms of a Kalman gain matrix. From the matrix inversion lemma, or the equivalence of formulas (3.15) and (3.16), it follows that

$$x^{j+1} = x_{k+1|k} + \left(1 - K^j_{k+1} \mathcal{D}G^j_{k+1}\right) \Gamma_{k+1|k} \left(y^j - \mathcal{D}G^j_{k+1}(x_{k+1|k} - x^j)\right),$$

where

$$K^j_{k+1} = \Gamma_{k+1|k} \mathcal{D}G^j_{k+1} \left(\mathcal{D}G^j_{k+1} \Gamma_{k+1|k} (\mathcal{D}G^j_{k+1})^T + \Gamma_{v_{k+1}}\right)^{-1}.$$

The iteration is repeated until convergence. Then one sets $x_{k+1|k+1} = x^j \approx x_*$. The covariance matrix $\Gamma_{k+1|k+1}$ is then obtained by formula (4.18), using x_* as a linearization point.

Finally, we remark that if the nonlinearities are weak, we may possibly consider the globally linearized problem in which the approximations $\mathcal{D}F(x_k) \approx \mathcal{D}F(\bar{x})$ and $\mathcal{D}G(x_k) \approx \mathcal{D}G(\bar{x})$ at some point $\bar{x}$ may yield practically feasible approximations.

4.3 Particle Filters

In applications, the observation and evolution models may be cumbersome or impossible to linearize. This is the case, for example, when the models are non-differentiable or not given in a closed form. As in the case of stationary inverse problems discussed in the previous chapter, one may try to use Monte Carlo methods to simulate the distributions by random samples. Such methods are known as *particle filters*. In this section, we discuss some of these simulation methods.

In principle, the goal in particle filter methods is to produce sequentially an ensemble of random samples $\{x^1_k, x^2_k, \ldots, x^N_k\}$ distributed according to the conditional probability distributions $\pi(x_{k+1} \mid D_k)$ (prediction) or $\pi(x_k \mid D_k)$ (filtering). The vectors x^j_k are called *particles* of the sample, thus the name particle filter. To this end, we have to consider two simulation steps: Having a sample simulating the probability distribution $\pi(x_k \mid D_k)$, we must be able to produce a new sample simulating $\pi(x_{k+1} \mid D_k)$, and, further, a sample simulating $\pi(x_{k+1} \mid D_{k+1})$. This is done by means of the updating formulas derived in Theorem 4.2.

The following rather straightforward particle filter method is known as the *sampling importance resampling* algorithm, or briefly SIR. We give the algorithm in a concise form below.

1. Draw a random sample $\{x_0^n\}_{n=1}^N$ from the initial distribution $\pi(x_0) = \pi(x_0 \mid D_0)$ of the random variable X_0 and set $k = 0$.
2. Prediction step: For $k \geq 0$ given, let $\{x_k^n\}_{n=1}^N$ be a sample distributed according to $\pi(x_k \mid D_k)$. Approximate the integral (4.5) as

$$\pi(x_{k+1} \mid D_k) = \int \pi(x_{k+1} \mid x_k)\pi(x_k \mid D_k)dx_k \approx \frac{1}{N}\sum_{n=1}^{N}\pi(x_{k+1} \mid x_k^n).$$

3. Sample from predicted density: Draw *one* new particle $\widetilde{x}_{k+1}^n$ from $\pi(x_{k+1} \mid x_k^n)$, $1 \leq n \leq N$.
4. Calculate relative likelihoods

$$w_{k+1}^n = \frac{1}{W}\pi(y_{k+1} \mid \widetilde{x}_{k+1}^n), \quad W = \sum_{n=1}^{N}\pi(y_{k+1} \mid \widetilde{x}_{k+1}^n).$$

5. Resample: Draw x_{k+1}^n, $1 \leq n \leq N$ from the set $\{\widetilde{x}_{k+1}^n\}$, where the probability of drawing the particle $\widetilde{x}_{k+1}^n$ is w_{k+1}^n. Increase $k \longrightarrow k+1$ and repeat from 2.

Before presenting applications and examples, a few comments are in order. The prediction step 2 is a straightforward Monte Carlo integration approximation that should be familiar already from the context of MCMC methods of the previous section. Indeed, it is but a realization of the approximation

$$\int f(x_k)\pi(x_k \mid D_k)dx_k \approx \frac{1}{N}\sum_{n=1}^{N}f(x_k^n),$$

with $f(x_k) = \pi(x_{k+1} \mid x_k)$.

The steps 3–5 can be understood as *importance sampling*. We use the density $\pi(x_{k+1} \mid D_k)$ as a proposal density and assign later a relative probability weight to each proposed particle by using the likelihood. Note that we draw one new particle from each one of the individual densities $x_{k+1} \mapsto \pi(x_{k+1} \mid x_k^n)$, $1 \leq n \leq N$. This updating strategy is called *layered sampling*. From the point of view of the algorithm, this choice is not essential. We might as well draw each new particle from the approximated density given in step 2. However, there is evidence that it is often advisable to use layered sampling to get better coverage of the density. Moreover, as we shall see in the forthcoming examples, the layered sampling is often very easy to accomplish.

To give a more concrete feel for the SIR algorithm, let us consider one updating round in a simple example.

Example 2: We consider a simple one-dimensional model. Assume that for some k, $\pi(x_k \mid D_k)$ is a Rayleigh distribution,

$$\pi(x_k \mid D_k) = x_k \exp\left(-\frac{1}{2}x_k^2\right), \quad x_k \geq 0.$$

Let $\{x_k^1, \ldots, x_k^N\}$ be a random sample drawn from this distribution.

Consider a simple random walk model,

$$X_{k+1} = X_k + W_{k+1}, \quad W_{k+1} \sim \mathcal{N}(0, \gamma^2),$$

corresponding to the transition density

$$\pi(x_{k+1} \mid x_k) = \frac{1}{\sqrt{2\pi\gamma^2}} \exp\left(-\frac{1}{2\gamma^2}(x_k - x_{k+1})^2\right).$$

Now we perform the layered sampling: For each n, draw w^n from the density $\mathcal{N}(0, \gamma^2)$ and set

$$\tilde{x}_{k+1}^n = x_k^n + w^n.$$

Thus we have produced the prediction cloud of particles. Observe that due to our assumptions, the particles should be distributed according to the density

$$\pi(x_{k+1} \mid D_k) = \int \pi(x_k \mid D_k) \pi(x_{k+1} \mid x_k) dx_k$$

$$= \frac{1}{\sqrt{2\pi\gamma^2}} \int_0^\infty x_k \exp\left(-\frac{1}{2}x_k^2 - \frac{1}{2\gamma^2}(x_k - x_{k+1})^2\right) dx_k.$$

Assume now that the observation model is again the simplest imaginable,

$$Y_{k+1} = X_{k+1} + V_{k+1}, \quad V_{k+1} \sim \mathcal{N}(0, \sigma^2),$$

that is, the likelihood density is

$$\pi(y_{k+1} \mid x_{k+1}) = \frac{1}{\sqrt{2\pi\sigma^2}} \exp\left(-\frac{1}{2\sigma^2}(y_{k+1} - x_{k+1})^2\right),$$

and the outcome of our measurement is $y_{k+1} = y_{\text{meas}}$. Then the relative likelihoods are simply

$$w_{k+1}^n = \frac{1}{W} \exp\left(-\frac{1}{2\sigma^2}(y_{\text{meas}} - \tilde{x}_{k+1}^n)^2\right),$$

where W is a norming constant. Finally, we do the resampling: For every n, $1 \leq n \leq N$, draw a random number $t \sim \mathcal{U}([0, 1])$ and set

132   4 Nonstationary Inverse Problems

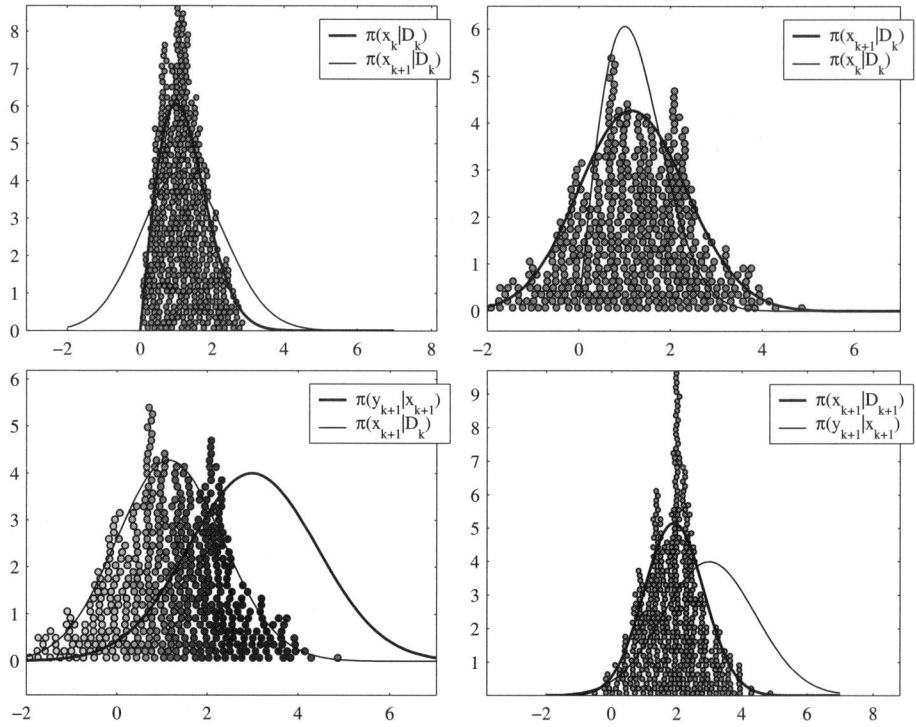

Figure 4.3. Schematic representation of the SIR filtering. Top left: Particles drawn from $\pi(x_k \mid D_k)$. Top right: Particles propagated to approximate the predicted distribution $\pi(x_{k+1} \mid D_k)$. Bottom left: Relative likelihoods coded by a gray scale. Bottom right: Resampled particles approximating $\pi(x_{k+1} \mid D_{k+1})$.

$$x_{k+1}^n = \tilde{x}_{k+1}^\ell, \quad \text{when} \quad \sum_{j=1}^{\ell-1} w_{k+1}^j < t \leq \sum_{j=1}^{\ell} w_{k+1}^j.$$

In Figure 4.3, the above steps have been illustrated graphically.   ◇

Having defined the particle filtering procedure, let us look at some of its properties.

Consider the problem of calculating an estimate for an integral of a given function $g(x_{k+1})$ with respect to the measure $\pi(x_{k+1} \mid D_{k+1})dx_{k+1}$, that is, the conditional expectation of $g(x_{k+1})$ conditioned on D_{k+1}. At the time slice $t = k + 1$, we have two samples at our disposal, namely the prediction sample $\{\tilde{x}_{k+1}^n\}$ and the resampled one, $\{x_{k+1}^n\}$. These two samples yield two different approximations, namely

$$\int g(x_{k+1})\pi(x_{k+1} \mid D_{k+1})dx_{k+1} \approx \sum_{n=1}^{N} w_{k+1}^n g(\tilde{x}_{k+1}^n) = \tilde{\theta}g, \qquad (4.19)$$

or alternatively,

$$\int g(x_{k+1})\pi(x_{k+1} \mid D_{k+1})dx_{k+1} \approx \frac{1}{N}\sum_{n=1}^{N} g(x_{k+1}^n) = \theta g. \quad (4.20)$$

The question arises, which one of these estimates is preferable. To settle this question, observe that both $\tilde{\theta}g$ and θg are realizations of a random variable, since the samples are generated via a random process. It is therefore advisable to select that variable which has a smaller variance. It can be shown that the resampling phase increases the variance, so the estimate (4.19) could produce more accurate estimates.

From the practical point of view, an important question is how large particle samples should be. The convergence of the estimation process can be assessed by estimating the variance of the estimate $\tilde{\theta}g$ in (4.19). Indeed, by repeating the filtering with the same data M times and computing the corresponding estimates (4.19), denoted here as $\tilde{\theta}_m g$, $1 \le m \le M$, we may estimate

$$\mathrm{var}(\tilde{\theta}g) \approx \frac{1}{M}\sum_{m=1}^{M}(\tilde{\theta}_m g)^2 - (\overline{\theta}g)^2, \quad \overline{\theta}g = \frac{1}{M}\sum_{m=1}^{m}\tilde{\theta}_m g. \quad (4.21)$$

This quantity gives often a good understanding of the adequacy of the number of particles.

The second comment concerns the so-called *thinning*, or *impoverishment* of the sample: It may happen that the relative likelihoods w_{k+1}^n of the prediction particles $\tilde{x}_{k+1}^n$ are very unevenly distributed, only a few of them having a relative likelihood of significant size. As a consequence, after resampling, the sample consists of copies of very few prediction particles. Such a situation may occur if the likelihood $\pi(y_{k+1} \mid x_{k+1})$ is very narrow or if the true time evolution of the system differs significantly from the evolution model used in the filtering so that the prediction sample goes badly astray. Remedies to this phenomenon have been suggested in the literature.

These problems are present also in the stationary Markov chain Monte Carlo algorithms. Indeed, a grossly infeasible prior makes reliable sampling always technically very difficult. In the dynamical case, the use of well-justified prior models (that is, reliable time evolution models) becomes even more important.

4.4 Spatial Priors

The time evolution model in the Bayes filtering can be seen as a form of prior information. Indeed, in the obsevation updating step, the evolution-based predicted density appears as a prior density for the subsequent measurement.

In several applications, the state vectors X_k represent discretized versions of spatially distributed parameters. We may have prior information concerning these parameters outside the context of the time evolution. Indeed, if we were to estimate X_k purely based on the observation Y_k, we might know that X_k

is a priori distributed according to a prior density $\pi_{\mathrm{pr}}(x_k)$. As an example, we may assume that the spatial prior is a smoothness prior. The question then becomes, how to incorporate such *spatial prior* density into the time evolution model.

The key to this is, naturally, Bayes' formula. Indeed, let us write

$$\pi(x_{k+1} \mid x_k) = \frac{\pi_{\mathrm{pr}}(x_{k+1})\pi(x_k \mid x_{k+1})}{\pi(x_k)}, \qquad (4.22)$$

where

$$\pi(x_k) = \int \pi_{\mathrm{pr}}(x_{k+1})\pi(x_k \mid x_{k+1})dx_{k+1}.$$

Here, the transition probability density $\pi(x_k \mid x_{k+1})$ can be seen as a *back transition density* from x_{k+1} to x_k.

Example 3: As an example of the above discussion, let us consider the random walk time evolution model augmented with a Gaussian smoothness prior. Let

$$\pi_{\mathrm{pr}}(x_{k+1}) \propto \exp\left(-\frac{1}{2}\|Lx_{k+1}\|^2\right)$$

be a smoothness prior, where L denotes a discrete approximation matrix of a differential operator. If no Markov model for the process $\{X_n\}_{n=0}^{\infty}$ is given, we could estimate X_k from the observation Y_k simply by using the above density as a prior. Let us assume, on the other hand, that in addition to the above prior information of X_{k+1}, we want to enforce a random walk time evolution,

$$X_{k+1} = X_k + W_{k+1}, \quad W_{k+1} \sim \mathcal{N}(0, \gamma^2 I).$$

From this model, we infer that

$$\pi(x_k \mid x_{k+1}) \propto \exp\left(-\frac{1}{2\gamma^2}\|x_k - x_{k+1}\|^2\right).$$

Hence, the forward transition kernel becomes

$$\pi(x_{k+1} \mid x_k) \propto \frac{1}{\pi(x_k)}\exp\left(-\frac{1}{2\gamma^2}\|x_k - x_{k+1}\|^2 - \frac{1}{2}\|Lx_{k+1}\|^2\right),$$

the denominator being the integral of the numerator. To understand how the random walk time evolution step should be modified, let us rewrite the above formula in a more suggestive form.

Rearranging the terms in the exponential, we have

$$\frac{1}{2\gamma^2}\|x_k - x_{k+1}\|^2 + \frac{1}{2}\|Lx_{k+1}\|^2 = \frac{1}{2\gamma^2}(x_{k+1} - G^{-1}x_k)^\mathrm{T} G(x_{k+1} - G^{-1}x_k)$$

$$+ \frac{1}{2\gamma^2}\left(\|x_k\|^2 - \|G^{-1/2}x_k\|^2\right),$$

4.4 Spatial Priors

where
$$G = I + \gamma^2 L^T L.$$

Hence,
$$\pi(x_{k+1} \mid x_k) \propto \exp\left(-\frac{1}{2\gamma^2}(x_{k+1} - G^{-1}x_k)^T G(x_{k+1} - G^{-1}x_k)\right).$$

This transition probability corresponds to a modified time evolution model, namely
$$X_{k+1} = G^{-1}X_k + U_{k+1}, \quad U_{k+1} \sim \mathcal{N}(0, \gamma^2 G^{-1}).$$

This formula is quite easy to interpret: The operator G^{-1} is a smoothing operator, thus removing nonsmoothness of X_k, while the noise U_{k+1} is drawn from a Gaussian smoothness density. ◇

Let us now see how the updating formulas (4.5) and (4.6) of Theorem 4.2 change. Substitution of the equation (4.22) into the evolution updating equation (4.5) yields

$$\pi(x_{k+1} \mid D_k) = \int \pi(x_{k+1} \mid x_k)\pi(x_k \mid D_k)dx_k \tag{4.23}$$

$$= \pi_{\text{pr}}(x_{k+1}) \int \frac{\pi(x_k \mid x_{k+1})}{\pi(x_k)} \pi(x_k \mid D_k)dx_k.$$

The observation updating (4.6) remains as it was before.

It turns out that for practical reasons, it is sometimes useful to regroup the various densities. Assume that we want to perform the SIR algorithm using the above updating scheme. Let $\{x_k^n\}_{n=1}^N$ denote a sample distributed according to the conditional density $\pi(x_k \mid D_k)$. Then, the layered approximation of the prediction density $\pi(x_{k+1} \mid D_k)$ becomes

$$\pi(x_{k+1} \mid D_k) \approx \frac{1}{N} \sum_{n=1}^N \pi_{\text{pr}}(x_{k+1}) \frac{\pi(x_k^n \mid x_{k+1})}{\pi(x_k^n)}.$$

Hence, the layered sampling of the prediction sample $\{\tilde{x}_{k+1}^n\}$ requires that we draw $\tilde{x}_{k+1}^n$ from the density $x_{k+1} \mapsto \pi_{\text{pr}}(x_{k+1})\pi(x_k^n \mid x_{k+1})/\pi(x_k^n)$. In particular in higher space dimensions this may be a cumbersome task if the prior density $\pi_{\text{pr}}(x_{k+1})$ is complicated. For this reason, we consider an alternative way of grouping the updating steps. Let us substitute (4.23) into the observation updating formula (4.6). We obtain

$$\pi(x_{k+1} \mid D_{k+1}) = \frac{\pi(y_{k+1} \mid x_{k+1})\pi(x_{k+1} \mid D_k)}{\pi(y_{k+1} \mid D_k)}$$

$$= \frac{\pi(y_{k+1} \mid x_{k+1})\pi_{\text{pr}}(x_{k+1}) \int \frac{\pi(x_k \mid x_{k+1})}{\pi(x_k)} \pi(x_k \mid D_k)dx_k}{\pi(y_{k+1} \mid D_k)}.$$

Let us now define a *weighted evolution updating* formula as

$$\widetilde{\pi}(x_{k+1} \mid D_k) = \int \frac{\pi(x_k \mid x_{k+1})}{\pi(x_k)} \pi(x_k \mid D_k) dx_k,$$

and the *augmented observation updating* as

$$\pi(x_{k+1} \mid D_{k+1}) = \frac{\pi(y_{k+1} \mid x_{k+1})\pi_{\mathrm{pr}}(x_{k+1})\widetilde{\pi}(x_{k+1} \mid D_k)}{\pi(y_{k+1} \mid D_k)}.$$

We collect these formulas for later reference into a theorem, which is a special version of the earlier and more fundamental Theorem 4.2

Theorem 4.4. *Assume that $\{X_k\}_{k=0}^{\infty}$ and $\{Y_k\}_{k=1}^{\infty}$ are processes that form an evolution–observation model. Further, let the transition probability $\pi(x_{k+1} \mid x_k)$ be given as*

$$\pi(x_{k+1} \mid x_k) = \frac{\pi_{\mathrm{pr}}(x_{k+1})\pi(x_k \mid x_{k+1})}{\pi(x_k)}.$$

Then either of the following updating formulas can be used to incorporate the spatial prior model:

1. *Weighted evolution updating: We have*

$$\widetilde{\pi}(x_{k+1} \mid D_k) = \int \frac{\pi(x_k \mid x_{k+1})}{\pi(x_k)} \pi(x_k \mid D_k) dx_k. \qquad (4.24)$$

2. *Augmented observation updating: We have*

$$\pi(x_{k+1} \mid D_{k+1}) = \frac{\pi(y_{k+1} \mid x_{k+1})\pi_{\mathrm{pr}}(x_{k+1})\widetilde{\pi}(x_{k+1} \mid D_k)}{\pi(y_{k+1} \mid D_k)}. \qquad (4.25)$$

We consider the above result from two different points of view. First, we give a modified particle filter algorithm including the spatial prior. Second, we use it to give an alternative interpretation of the spatial prior in the case of Gaussian linear case.

A modification of the sampling importance resampling algorithm can be obtained in a rather straightforward manner from the above theorem:

1. Draw a random sample $\{x_0^n\}_{n=1}^N$ from the initial distribution $\pi(x_0) = \pi(x_0 \mid D_0)$ of X_0 and set $k = 0$.
2. Prediction step: For $k \geq 0$ given, assume that $\{x_k^n\}_{n=1}^N$ is a sample distributed according to the density $\pi(x_k \mid D_k)$. Approximate (4.24) as

$$\widetilde{\pi}(x_{k+1} \mid D_k) \approx \frac{1}{N} \sum_{n=1}^{N} \frac{\pi(x_k^n \mid x_{k+1})}{\pi(x_k^n)}.$$

4.4 Spatial Priors

3. Layered sampling: for $1 \leq n \leq N$, draw $\widetilde{x}_{k+1}^n$ from the density $x_{k+1} \to \pi(x_k^n \mid x_{k+1})/\pi(x_k^n)$.
4. Calculate the relative augmented likelihoods

$$w_{k+1}^n = \frac{1}{W}\pi_{\mathrm{pr}}(\widetilde{x}_{k+1}^n)\pi(y_{k+1} \mid \widetilde{x}_{k+1}^n), \quad W = \sum_{n=1}^N \pi_{\mathrm{pr}}(x_{k+1}^n)\pi(y_{k+1} \mid x_{k+1}^n).$$

5. Resample by drawing $\{x_{k+1}^n\}_{n=1}^N$ from the discrete set $\{\widetilde{x}_{k+1}^n\}_{n=1}^N$, with $P(\widetilde{x}_{k+1}^n) = w_{k+1}^n$.

As explained before, the above form of the SIR algorithm may be convenient when the back transition probability density $\pi(x_k \mid x_{k+1})$ is simple (for example, random walk) while the spatial prior density is such that drawing from the density $\pi_{\mathrm{pr}}(x_{k+1})\pi(x_k \mid x_{k+1})$ would be cumbersome.

As a follow-up to Theorem 4.4, consider the special case in which the observation model has additive Gaussian noise and the spatial prior is, for simplicity, a Gaussian density. Hence, let the observation model at time t_{k+1} be of the form

$$Y_{k+1} = G(X_{k+1}) + V_{k+1}, \quad V_{k+1} \sim \mathcal{N}(0, S). \tag{4.26}$$

Further, let us assume that the spatial prior is of the form

$$\pi_{\mathrm{pr}}(x_{k+1}) \propto \exp\left(-\frac{\alpha^2}{2}\|Lx_{k+1}\|^2\right).$$

Consider the formula (4.25). We have

$$\pi(x_{k+1} \mid D_{k+1}) = \frac{\pi(y_{k+1} \mid x_{k+1})\pi_{\mathrm{pr}}(x_{k+1})\widetilde{\pi}(x_{k+1} \mid D_k)}{\pi(y_{k+1} \mid D_k)}$$

$$= \frac{\widetilde{\pi}(y_{k+1} \mid x_{k+1})\widetilde{\pi}(x_{k+1} \mid D_k)}{\pi(y_{k+1} \mid D_k)},$$

where the modified likelihood function is defined as

$$\widetilde{\pi}(y_{k+1} \mid x_{k+1}) = \pi(y_{k+1} \mid x_{k+1})\pi_{\mathrm{pr}}(x_{k+1})$$

$$\propto \exp\left(-\frac{1}{2}\|S^{-1/2}(G(x_{k+1}) - y_{k+1})\|^2 - \frac{\alpha^2}{2}\|Lx_{k+1}\|^2\right)$$

$$= \exp\left(-\frac{1}{2}\left\|S^{-1/2}\left(\begin{bmatrix}G(x_{k+1})\\ \alpha S^{1/2}Lx_{k+1}\end{bmatrix} - \begin{bmatrix}y_{k+1}\\ 0\end{bmatrix}\right)\right\|^2\right).$$

By comparing the resulting formulas to those appearing in the original formulation of Theorem 4.2, we may now give the following interpretation. Assume that instead of (4.26), the observation model is

$$\begin{bmatrix} Y_{k+1} \\ Z_{k+1} \end{bmatrix} = \begin{bmatrix} G(X_{k+1}) \\ \alpha S^{1/2} L X_{k+1} \end{bmatrix} + \begin{bmatrix} V_{k+1} \\ \tilde{V}_{k+1} \end{bmatrix},$$

where

$$V_{k+1},\ \tilde{V}_{k+1} \sim \mathcal{N}(0, S)$$

are mutually independent. Then, $\tilde{\pi}(y_{k+1} \mid x_{k+1})$ is the likelihood function of this observation with the realized data

$$\begin{bmatrix} Y_{k+1} \\ Z_{k+1} \end{bmatrix} = \begin{bmatrix} y_{k+1} \\ 0 \end{bmatrix}.$$

This interpretation appears rather natural: We may assume that we make a fictitious observation concerning the "nonsmoothness" LX_{k+1} of the state X_{k+1}, and, according to our prior information, we expect this random variable to be small.

4.5 Fixed-lag and Fixed-interval Smoothing

Kalman filtering is an example of *on-line* estimation schemes in which the computational load and memory requirements do not increase with time. In automatic control problems, the estimates of X_k are needed immediately after the observation Y_k; such applications require *real-time* estimation, in which the requirements for computational speed are strict.

In some situations the estimates can be computed completely off-line, that is, only after all observations have been obtained. For nonstationary inverse problems, the relevant off-line schemes are the fixed-lag smoother and the fixed-interval smoother, defined in Definition 4.1. We treat here only the Gaussian linear case with first-order Markov assumptions. The extended smoothers for nonlinear evolution–observation models are similar to the extended Kalman filter explained above. There are also particle filter versions for smoothers, but for common inverse problems these are usually computationally much too complex to implement in practice.

The term "smoothing" is historic and was coined for the first applications in which the observation model was simply $Y_k = X_k + V_k$, $Y_k \in \mathbb{R}$. By assuming, for example, a random walk for the evolution model, the estimates $x_{k|k+p}$ clearly resembled a "smoothed" observation sequence.

Consider a linear evolution–observation model (4.8–4.9) with mutually independent Gaussian state noise and observation noise. As stated in Definition 4.1, the fixed-lag smoothing problem is to calculate $\pi(x_k \mid D_{k+p})$, $p \geq 1$. Clearly, in the Gaussian case, it suffices to calculate the midpoints and the covariances, $x_{k|k+p}$ and $\Gamma_{k|k+p}$. To emphasize the instant when the estimate is computed, we consider instead the quantities $x_{k-p|k}$ and $\Gamma_{k-p|k}$.

We define an augmented state vector

4.5 Fixed-lag and Fixed-interval Smoothing   139

$$Z_k = \begin{bmatrix} X_k \\ X_{k-1} \\ \vdots \\ X_{k-p} \end{bmatrix},$$

and write an evolution–observation model for it. By augmenting the equation (4.8) by identities $X_{k-\ell} = X_{k-\ell}$, we obtain

$$Z_{k+1} = \begin{bmatrix} F_{k+1} & 0 & \cdots & 0 \\ I & 0 & & \\ 0 & I & & \vdots \\ \vdots & & \ddots & \\ 0 & & \cdots & I & 0 \end{bmatrix} Z_k + \begin{bmatrix} W_{k+1} \\ 0 \\ 0 \\ \vdots \\ 0 \end{bmatrix}, \qquad (4.27)$$

$$Y_k = [G_k\ 0\ \cdots\ 0] Z_k + V_k. \qquad (4.28)$$

The Markov properties of the pair $\{Z_k\}$, $\{Y_k\}$ required in Kalman filtering are clearly fulfilled. The representation (4.27)–(4.28) is of the form (4.8)–(4.9).

By applying the Kalman filter to the system (4.27-4.28) we obtain the estimates $z_{k|k}$. But clearly the last block of the Kalman filter estimate $z_{k|k}$ is $x_{k-p|k}$ as desired. Of course, the above yields all fixed-lag estimates up to lag p. The above calculation shows the versatility of the evolution–observation models: Given one model, one can easily generate new ones.

While the pair (4.27)–(4.28) could be used to obtain the fixed-lag estimates, this is not computationally efficient unless the the sparsity of the associated matrices is exploited. Also, we are not usually interested in all possible cross-covariances between the estimates with different lags. There are a number of variations that differ in computational cost and stability. We show below a version in which we replace the observation update of the standard Kalman filter recursion with the following: At each time k, compute recursively for $\ell = 0, \ldots, p$,

$$x_{k-\ell|k} = x_{k-\ell|k-1} + K_{k-\ell}(y_k - G_k x_{k|k-1}), \qquad (4.29)$$

$$K_{k-\ell} = \Gamma^{(\ell,0)}_{k|k-1} G_k^T (G_k \Gamma_{k|k-1} G_k^T + \Gamma_{v_k})^{-1}, \qquad (4.30)$$

$$\Gamma^{(\ell+1,0)}_{k+1|k} = \Gamma^{(\ell,0)}_{k|k-1} (I - K_k G_k)^T F_k^T, \qquad (4.31)$$

where

$$\Gamma^{(\ell,0)}_{k|k-1} = \mathrm{E}\left\{(x_{k-\ell} - x_{k-\ell|k-1})(x_k - x_{k|k-1})^T\right\},$$

and $\Gamma^{(0,0)}_{k|k-1} = \Gamma_{k|k-1}$ is the one-step predictor covariance which initializes the lag iteration.

To derive the recursion (4.29)–(4.31), one starts from the generic form (4.27)–(4.28) and effectively takes into account the structure of the matrices, which yields the recursions. The detailed derivation is omitted.

Depending on p, we may have a very large number of matrices and vectors that we have to store. However, note that the overall computational complexity is less that p-fold when compared to the Kalman filter.

The fixed-interval smoother consists of the conventional computation of the Kalman filter and predictor estimates and storing these estimates. Then, by exploiting the fact that the Kalman filter can be written in a form that orthogonalizes the observation sequence, and taking into account that the estimates can be written as a projection with respect to this orthogonal sequence, one can show that the fixed-interval smoother estimates and the respective covariances can be obtained by the backward iteration

$$x_{k-1|T} = x_{k-1|k-1} + A_{k-1}(x_{k|T} - x_{k|k-1}), \qquad (4.32)$$

$$\Gamma_{k-1|T} = \Gamma_{k-1|k-1} + A_{k-1}\left(\Gamma_{k|T} - \Gamma_{k|k-1}\right)A_{k-1}^{\mathrm{T}}, \qquad (4.33)$$

for $k = T, T-1, \ldots, 2$ where the backward gain matrices A_k are given by

$$A_k = \Gamma_{k|k} F_{k+1} \Gamma_{k+1|k}^{-1}, \qquad k = 1, \ldots, T-1. \qquad (4.34)$$

Note again that the storage requirement can be significant. However, the backward gain matrices are not needed simultaneously and so the requirement is not on the core memory.

How much better the smoothed estimates are when compared to the on-line Kalman filtered estimates depends on the evolution–observation model in a nontrivial way. In some cases the filtered estimate errors possess a delay type structure which is largely absent in the smoothed estimates. In some other cases the decrease of estimation error can be practically negligible and costly in view of the increased computational complexity.

4.6 Higher-order Markov Models

The discussion of nonstationary inverse problems so far is based on first-order Markov properties of the processes $\{X_n\}_{n=0}^{\infty}$ and $\{Y_n\}_{n=1}^{\infty}$. In this section, we generalize the discussion to the case in which some of these properties are no longer valid. In particular, we consider the cases in which the state process is a p–Markov process, or the observation has memory, that is, the measured output depends on the past. Also, a combination of these cases is studied.

Consider first the case in which the pair $\{X_n\}_{n=0}^{\infty}$ and $\{Y_n\}_{n=1}^{\infty}$ satisfies the following properties:

1. The process $\{X_k\}_{k=0}^{\infty}$ is a p-Markov process, that is,

$$\pi(x_{k+1} \mid x_0, x_1, \ldots, x_k) = \pi(x_{k+1} \mid x_{k-p+1}, x_{k-p+2}, \ldots, x_k).$$

2. The process $\{Y_k\}_{k=1}^{\infty}$ is a Markov process with respect to the history of $\{X_k\}$, that is,

$$\pi(y_k \mid x_0, x_1, \ldots, x_k) = \pi(y_k \mid x_k).$$

4.6 Higher-order Markov Models

3. The process $\{X_k\}_{k=0}^{\infty}$ depends of the past observations only through its own history, that is,

$$\pi(x_{k+1} \mid x_k, y_1, y_2, \ldots, y_k) = \pi(x_{k+1} \mid x_k).$$

It is understood above that for negative index values k, we set $X_k = 0$ identically, that is, the process is trivial. Compared to the earlier discussion, only the first condition is altered, saying that the current value depends on the past only via the p previous values.

There is a very simple method to transform this problem such that the previous results can be applied. Similarly as in the previous section, let us define a new process $\{Z_k\}_{k=0}^{\infty}$ as

$$Z_k = \begin{bmatrix} X_k \\ X_{k-1} \\ \vdots \\ X_{k-p+1} \end{bmatrix}, \quad k \geq 0.$$

The realizations of Z_k are denoted by z_k. It is easy to see that for this process, we have

$$\pi(z_{k+1} \mid z_0, \ldots, z_k) = \pi(x_{k-p+2}, \ldots, x_{k+1} \mid x_0, x_1, \ldots, x_k)$$
$$= \pi(x_{k-p+2}, \ldots, x_{k+1} \mid x_{k-p+1}, \ldots, x_k)$$
$$= \pi(z_{k+1} \mid z_k),$$

that is, the process $\{Z_k\}_{k=0}^{\infty}$ is a Markov process. Furthermore, we have

$$\pi(y_k \mid z_0, z_1, \ldots, z_k) = \pi(y_k \mid x_0, x_1, \ldots, x_k) = \pi(y_k \mid x_k)$$
$$= \pi(y_k \mid x_{k-p+1}, \ldots, x_k) = \pi(y_k \mid z_k).$$

Finally, we deduce that

$$\pi(z_{k+1} \mid z_k, y_1, y_2, \ldots, y_k) = \pi(x_{k-p+2}, \ldots, x_{k+1} \mid x_{k-p+1}, \ldots, x_k, y_1, \ldots, y_k)$$
$$= \pi(x_{k-p+2}, \ldots, x_{k+1} \mid x_{k-p+1}, \ldots, x_k)$$
$$= \pi(z_{k+1} \mid z_k),$$

as a moments reflection shows. In conclusion, the pair $\{Z_k\}_{k=0}^{\infty}$, $\{Y_k\}_{k=1}^{\infty}$ is an evolution–observation model. Thus we may use directly all the results to estimate Z_k and hence X_k.

Consider the following evolution model. Assume that we have a linear p-Markov evolution model,

$$X_{k+1} = F_1 X_k + F_2 X_{k-1} + \cdots + F_p X_{k-p+1} + W_{k+1},$$

where F_j, $1 \leq j \leq p$, are known matrices. By definition of the process Z_k, we obtain immediately the Markov-type evolution equation,

$$\begin{bmatrix} X_{k+1} \\ X_k \\ \vdots \\ X_{k-p+2} \end{bmatrix} = \begin{bmatrix} F_1 & F_2 & \cdots & F_{p-1} & F_p \\ I & 0 & \cdots & 0 & 0 \\ \vdots & \vdots & & \vdots & \\ 0 & 0 & \cdots & I & 0 \end{bmatrix} \begin{bmatrix} X_k \\ X_{k-1} \\ \vdots \\ X_{k-p+1} \end{bmatrix} + \begin{bmatrix} W_{k+1} \\ 0 \\ \vdots \\ 0 \end{bmatrix},$$

that is, Z_k satisfies an evolution model of the form

$$Z_{k+1} = \mathcal{F} Z_k + \mathcal{W}_{k+1}.$$

which is a linear first-order Markov model for Z_k. The linear observation model is written as (4.28). This allows one to apply the standard Kalman filtering.

Quite similarly, we may treat the case in which the evolution model is Markovian while the observation equation has memory, that is, instead of the simple equation (4.4), we have an observation having q-timestep memory,

$$Y_n = G_n X_n + G_{n-1} X_{n-1} + G_{n-q+1} X_{n-q+1} + V_n, \quad q > 1.$$

With the same definition for the multistate $Z_n = [X_n; X_{n-1}; \ldots; X_{n-q+1}]$ as before, we have now the observation equation

$$Y_n = [G_n \; G_{n-1} \; \cdots \; G_{n-q+1}] Z_n + V_n,$$

the state evolution in this case being similar to (4.27).

Finally, we may naturally combine the above ideas and consider p-Markov processes with observation having q-timestep memory. The multistate vector is then defined as $Z_n = [X_n; X_{n-1}, \ldots, X_{n-r+1}]$, where $r = \max(p, q)$.

Higher-order Markov models appear naturally when the continuous time evolution is described by higher-order differential equations. As an example, consider an evolution model

$$\frac{\partial^2 u(r, t)}{\partial t^2} = p(r, D) u(r, t) + e(r, t).$$

Here, $p(r, D)$ is a spatial differential operator, and the term $e(r, t)$ represents the modelling error. An explicit discretization in the spatial direction leads to a semidiscrete vector differential equation

$$\frac{d^2 x(t)}{dt^2} = P x(t) + w(t), \quad x(t) \in \mathbb{R}^N.$$

Similarly to the discussion of Example 1, we may discretize the time variable. Setting $t_k = k\tau$, $k = 0, 1, 2, \ldots$, we write an approximate equation

$$\tau^{-2} \big(x(t_{k-1}) - 2x(t_k) + x(t_{k+1}) \big) = P x(t_{k+1}) + w(t_{k+1}).$$

By solving this equation for $x(t_{k+1})$, we obtain a 2–Markov evolution model.

It is left to the reader to verify that in fact, the reduction of the 2-Markov model into a 1-Markov model in this section corresponds to writing the second-order equation above as a first-order system.

4.7 Notes and Comments

The original idea of Kalman filters goes back to the classic work [70] of Kalman. The idea of using Kalman filters in parameter estimation, or in inverse problems, is not new in the engineering literature. The first instance known to the authors is [77] in which Kalman filtering is mentioned as a possible algorithm for inverse thermal problems. A classical book on Kalman filters and smoothers is [4]. For particle filters, or *bootstrap filters*, as they are sometimes called, see [85]. The discussion of the particle filters is also based on numerous articles; see [21], [29], [46], and those in the book [33].

The derivation of discrete evolution models from continuous ones should be based on the use of stochastic differential equations. The reference [98] contains a good introduction to this topic. The book discusses also some interesting applications with potential sources of new inverse problems. Observe also that there is a continuous time version of the Kalman filter which relies on continuous time measurements, known as the Kalman–Bucy filter; see, for example, [69] for further details.

The Kalman filter recursions are obtained also by considering any finite variance Markov problem with possibly non-Gaussian noise processes as the optimal estimator with linear structural constraint for the estimator; see [17]. This treatment in this reference is based on Hilbert spaces induced by time series.

The computational complexity of the Kalman filter and smoothers may seem prohibitive for many real-time applications. However, in the linear Gaussian case, note that the covariances and the Kalman gain do not depend on the observations, hence these can be precomputed. Also, if the evolution and observation models are time invariant, that is, $F_k \equiv F$ and $G_k \equiv G$, the covariances and the Kalman gain may converge in time so that we only need to compute

$$x_{k+1|k+1} = Fx_{k|k} + K(y_{k+1} - GFx_{k|k})$$
$$= Ky_{k+1} + (I - G)Fx_{k|k} = Ky_{k+1} + Bx_{k|k}$$

at each time.

The treatment in this chapter assumed that the state space is continuous. In fact, the presented evolution–observation model is an example of a *hidden Markov model*. This larger class includes also models for which the state is discrete and finite, that is, it assumes values only in a finite set. These models and the respective estimation algorithms are discussed, for example, in [34]. Also, we have assumed that the state variable has finite dimensions. For a treatment on infinite-dimensional system theory, see [26].

The idea of spatial regularization in Bayes filtering by augmented fictitious observations was presented in the articles [10] and [68].

5
Classical Methods Revisited

This chapter contains a refined discussion of various topics in inverse problems. In addition to a deeper look at the classical methods, we also review briefly the fundamentals of estimation theory both in the standard and the general linear estimator form. First, we consider some of the classical regularization methods from the point of view of the statistical approach. We set up a few test cases of linear inverse problems of different levels of difficulty. These test cases are spatial imaging problems with either blurred or tomography observations. We shall then use the most common variants of truncated singular value decomposition, truncated conjugate gradient iterations and Tikhonov regularization and see how these methods perform statistically. In particular, we compare the performance of these methods with conditional mean or maximum a posteriori estimators employing feasible prior models.

We emphasize that the main purpose is to show the importance of prior modelling rather than actual comparison. We also demonstrate how the statistical techniques can be used to assess the performance of classical methods that are derived without a reference to statistics. Rather than considering single test cases, we set up distributions and carry out the performance analysis over ensembles. Another question discussed here is the sensitivity problem of mismodelling the prior and the likelihood. Of course, this is an issue with both the classical and statistical approaches.

In prior modelling, the discussion focuses on effects due to discretization. Discretization also naturally affects the likelihood model. Another topic is the proper modelling of the measurement noise. We demonstrate that the knowledge of the noise level alone is often not sufficient for satisfactory estimation of the unknowns. Hence, a measurement noise with a special covariance structure may be a severe problem for methods using the discrepancy principle for selecting regularization parameters.

Last but not least, we discuss the infamous inverse crimes, mentioned occasionally in the preceeding chapters. Inverse crimes are closely related to discretization of the likelihood models. We demonstrate that especially in the low noise case - not to mmention the asymptotic case in which the noise level tends to zero - numerical tests performed with inverse crimes may yield completely misleading conclusions regarding the feasibility and performance of a method in practice. Several of the topics discussed in this chapter are recurrent themes that show up also in the last chapter of the book.

5.1 Estimation Theory

The solution of an inverse problem, from the point of view of Bayesian statistics, is the posterior probability density. However, rather than the density, a practitioner usually wants a single estimate of the unknown. The classical methods satisfy this demand typically by setting up a regularized optimization problem whose solution is a feasible estimate. The regularization is often carried out by penalizing the solution for unwanted properties. Tikhonov regularization is the model method of such thinking. However, as noted earlier, adhering to classical methods neglecting feasible statistical modelling makes it difficult, for example, to assess the errors in the solutions.

We discuss first briefly the statistical estimation theory and show that the above reasoning can also lead to the typical Bayesian estimators - the maximum a posteriori and conditional mean estimates. We discuss the estimation error of affine estimators. This part of the work is used later when we analyze the *average performance* of classical inversion methods over ensembles rather than in a single example.

5.1.1 Maximum Likelihood Estimation

The maximum likelihood estimation is by far the most popular estimation method, for example, in engineering literature. Recall that given the likelihood density $\pi(y \mid x)$, the *maximum likelihood estimator* x_{ML} of x is characterized by the condition

$$\pi(y \mid x_{\mathrm{ML}}) \geq \pi(y \mid \widehat{x}) \tag{5.1}$$

for any estimator $\widehat{x}$ of x. From the point of view of ill-conditioned inverse problems the problem is the sensitivity of the maximum likelihood estimates to noise and other errors. In the case of the linear Gaussian observation model

$$Y = AX + E, \qquad E \sim \mathcal{N}(0, \Gamma_{\mathrm{n}}),$$

where $A : \mathbb{R}^n \to \mathbb{R}^m$ and Γ_{n} is positive definite, the maximum likelihood estimation is reduced to the (weighted) *output least squares* (OLS) problem,

$$\min_x \left((y - Ax)^{\mathrm{T}} \Gamma_{\mathrm{n}}^{-1}(y - Ax)\right) = \min_x \|L_{\mathrm{n}}(y - Ax)\|^2, \tag{5.2}$$

where $L_{\mathrm{n}}^{\mathrm{T}} L_{\mathrm{n}} = \Gamma_{\mathrm{n}}^{-1}$. This problem is unstable for ill-conditioned matrices A and the minimizer is meaningless.

In the *regularized output least squares* (ROLS) approaches, a penalty term is added to the object functional. Such methods are essentially the same as the Tikhonov regularization. As pointed out earlier in this book, Tikhonov regularization can be seen as solving a maximum a posteriori estimator with an appropriately defined prior model. Hence, it acts as a bridge between non-Bayesian and Bayesian estimation problems.

5.1.2 Estimators Induced by Bayes Costs

In the sequel, consider an indirect observation $Y = y \in \mathbb{R}^m$ of a random variable $X \in \mathbb{R}^n$. Let $\widehat{x} = \widehat{x}(y)$ denote an estimator of x that depends on the observed y. To set up a Bayesian framework for estimation, let us define a *cost function* $\Psi : \mathbb{R}^n \times \mathbb{R}^n \to \mathbb{R}$ so that $\Psi(x, \widehat{x})$ gives a measure for the desired and undesired properties of the estimator $\widehat{x}$ of x. The *Bayes cost* is defined as

$$B(\widehat{x}) = \mathrm{E}\{\Psi(X, \widehat{x}(Y))\} = \int \int \Psi(x, \widehat{x}(y))\pi(x, y) dx\, dy.$$

Further, we can write

$$B(\widehat{x}) = \int \int \Psi(x, \widehat{x})\pi(y \mid x) dy\, \pi_{\mathrm{pr}}(x) dx$$
$$= \int B(\widehat{x} \mid x)\pi_{\mathrm{pr}}(x) dx = \mathrm{E}\{B(\widehat{x} \mid x)\},$$

where

$$B(\widehat{x} \mid x) = \int \Psi(x, \widehat{x})\pi(y \mid x) dy$$

is the *conditional Bayes cost*.

In the *Bayes cost method* we fix the cost function Ψ and define the estimator $\widehat{x}_{\mathrm{B}}$ so that

$$B(\widehat{x}_{\mathrm{B}}) \leq B(\widehat{x})$$

for all estimators $\widehat{x}$ of x.

By using the Bayes formula, we write the Bayes cost in the form

$$B(\widehat{x}) = \int \int \Psi(x, \widehat{x})\pi(x \mid y) dx\, \pi(y) dy.$$

Since the marginal density $\pi(y)$ satisfies $\pi(y) \geq 0$ and the estimator $\widehat{x}(y)$ depends only on y, the minimizer of the Bayes cost is found by solving

$$\widehat{x}_{\mathrm{B}}(y) = \arg\min\left\{\int \Psi(x, \widehat{x})\pi(x \mid y) dx\right\} = \arg\min\left\{\mathrm{E}\{\Psi(x, \widehat{x}) \mid y\}\right\}.$$

The most common choice for the cost function is $\Psi(x, \widehat{x}) = \|x - \widehat{x}\|^2$ which induces the *mean square error criterion*,

$$B(\widehat{x}) = \mathrm{E}\{\|X - \widehat{X}\|^2\} = \mathrm{Tr}\left(\mathrm{corr}(X - \widehat{X})\right),$$

where we denoted $\widehat{X} = \widehat{x}(Y)$, and $\mathrm{corr}(X - \widehat{X})$ denotes the correlation matrix,

$$\mathrm{corr}(X - \widehat{X}) = \mathrm{E}\{(X - \widehat{X})(X - \widehat{X})^{\mathrm{T}}\} \in \mathbb{R}^{n \times n}.$$

In this case, the Bayes estimator is called the *mean square estimator*, denoted by x_{MS}. It turns out that in fact,

$$x_{\mathrm{MS}} = \int x \pi(x \mid y) \, dx = x_{\mathrm{CM}}, \qquad (5.3)$$

that is, the mean square estimator and the conditional mean of x given y coincide. To see the equality (5.3), we observe that for any estimator $\widehat{x}$,

$$\mathrm{E}\{\|X - \widehat{x}\|^2 \mid y\} = \mathrm{E}\{\|X\|^2 \mid y\} - \|\mathrm{E}\{X \mid y\}^2\| + \|\mathrm{E}\{X \mid y\} - \widehat{x}\|^2$$
$$\geq \mathrm{E}\{\|X\|^2 \mid y\} - \|\mathrm{E}\{X \mid y\}^2\|,$$

and the equality holds only if $\widehat{x}(y) = \mathrm{E}\{X \mid y\} = x_{\mathrm{CM}}$. Furthermore, the expectation of the estimation error of the mean square estimator vanishes, i.e.,

$$\mathrm{E}\{X - x_{\mathrm{CM}}\} = \mathrm{E}\{X - \mathrm{E}\{X \mid y\}\} = 0.$$

Hence, the mean square estimator minimizes the trace of the covariance of the estimation error matrix and thus we have yet another equivalent term, *the minimum (error) variance estimator*, for the mean square estimator.

Summarizing, although the computation of the conditional mean is technically an integration problem, it is thus also a solution to an optimization problem which needs little persuasion.

Consider now the other commonly used Bayes estimator, the maximum a posteriori estimator. It turns out that the MAP estimate is asymptotically a solution to the Bayes cost optimization with a specific cost function. Define

$$\Psi(x, \widehat{x}) = \begin{cases} 0, & |x_k - \widehat{x}_k| < \varepsilon \text{ for all } k, \quad 1 \leq k \leq n \\ 1 & \text{otherwise} \end{cases}$$

where $\varepsilon > 0$ is a small constant. We have

$$B(\widehat{x} \mid y) = \int_{|x_k - \widehat{x}_k| > \varepsilon} \pi(x \mid y) \, dx$$
$$= 1 - \prod_{k=1}^{n} \int_{\widehat{x}_k - \varepsilon}^{\widehat{x}_k + \varepsilon} \pi(x \mid y) \, dx_k$$
$$\approx 1 - (2\varepsilon)^n \pi(\widehat{x} \mid y)$$

by the mean value theorem. Thus we can minimize $B(\widehat{x} \mid y)$ by maximizing $\pi(\widehat{x} \mid y)$ which is equivalent to computing the MAP estimate. Note that as

$\varepsilon \to 0$, $\Psi \to 1$ uniformly and thus this cost is called the *uniform cost*. Loosely speaking, while the CM estimators penalized heavily large errors that are due to some subsets of the support of π_{pr}, for the MAP estimator small and large errors are weighted equivalently. The important thing is to understand the following: both estimates are optimal over their own criteria.

Statistical estimation theory is closely related to statistical decision theory. Bayesian decision theory employs Bayes costs that assign appropriate weights to decisions. For instance, if in a medical application, one needs to avoid a false negative decision in patient treatment, one sets a heavy penalty for such an outcome. The importance of careful Bayesian modelling is emphasized in decision making since ambiguous decision instances are often related to infrequent events that occur far from the maxima of the probability densities.

5.1.3 Estimation Error with Affine Estimators

In this section we restrict ourselves to the linear Gaussian likelihoods and priors. Hence, consider a linear model with additive noise,

$$Y = AX + E, \tag{5.4}$$

where $A \in \mathbb{R}^{m \times n}$ is known, $X \in \mathbb{R}^n$ and $E \in \mathbb{R}^m$ are independent and Gaussian, $X \sim \mathcal{N}(x_0, \Gamma_{\mathrm{pr}})$, $E \sim \mathcal{N}(0, \Gamma_{\mathrm{n}})$. We wish to obtain an estimate for X using the affine estimation rule

$$\widehat{x} = \varphi + \Phi y \in \mathbb{R}^n, \quad \varphi \in \mathbb{R}^n, \ \Phi \in \mathbb{R}^{n \times m}. \tag{5.5}$$

Note that estimates obtained by some classical methods, such as Tikhonov regularization, TSVD, Landweber and Kaczmarz iteration can be written in this form. Similarly, the Bayesian CM and MAP estimates are included in the discussion. On the other hand, the estimates obtained by truncated Krylov subspace methods do not fall into this category. For instance, the truncated CG method leads to estimates of the type

$$\widehat{x}(y) = \sum_{j=1}^{k} a_j A^j (y - Ax_0),$$

but the coefficients a_j depend on the data y.

The estimate error correlation[1] is

$$\mathrm{corr}(X - \widehat{X}) = \mathrm{E}\{(X - \widehat{X})(X - \widehat{X})^{\mathrm{T}}\} \tag{5.6}$$
$$= \mathrm{E}\{(X - \Phi AX - \Phi E - \varphi)(X - \Phi AX - \Phi E - \varphi)^{\mathrm{T}}\}$$
$$= (1 - \Phi A)\Gamma_{\mathrm{pr}}(1 - \Phi A)^{\mathrm{T}} + \Phi \Gamma_{\mathrm{n}} \Phi^{\mathrm{T}} + C,$$

---

[1] In our terminogy the correlation and covariance matrices coincide when the mean of the error vanishes. In statistics, correlation is sometimes used to refer to scaled covariance.

where
$$C = \bigl((1 - \Phi A)x_0 - \varphi\bigr)\bigl((1 - \Phi A)x_0 - \varphi\bigr)^{\mathrm{T}} \qquad (5.7)$$
is a rank-one matrix. Observe that
$$\mathrm{E}\{X - \widehat{X}\} = x_0 - \varphi - \Phi A x_0.$$

Thus, in the frequentist terminology, C would be said to be due to the *bias*, that is, $\mathrm{E}\{X - \widehat{X}\}$ being possibly nonvanishing. For the conditional mean estimate, $C = 0$. This result requires that the conditional mean estimate be constructed using the actual prior density of the variable X. In practice, however, since the estimators are always based on computational models of the priors, the condition $C = 0$ may not be satisfied. The modelling errors of prior densities will be discussed later in this chapter.

In this chapter we use the trace of the estimate error correlation matrix as a measure of the quality of an estimator. Let us define the *normalized estimate error* of an estimate $\widehat{x}$ by

$$\mathcal{D} = \mathcal{D}(\widehat{x}) = \frac{\mathrm{Tr}\bigl(\mathrm{corr}(X - \widehat{X})\bigr)}{\mathrm{Tr}\bigl(\mathrm{corr}(X)\bigr)} = \frac{\mathrm{Tr}\bigl(\mathrm{corr}(X - \widehat{X})\bigr)}{\bigl(\mathrm{Tr}\,(\Gamma_{\mathrm{pr}}) + \|x_0\|^2\bigr)}. \qquad (5.8)$$

The denominator does not depend on y and is used to normalize the error measure. Observe that with this normalization, for the trivial estimator $\widehat{x} = 0$ we have $\mathcal{D}(0) = 1$. Thus, for an affine estimator $\widehat{x}$, the condition $\mathcal{D}(\widehat{x}) = 1$ means that the estimate is as informative as the trivial one, while for a perfect estimator $\mathcal{D} = 0$. In the case of severely ill-posed inverse problems we can easily have $\|X - \widehat{X}\| > \mathrm{Tr}\bigl(\mathrm{corr}\,(X)\bigr)$, or $\mathcal{D} > 1$, as we shall see later in this chapter. We remark that if we forget about the issues related to modelling, $\mathcal{D}$ is precisely the measure that the mean square, or conditional mean estimator seeks to minimize.

The sensitivity of $\mathcal{D}$ to typical modelling errors of either the prior density or the likelihood will be discussed later in this and the last chapter.

5.2 Test Cases

In this section we specify first the numerical models that we use as test problems later in this chapter. We discuss only additive noise and linear observation models with the assumption that the unknown and the noise are independent. In the following three subsections, we specify the prior distributions, the observation operators and the additive noise distributions.

5.2.1 Prior Distributions

Consider two different Gaussian distributions that are used as prior models. In both cases, the object to be estimated is a pixelized image in $\Omega = [-1, 1] \times [-1, 1]$ discretized into n identical elements.

5.2 Test Cases   151

The first prior density, denoted by π_{smooth}, is homogeneously smooth and is constructed by convolving a Gaussian white noise field with a truncated Gaussian blurring kernel. Let $g_\mu : \mathbb{R}^2 \to \mathbb{R}$ be a blurring kernel,

$$g_\mu(p) = \begin{cases} \exp(-\mu|p|^2), & |p|^2 < (1/\mu)\log(1/a), \\ 0, & \text{otherwise}, \end{cases} \tag{5.9}$$

where $\mu = 200$ and $a = 10^{-4}$. Hence, we have truncated the values below the cut-off level a. Further, if we let $G \in \mathbb{R}^{n \times n}$ be a discrete approximation matrix for convolution with the kernel g_μ, we have

$$\pi_{\text{smooth}} \sim \mathcal{N}(0, GG^{\mathrm{T}}).$$

If $X \sim \pi_{\text{smooth}}$, we may easily generate samples of X by setting

$$x_{ij} = \sum_k \sum_\ell g_\mu(p_{1,i-k}, p_{2,j-\ell}) \nu_{k,\ell},$$

where $(p_{1,i}, p_{2,j})$ is the center of the pixel (i, j) and ν is white noise. Examples of draws from this distribution are given in Figure 5.1.

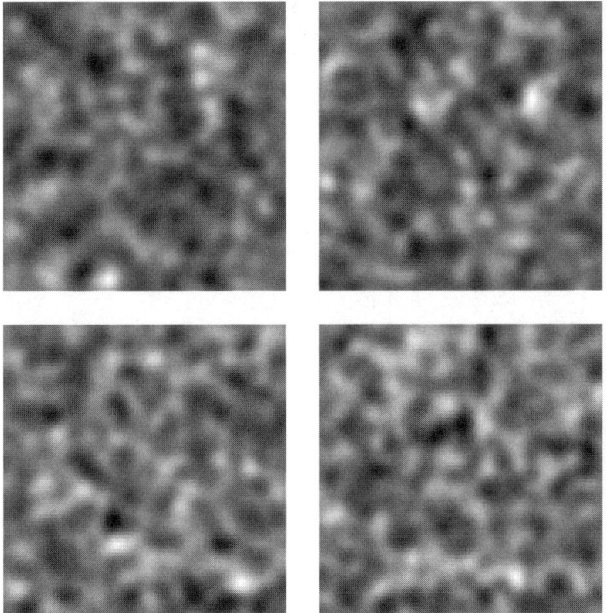

Figure 5.1. Four draws from the smooth Gaussian distribution π_{smooth}.

The second, more complicated distribution is based on external information on structural properties of the image. We denote this prior by π_{struct}.

152   5 Classical Methods Revisited

The basic setup has already been discussed in Section 3.4. Qualitatively, the images are known to consist of an annulus, its exterior and interior, and the the variable is smooth within all three subdomains. The correlation across the boundaries of the three subdomains is almost vanishing.

The boundaries of the annulus are concentric circles with radii 0.62 and 0.9. The prior is constructed similarly to that in Example 8 in Section 3.4. The eigenvalues of the anisotropy matrix $A(p) \in \mathbb{R}^2$ are $\lambda_1(p) = \lambda_2(p) = c$ for the pixels p in the interior of the smooth domains. The constant c is different inside and outside of the annulus. In a layer of approximately two pixels close to the discontinuities, the ratio of the eigenvalues is $\lambda_1/\lambda_2 = 10^{15}$, the smaller eigenvalue corresponding to an eigenvector perpendicular to the circle of discontinuity. Such a choice leads to a very low correlation between the subdomain pixels. This construction leads to an improper preliminary prior density of the form

$$\pi_{\text{pre}}(x) \propto \exp\left(-\frac{1}{2}x^{\text{T}}Bx\right),$$

where the matrix B is costructed similarly to that in Example 8 in Section 3.4. Drawing random samples of the density π_{pre} is of course impossible since the matrix B has at least one vanishing eigenvalue and is therefore not invertible. However, we can construct a proper prior distribution that has all the desired properties by a conditioning trick similar to that in Example 7 of Section 3.4: We select a number of pixels and condition the remaining ones on them. Thus, if $I = [1, 2, \ldots, n] = I' \cup I''$, $I' \cap I'' = \emptyset$, we write

$$\pi_{\text{struct}}(x) = \pi_{\text{pre}}(x_{I'} \mid x_{I''})\pi_{\text{marg}}(x_{I''}).$$

In other words, we assume that we can pose a marginal distribution on the variables $x_{I''}$ and that the set I'' is such that the conditional distribution is proper. In this example we assume that the selected pixels $x_{I''}$ are independent and Gaussian.

Figure 5.2 shows the geometric setting. The pixels with indices in I'' are marked by circles. Figure 5.3 shows four random draws from this density. The relevant parameter values used in the generation of these draws are given in Table 5.1.

5.2.2 Observation Operators

Consider different types of observation operators, a conventional truncated Gaussian blur with various widths and limited angle tomography operators. The limited angle tomography problems are numerically more challenging.

As blurring operator we employ a discretized convolution operator which is similar to the smoothing operator that was used in the construction of the smooth prior distribution. Thus the kernel g_μ is as in (5.9). The truncation parameter is $a = 10^{-4}$. For the width parameter, we use two different values, $\mu = 80$ or $\mu = 40$. The former value gives a faster decaying kernel than the

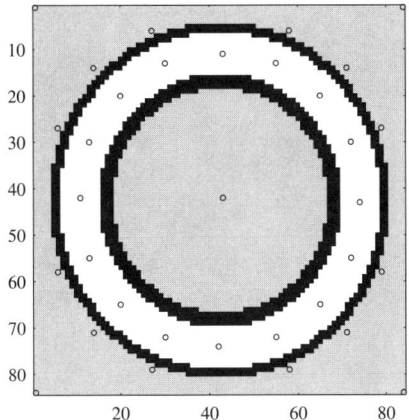

Figure 5.2. The overall setting of the second Gaussian example, the anisotropic smoothness Gaussian distribution π_{struct}, showing the 16+16+1 marginalized pixels.

Table 5.1. The parameters used to create the second Gaussian ensemble with an annulus. The eigenvalues refer to the eigenvalues $\lambda_j(p)$, the mean and the variance to those pixels with indices in I''. We have $c_0 = 200/84$.

	eigenvalues	mean	variance
Outer subdomain			
corners	c_0	0	$100\,c_0^{-2}$
near annulus	c_0	0	$10\,c_0^{-2}$
Inner subdomain	c_0	1	$10\,c_0^{-2}$
Annulus	$100\,c_0$	6	$10\,c_0^{-2}$
Annulus boundary			
tangential	c_0		
radial	$10^{-15}\,c_0$		

latter. The discretized convolution matrices are denoted by $A_{\text{lb}} \in \mathbb{R}^{n \times n}$, "lb" for low blur, and $A_{\text{mb}} \in \mathbb{R}^{n \times n}$, "mb" for medium blur, respectively.

The tomography problem was already introduced in Subsection 2.4.2. The tomography data consists of projections of the image in various directions. Conventional tomographic imaging is based on obtaining the projection data of the image to directions that cover at least half a revolution. If the projection directions θ are limited between $\theta_{\max}$ and $\theta_{\min}$, the angle between these limits is called the *illumination angle*. Furthermore, the angle separations $\theta_{k+1} - \theta_k$ between single projection directions should be small and the measurement density on the transversal line should be sufficiently dense. In some applications these requirements are not met. We investigate two such cases.

In the first one, although the illumination angle is relatively large, $3\pi/4$, there are only 7 projection angles so that the angle separations are large. The

154   5 Classical Methods Revisited

Figure 5.3. Four draws from the anisotropic smoothness Gaussian distribution π_{struct}.

measurement density on the transversal line is large enough to be considered as full.

In the second case we have a small illumination angle of $3\pi/8$. Moreover, there are again only 7 projection angles, and in this case the measurement density on the transversal plane is very sparse, one-sixth of the previous case. The measurement geometry in this case for a single angle is depicted in Figure 5.4. We denote the corresponding observation matrices by A_{wa}, "wa" for wide angle, and A_{na}, "na" for narrow angle, respectively.

A good overall picture of the sparsity of the data can be obtained by viewing an example sinogram in the conventional full angle dense data case and the two considered limited angle cases; see Figure 5.5. In a sinogram each column contains projection data for a single angle. Neighboring columns

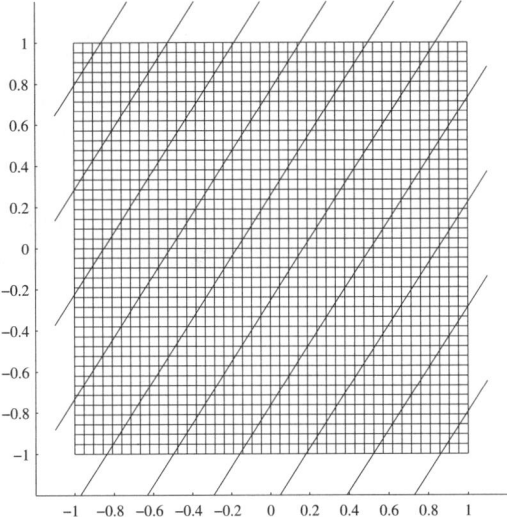

Figure 5.4. The geometry in the second case of limited angle tomography showing the 9 projections per angle.

correspond to neighboring angles. Thus, for example, a single pixel will be seen as a sinusoidal trace in the sinogram, the amplitude and phase determining the location of the pixel in the image.

5.2.3 The Additive Noise Models

The modelling of noise is very often neglected and the standard Gaussian white noise model is employed. In many cases it can be shown that the noise is actually Gaussian and in some cases the noise variance can also be calculated. However, with complex measurement systems the modelling of the noise may not be particularly simple. There are several kinds of noise sources such as electromagnetic interference in industrial environments, disturbances in power supply and contact problems with leads, for example, for which the white noise model is completely inappropriate even as an approximation.

Furthermore, many error models are not additive. For example, Poisson distributed observation models are common in emission tomography. In this case one has to consider the observation itself to be a random draw from a distribution rather than having any additive error. Also, in many cases the errors might be multiplicative such as when the observations have been obtained by analog demodulation. The most common noise models were discussed in Chapter 3.

We shall consider mostly Gaussian noise in this chapter. An example of mismodelling of non-Gaussian noise will be discussed later in this chapter. If $E \sim \mathcal{N}(0, \Gamma_\mathrm{n})$, we define the *noise level* of E to be

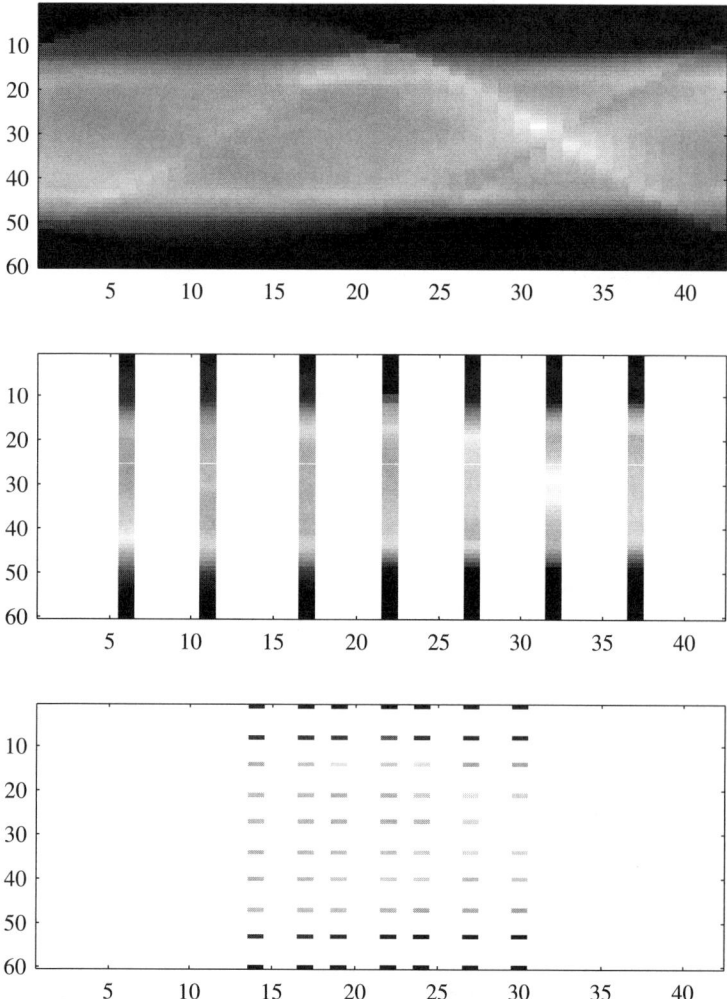

Figure 5.5. The sinograms of a draw from the anisotropic nonhomogeneous smooth Gaussian distribution. Top: Full sinogram. The horizontal axis refers to the number of projection angle and the vertical axis to a single projection for the associated projection angle. Middle: The first limited angle case (sparse angle, full transversal density). Bottom: The second limited angle case (sparse angle, low transversal density).

$$\text{Noise level} = \left(\frac{\text{Tr}\,(\Gamma_n)}{\max \text{Tr}\,(A\Gamma_{\text{pr}} A^T)} \right)^{1/2}. \quad (5.10)$$

Classical regularization techniques rely on some principle for choosing the regularization parameter, eigenvalue truncation level or iteration stopping. The usual criterion is the discrepancy principle, which does not take into account the statistical noise structure. To investigate how severely this affects the estimation error, we consider the following observation model,

$$Y = AX + E_\nu, \quad E_\nu = \sqrt{1-\nu}\,E_0 + \sqrt{\nu}\,E_1,$$

where E_0 and E_1 constitute the observation error and $\nu \in [0,1]$. Both E_0 and E_1 are Gaussian, zero mean and mutually independent with covariances Γ_{n0} and Γ_{n1}, respectively, so that

$$\text{cov}\,(E_\nu) = \text{cov}\,(\sqrt{1-\nu}\,E_0 + \sqrt{\nu}\,E_1) = (1-\nu)\Gamma_{n0} + \nu\Gamma_{n1}.$$

By assuming that
$$\text{Tr}\,(\Gamma_{n0}) = \text{Tr}\,(\Gamma_{n1}) = \delta^2, \quad (5.11)$$

we observe that
$$\text{Tr}\,(\text{cov}\,(E_\nu)) = \delta^2.$$

Thus, when the Morozov criterion is used, the discrepancy is independent of the balance parameter ν. This means that classical regularization methods that rely on the discrepancy principle yield the same estimators for all ν. However, it turns out that the estimation error is sensitive to the value of the parameter.

5.2.4 Test Problems

Using the above models for observation operators, priors and noise, we set up four problems that we seek to solve and analyze by classical methods and by Bayesian estimation. The test problems consist of combinations of prior models, observation models and noise models that are listed in Table 5.2. In addition, we consider separately problems related to noise mismodelling and prior mismodelling.

Table 5.2. The test problems. The noise in these cases is white noise. The noise levels are computed according to formula (5.10).

	prior density	observation	noise level
Case I	π_{smooth}	A_{lb}	2%
Case II	π_{struct}	A_{mb}	10%
Case III	π_{struct}	A_{wa}	5%
Case IV	π_{struct}	A_{na}	5%

5.3 Sample-Based Error Analysis

Having defined our test models in Table 5.2, we apply the statistical error analysis to the classical inversion methods discussed in Chapter 2. In particular, we are interested in calculating the estimation errors $\mathcal{D}(\widehat{x})$ defined in (5.8) and its dependence on the various parameters appearing in the estimators.

Here, we return to the recurrent theme of modelling errors. When deriving formula (5.6), we assumed that the random variable Y representing the data came from the linear model (5.4) and the linear estimator (5.5) is *based on the same model*. This assumption, however, is at the core error of an inverse crime. In reality, estimators are *always* based on an incomplete model while the data comes from the real world. If we base our error analysis on closed formulas derived with the flawed assumption of identifiability of an object with its incomplete model, the results are bound to be overly optimistic. For this reason, we simulate the reality as follows.

Assume that the affine estimator $\widehat{x}$ is calculated based on the model (5.4). Then we generate a more accurate model

$$\widetilde{Y} = \widetilde{A}\widetilde{X} + \widetilde{E}, \qquad (5.12)$$

where $\widetilde{X}$ represents the true unkown. Typically, we assume that $\widetilde{X}$ is represented in a finer discretization mesh than the model vector X. Accordingly, the matrix $\widetilde{A}$ corresponds to a finer mesh than the model matrix A. Also, the noise vector $\widetilde{E}$ may be different from the noise E used in constructing the estimator. The dimension of $\widetilde{Y}$, however, is the same as the dimension of Y, since we are modelling data of known dimension.

Assume that the model vector X is obtained from $\widetilde{X}$ by a linear operator P, i.e.,

$$X = P\widetilde{X}.$$

Typically, P is a projection operator from a finer mesh onto the coarser mesh where the estimation is done. If $\widetilde{X}$ is a Gaussian random variable, $\widetilde{X} \sim \mathcal{N}(\widetilde{x}_0, \widetilde{\Gamma}_{\mathrm{pr}})$, then X is a random variable with $X \sim \mathcal{N}(P\widetilde{x}_0, P\widetilde{\Gamma}_{\mathrm{pr}}P^{\mathrm{T}})$. In our numerical simulations, we assume that the prior distribution of X is the projected one, i.e.,

$$x_0 = P\widetilde{x}_0, \quad \Gamma_{\mathrm{pr}} = P\widetilde{\Gamma}_{\mathrm{pr}}P^{\mathrm{T}}.$$

This assumption is not necessary, and in fact we will show an example where this assumption is violated.

The error estimation strategy in the ensuing chapters is now the following: Having established an affine estimator $\widehat{x}$ based on the incomplete model (5.4), we generate a large sample

$$\{(\widetilde{x}_1, \widetilde{y}_1), \ldots, (\widetilde{x}_N, \widetilde{y}_N)\}$$

based on the model (5.12). From this sample, we calculate a sample of norm squared errors,

$$\{\|P\widetilde{x}_1 - \widehat{x}(\widetilde{y}_1)\|^2, \ldots, \|P\widetilde{x}_N - \widehat{x}(\widetilde{y}_N)\|^2\},$$

and consider the distribution of these errors. The estimate error $\mathcal{D}$ can be computed as a sample average, i.e., we compute

$$\operatorname{corr}(X - \widetilde{X}) \approx \frac{1}{N} \sum_{j=1}^{N} \left((P\widetilde{x}_j - \widehat{x}(\widetilde{y}_j))(P\widetilde{x}_j - \widehat{x}(\widetilde{y}_j))^{\mathrm{T}} \right),$$

and the scaled trace of this matrix.

We use the estimation error (5.8) as a measure for the modelling error. Since the conditional mean was shown to coincide with the mean square estimator that minimizes the estimation error, it may seem unfair to compare the performance of the classical estimators to that of the CM estimator. However, we emphasize that since the CM estimator that we use is also constructed by using the approximate model (5.4), the a priori optimality of this estimator is no longer true. More than comparing the estimator performances, the emphasis in this section is on demonstrating the importance of prior modelling when the overall difficulty that is related to the ill-posedness of A and the noise covariance Γ_e, gets heavier. Thus, the CM estimator which is not based on an inverse crime is a natural reference estimator.

5.4 Truncated Singular Value Decomposition

Consider the model (5.4). Let $\{(u_n, v_n, \lambda_n)\}$ be the singular system of the matrix A, i.e., $A = U\Lambda V^{\mathrm{T}}$ and r the TSVD truncation index; see Section 2.2. If r is determined by the discrepancy principle, the truncation level depends on the data. Statistically, the TSVD method can be thought of as a MAP estimator, where the prior information corresponds to having

$$X \in \operatorname{span}\{v_1, \ldots, v_r\}$$

with probability one. Strictly speaking, such information is not in line with the Bayesian interpretation, where the prior information is independent of the observation. In the presence of independent prior information, the performance of the TSVD method, from the Bayesian viewpoint, depends how well the eigenvectors of the observation matrix and the prior covariance matrix correspond to each other.

Letting

$$V_r = [v_1, \ldots, v_r], \quad U_r = [u_1, \ldots, u_r], \quad \Lambda_r = \operatorname{diag}(1/\lambda_1, \ldots, 1/\lambda_r),$$

we can write the affine estimator corresponding to TSVD as

$$\widehat{x}_{\mathrm{TSVD}}(y) = \Phi_{\mathrm{TSVD}} y = V_r \Lambda_r^{-1} U^{\mathrm{T}} y.$$

160   5 Classical Methods Revisited

If we denote the dependency of the TSVD estimator on the truncation level r explicitly by $\widehat{x}_{\mathrm{TSVD}}(y) = \widehat{x}_{\mathrm{TSVD},r}(y)$, we select the truncation according to the Morozov discrepancy principle as

$$r = \max\left\{r' \mid \|y - A\widehat{x}_{\mathrm{TSVD},r'}(y)\| \leq \tau\sqrt{m}\sigma\right\},$$

where the adjustment parameter τ is chosen $\tau = 1.1$ and the noise $E \in \mathbb{R}^m$ is white noise, $\mathrm{Tr}\,(\Gamma_{\mathrm{n}}) = \mathrm{Tr}\,(\sigma^2 I) = m\sigma^2$ (see Example 4 in Section 2.3).

We shall test the performance of the TSVD estimator in all four test cases of Table 5.2. The results are compared to the CM estimator,

$$\widehat{x}_{\mathrm{CM}}(y) = \varphi_{\mathrm{CM}} + \Phi_{\mathrm{CM}} y \tag{5.13}$$

$$= \left(x_0 - \Gamma_{\mathrm{pr}} A^{\mathrm{T}} \left(A\Gamma_{\mathrm{pr}} A^{\mathrm{T}} + \Gamma_{\mathrm{n}}\right)^{-1} A x_0\right) + \Gamma_{\mathrm{pr}} A^{\mathrm{T}} \left(A\Gamma_{\mathrm{pr}} A^{\mathrm{T}} + \Gamma_{\mathrm{n}}\right)^{-1} y.$$

Consider first the Case I, a smooth object with mild blurring. The overall ill-posedness of the problem is relatively low. Some true, finely meshed unknowns $\widetilde{x}_j$ drawn from the prior density $\mathcal{N}(\widetilde{x}_0, \widetilde{\Gamma}_{\mathrm{pr}})$ and the corresponding CM and TSVD estimates are shown in Figure 5.6.

To analyze the performance in terms of estimate error $\mathcal{D}$, we draw a sample of 500 pairs $(\widetilde{x}_j, \widetilde{y}_j)$, compute the estimates $\widehat{x}_{\mathrm{TSVD}}(\widetilde{y}_j)$ and $\widehat{x}_{\mathrm{CM}}(\widetilde{y}_j)$ and the corresponding norm squared errors $\|P\widetilde{x}_j - \widehat{x}(\widetilde{y}_j)\|^2$ for both estimators. We show the histograms of the norm-squared errors in Figure 5.7. The estimate errors are the normed averages of these errors.

Due to the mildly ill-posed nature of the problem, the TSVD estimates are relatively good although the CM estimates clearly carry more fine grain structural information. The errors are small for both estimators. We carry out the same analysis with sample size $N = 500$ with the test problems II–IV.

In Case II the distribution of the unknown is the structured distribution π_{struct} but the observation operator is the medium blur. The level (5.10) of the additive noise is 10% which means that although the ill-posedness of the problem is somewhat higher than in the previous example, the problem is still not too difficult to solve. Some true unknowns and the corresponding CM and TSVD estimates are shown in Figure 5.8. The distribution of the norm squared errors is shown in Figure 5.9.

With the TSVD estimates, the boundaries of the annulus are severely blurred and the artifacts due to the approximating subspaces are clearly visible as blobs in the corners of the images and as notches along the annulus. This case is much more difficult also for the CM estimates due to the model approximation error for the prior. The changes in the intensity along the annulus can, however, be traced. As for the quantitative measure, the estimate errors for the TSVD begin to escape from the CM errors; see Figure 5.9

In Case III, the unknown follows the nonsmooth distribution and the observation is the transversally densely sampled limited angle tomography data. The level of the additive noise is 5%, which means that the overall ill-posedness of the problem is not too high since the distribution of the projection angles is almost π.

5.4 Truncated Singular Value Decomposition 161

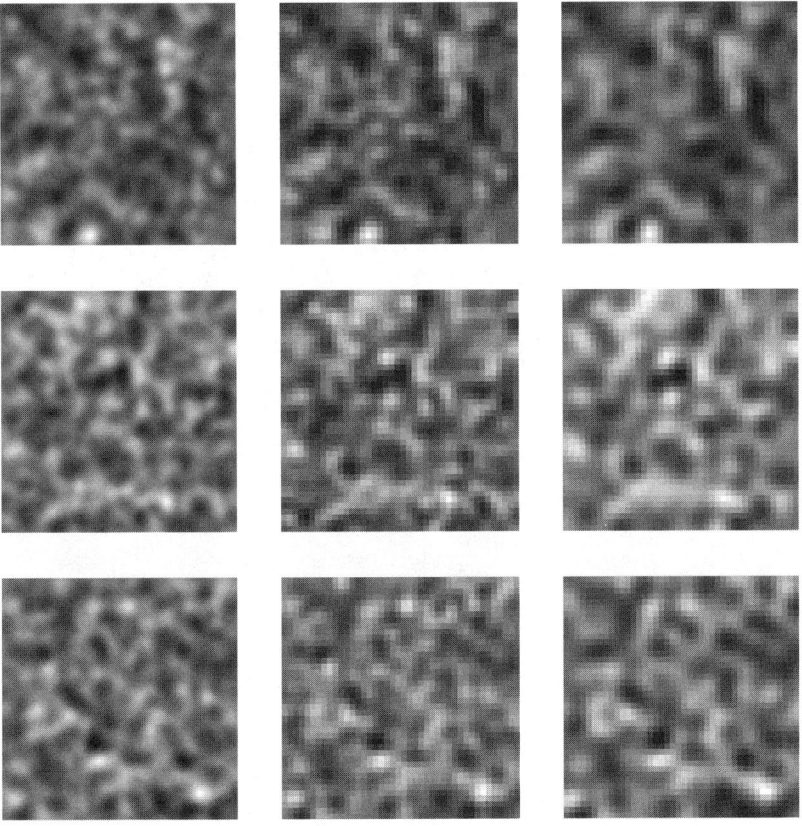

Figure 5.6. Case I. Left: draws form the smooth image distribution. Middle: CM estimates. Right: TSVD estimates.

Some true unknowns and the corresponding CM and TSVD estimates are shown in Figure 5.10. The distribution of the estimation errors is shown in Figure 5.11. In this case the approximating subspace of TSVD is different from the two previous cases. This is clearly visible in the overall visual quality of the TSVD estimates. The boundaries of the annulus are more pronounced, but the estimates are noisier than in the two preceding cases. Note that the streak artifacts typical for the limited angle tomography are clearly visible in both the TSVD and CM estimates. However, note that in the CM estimates the streaks are not visible within the annulus. This is due to the prior covariance structure: The annulus subregion has smaller variance between the neighboring pixels than the two other subregions; see Table 5.1. As for the

162   5 Classical Methods Revisited

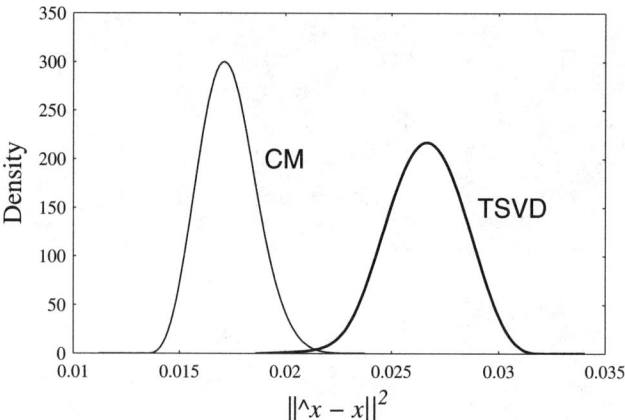

Figure 5.7. Case I. The norm squared error distributions for the CM and TSVD estimates. The estimate error norm here and in the figures to ensue mean the normalized estimate error.

errors, the TSVD errors have clearly surpassed the CM errors with maximum error of approximately 0.7. The CM errors are still concentrated near 0.01.

Finally, in Case IV the distribution of the unknown is again the nonsmooth distribution and the observation is the narrow angle tomography data with sparse tranversal sampling. The additive noise level is 5% and the overall ill-posedness of the problem is very high since the distribution of projection angles is less than $\pi/2$. Some true unknowns and the corresponding CM and TSVD estimates are shown again in Figure 5.12. The distribution of the estimation errors is shown in Figure 5.13.

This is the case in which the TSVD estimates finally blow up. The data is completely inadequate to allow the estimation of the unknown without a feasible prior model. As in the previous case, the structure of the approximating TSVD subspace is clearly visible. It is also to be noted that the data is so poor that the CM estimates begin to be *prior dominated*, thus becoming smoother than in the previous case. Note that for the CM estimates $\widehat{x}_{\mathrm{CM}}$ tends to the prior mean x_0 as either the number of measurements tends to zero or the likelihood covariance tends to infinity. However, note that the intensity variations along the annulus can be traced relatively well in the CM estimates. This information is not included in the prior model.

5.5 Conjugate Gradient Iteration

The analysis for the iterative methods depends on the method. We shall discuss here only one method, the basic conjugate gradient (CG) iterations with

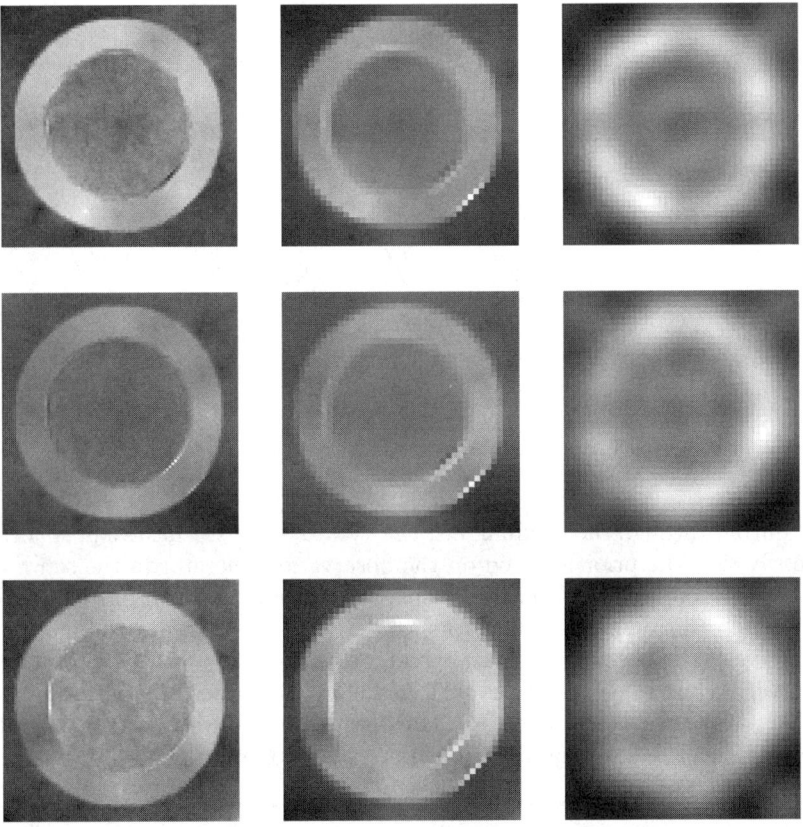

Figure 5.8. Case II. Left: Draws form the structured image distribution. Middle: CM estimates. Right: TSVD estimates.

discrepancy-based stopping criterion. In the case when the matrix A is nonsquare, we use the conjugate gradient normal residual (CGNR) iterations which solve the normal equations without explicitly forming them.

The affine estimator corresponding to the conjugate gradient iteration can be written explicitly in terms of the projector on the Krylov subspace, $\mathcal{K}_\ell = x_0 + \text{span}\{r, Ar, \ldots, A^{\ell-1}r\}$ where $r = y - Ax_0$ is the first residual and x_0 is the initial value of the iteration. Such a representation is not very informative due to the iterative nature of the algorithm.

It is worth mentioning that the statistical interpretation of the CG estimator is not straightforward. The stopping criterion of the iteration restricts the solution to an affine subspace. Hence, one would be tempted to say that the prior information here is a subspace prior, but the subspace itself depends on

164   5 Classical Methods Revisited

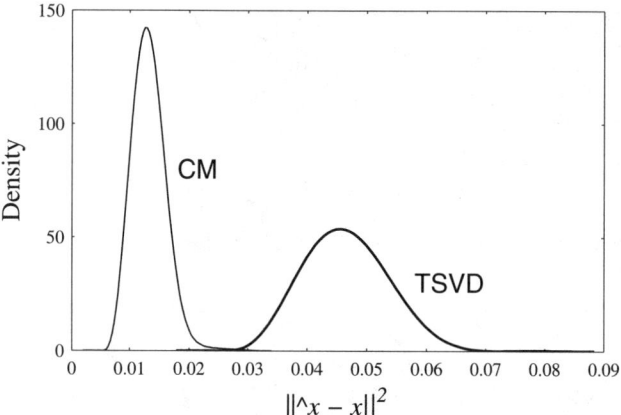

Figure 5.9. Case II. Estimate error distribution for the CM and TSVD estimates.

the output through the residual, i.e., the subspace is a stochastic subspace. We can say that the prior is based on the observation operator in the sense that the basis vectors of the Krylov subspace are linear combinations of the rows of the observation matrix. Thus the estimates are also linear combinations of the rows of the observation matrix and, of course, of x_0.

We run the same test cases I–IV as before to analyze the performance of the CG estimation. As in the TSVD examples above, we use the discrepancy principle for the stopping criterion. Let $x_{\text{init}} = 0$, and denote by x_j the jth CG iterate. We set

$$\widehat{x}_{\text{CG}}(y) = x_r, \quad r = \min\{j \mid \|y - Ax_j\| \leq \tau\sqrt{m}\sigma\},$$

where $\tau = 1.1$ as before and σ is the standard deviation of the additive white noise of dimension m.

Visually, the results are very similar to those obtained by TSVD estimation. For this reason, we do not display single estimates. Instead, we show the error distributions computed from an ensemble of 500 samples. In Figure 5.14, the error distributions of the estimates are shown in all four test cases. In comparison to the TSVD case, the CG method performs somewhat better in Cases I–III. In Case IV where the prior modelling becomes crucial the CG estimates become useless as expected.

5.6 Tikhonov Regularization

We consider here the basic form of the Tikhonov regularization,

$$\widehat{x}_{\text{Tikh}}(y) = \arg\min(\|Ax - y\|^2 + \alpha\|x\|^2),$$

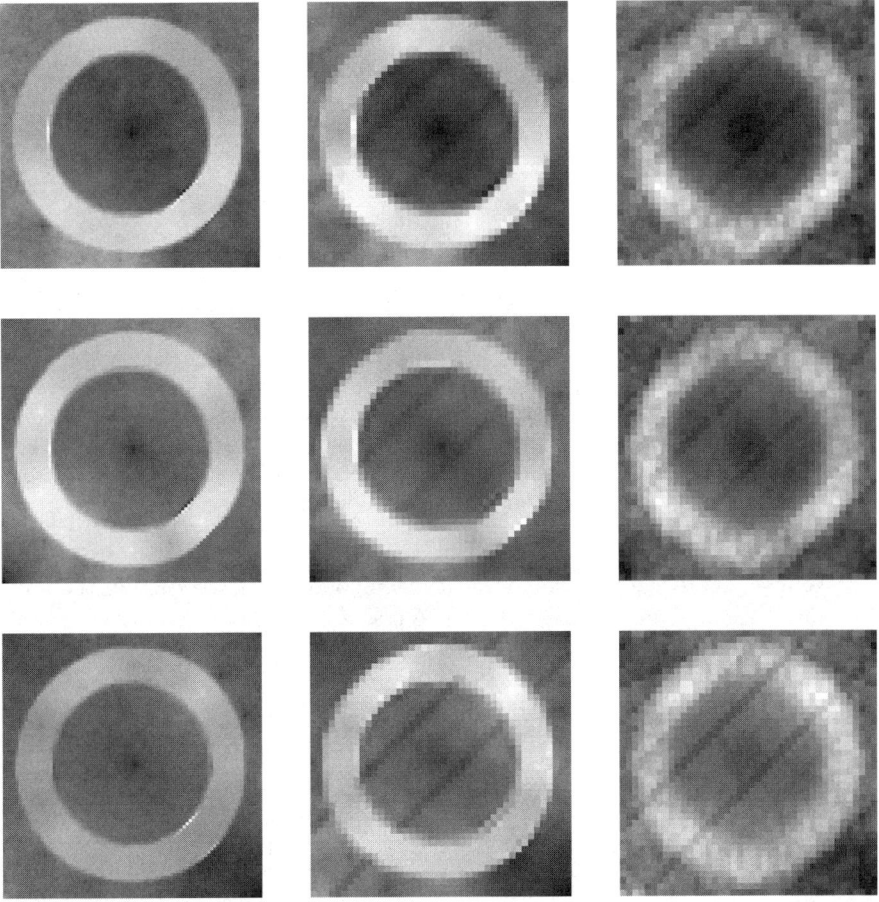

Figure 5.10. Case III. Left: Draws form the nonsmooth image distribution. Middle: CM estimates. Right: TSVD estimates. Limited angle tomography with dense tranversal density data.

or, explicitly,

$$\widehat{x}_{\text{Tikh}}(y) = \begin{bmatrix} A \\ \sqrt{\alpha}I \end{bmatrix}^{\dagger} \begin{bmatrix} y \\ 0 \end{bmatrix}$$
$$= \left(A^{\text{T}}A + \alpha I\right)^{-1} A^{\text{T}} y = \Phi_{\text{Tikh}} y.$$

The regularization parameter α is again determined by the Morozov discrepancy principle with the discrepancy $\delta = \tau\sqrt{m}\sigma$, $\tau = 1.1$.

As discussed in Chapter 3, the Tikhonov regularization estimate can be interpreted statistically as the CM or MAP estimate with a white noise prior.

166   5 Classical Methods Revisited

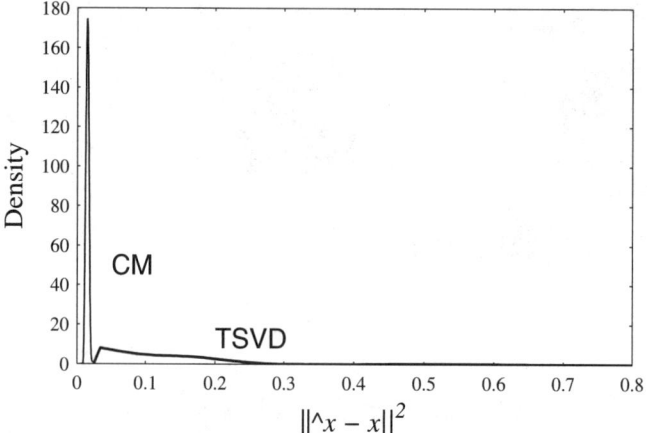

Figure 5.11. Case III. Estimate error distribution for the CM and TSVD estimates. Limited angle tomography with dense tranversal density data.

We investigate the performance of the Tikhonov regularization in the four test cases. Again, since the results with single estimates are visually very similar to those obtained by the TSVD and CGNR estimators, we show only the error distributions of each case; see Figure 5.15. The errors are somewhat smaller than for the CGNR estimates, especially in Case I. Also, the maximum error is less than half in Case III when compared to TSVD and CGNR estimates.

5.6.1 Prior Structure and Regularization Level

Often, the prior covariance design is based on qualitative information about the unknown. Consequently, the prior covariance matrix is only modelled up to a scaling factor. In Chapter 3, it was demonstrated that in such cases, the scaling factor can be left as an unknown and be determined a posteriori.

The following experiment is intended to show the importance of using qualitative information in the prior design. More precisely, let us consider the Case III problem, where the prior density is the structural prior π_{struct}. We consider two estimators, the Tikhonov estimator and the CM estimator with a variable scaling factor. To emphasize the dependence of the estimators on the parameters, let us write

$$\widehat{x}_\alpha(y) = (A^T A + \alpha I)^{-1} A^T y$$

for the Tikhonov regularized solution with regularization parameter α, and let

$$\widehat{x}_\kappa(y) = x_0 + \kappa \Gamma_{\text{pr}} A^T (\kappa A \Gamma_{\text{pr}} A^T + \Gamma_{\text{n}})^{-1} (y - Ax_0)$$

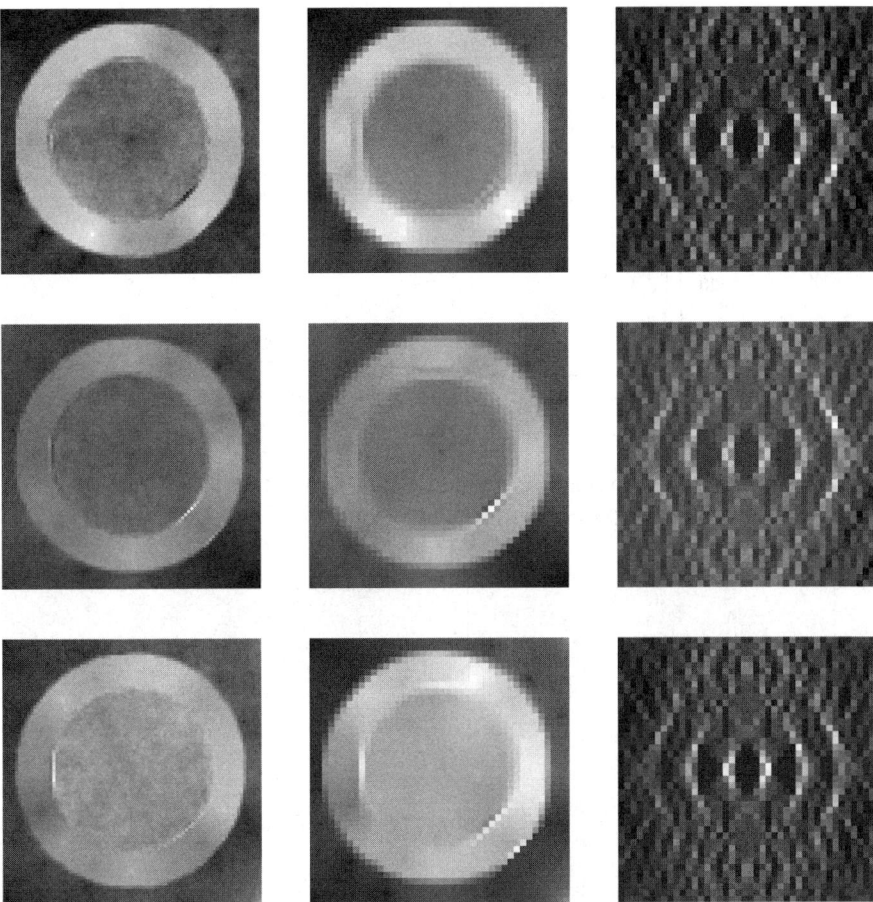

Figure 5.12. Case IV. Left: Draws form the nonsmooth image distribution. Middle: CM estimates. Right: TSVD estimates. Limited angle CT with sparse tranversal density data.

denote the CM estimator based on the prior $\mathcal{N}(x_0, \kappa\Gamma_{\mathrm{pr}})$, $\kappa > 0$. We remark, again, that is not the correct but rather the modelled prior distribution, that is, the matrix Γ_{pr} is given in the coarse mesh in which inverse problems are solved. Furthermore, we assume that we have been able to model the prior covariance only up to a scaling constant κ. If the matrix Γ_{pr} would be the true prior covariance and the data would be generated by committing an inverse crime, the value $\kappa = 1$ would be the choice corresponding to the minimum estimate error. With the present data, the value that minimizes the estimate error differs slightly from unity.

Again, we draw a sample of $N = 500$ pairs $(\widetilde{x}_j, \widetilde{y}_j)$ and plot the functions

168   5 Classical Methods Revisited

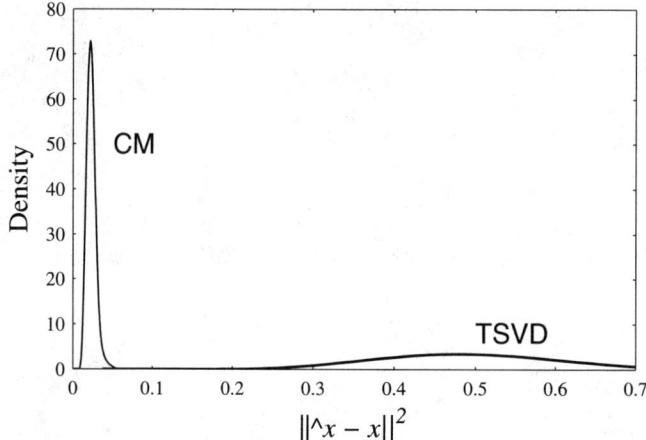

Figure 5.13. Case IV. Estimate error distribution for the CM and TSVD estimates. Limited angle CT with sparse tranversal density data.

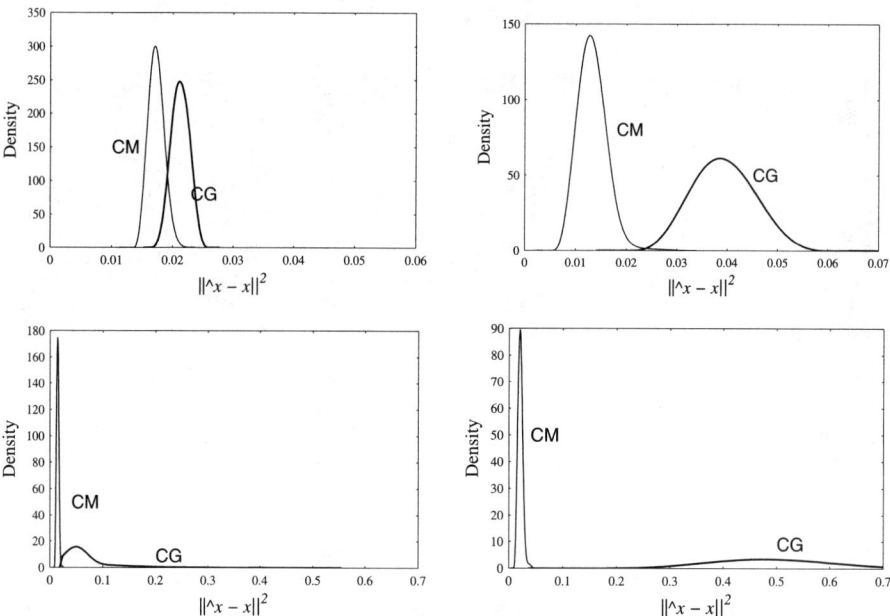

Figure 5.14. The conjugate gradient and CM estimate error distributions. Rowwise from top left: Case I to Case IV.

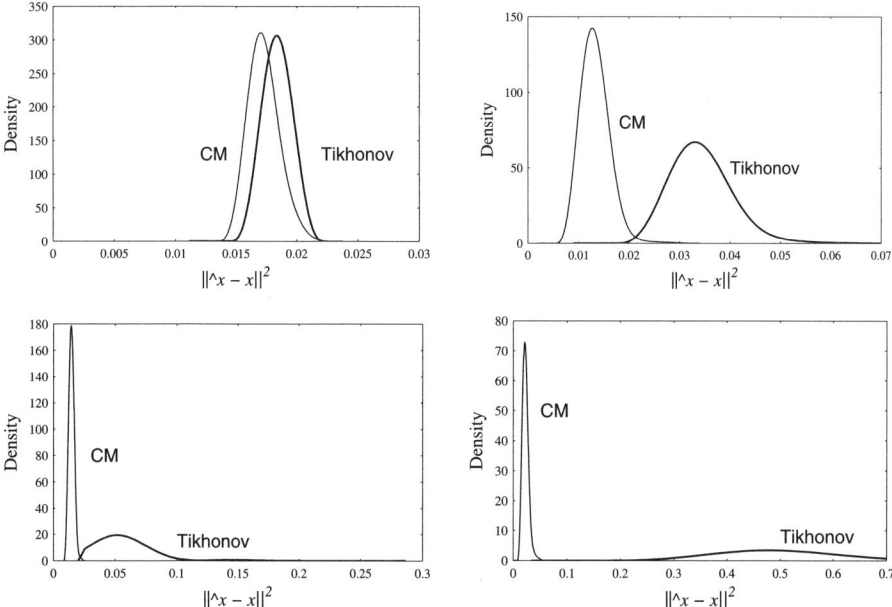

Figure 5.15. The Tikhonov and CM estimate error distributions. Rowwise from top left: Case I to Case IV.

$$\alpha \mapsto \mathrm{E}\{\|\widehat{X}_\alpha - X\|^2\} \approx \frac{1}{N} \sum_{j=1}^{N} \|\widehat{x}_\alpha(\widetilde{y}_j) - \widetilde{x}_j\|^2$$

and

$$\kappa \mapsto \mathrm{E}\{\|\widehat{X}_\kappa - X\|^2\} \approx \frac{1}{N} \sum_{j=1}^{N} \|\widehat{x}_\kappa(\widetilde{y}_j) - \widetilde{x}_j\|^2.$$

The results are shown in Figure 5.16. There are two things worth noticing. First, we observe that with the right covariance structure, one reaches lower estimation error levels than with Tikhonov regularization, which can be seen as a CM solution with an erroneous prior. Second, we observe that the error level with the correct structural prior is relatively stable for large values of the scaling factor κ. This suggests that if we have information about the structure of the prior covariance, the solution is not as sensitive to scaling as the Tikhonov regularized solution. Notice also that, due to prior modelling errors, minimum estimate error with respect to κ is reached at values slightly larger than one.

It may be argued that the above error behavior is not relevant if the Morozov or some equivalent scheme is employed. However, the following should be considered: In the linear fully Gaussian case the scaling of the regularization parameter and the covariance scale is equivalent to misspecifying the observation error norm δ. The determination of δ - or even its order of magnitude - is not always an easy task. While in many cases it can be argued that one can visually see from the data whether the noise level is 5% or 50%, this is not necessarily the case with levels of the order 0.05% and 0.5%.

It turns out that the small noise cases, in which most proposed inversion algorithms are evaluated, are most sensitive to the approximation errors, studied later in this chapter. In these cases it is easy to make an order of magnitude misspecification of δ.

Another observation is in order. In our examples, we calculated the conventional Tikhonov regularized solutions using the Morozov principle. Over the sample, the parameter α varies from specimen to another, and we can consider its distribution over the sample. Typically, the variation is not large, and we can choose a constant α that produces a minimum squared error almost as small as the sample average $N^{-1}\sum_j \|\widehat{x}_\kappa(\widetilde{y}_j) - \widetilde{x}_j\|^2$. It also means that if there is a feasible prior model but, for some reason, Tikhonov regularization is the method of choice, for example, for numerical reasons, this α could be precomputed based on the affine estimator formulas.

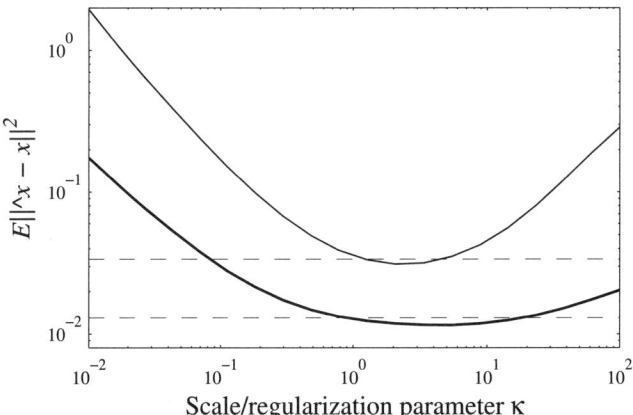

Figure 5.16. The sensitivity of the modelled CM estimate to the covariance scale parameter (bold line) and the Tikhonov estimate to the regularization parameter (weak line). The upper horizontal line corresponds to the mean error when the regularization has been chosen independently for each sample with the Morozov principle. The lower horizontal line corresponds to the CM estimate error with $\kappa = 1$.

5.6.2 Misspecification of the Gaussian Observation Error Model

So far, in our examples we have assumed that the additive noise is Gaussian white noise and that it is correctly modelled. In this section we consider the

5.6 Tikhonov Regularization

additive noise models discussed in Section 5.2.3 to appreciate the importance of the noise *structure* in addition to the noise *level* in the discrepancy principle.

Consider the model

$$Y = AX + E_\nu, \quad A = A_{\mathrm{mb}}, \quad X \sim \pi_{\mathrm{struct}}.$$

Set the noise level (5.10) at 5% and assume that it is known exactly. However, assume that the practitioner adheres to the assumption of an independent identically distributed noise model, that is,

$$Y = AX + E_0, \quad \Gamma_{\mathrm{n}} = \sigma^2 I,$$

and constructs the affine estimator based on this model.

In this example, we consider measurement errors of a particular type. Assume that E_1 is an *offset error*, defined as

$$E_1 = \eta v, \quad v \in \mathbb{R}^m, \quad \|v\| = 1, \quad \eta \sim \mathcal{N}(0, \sigma^2).$$

Then the covariance matrix $\Gamma_{\mathrm{n}1}$ is a rank one matrix,

$$\Gamma_{\mathrm{n}1} = \sigma^2 v v^{\mathrm{T}}, \quad \mathrm{Tr}\left(\Gamma_{\mathrm{n}1}\right) = \sigma^2.$$

Several measurement systems may have offset errors which are of this form. For example, when simultaneous electrical measurements are done over long unshielded cables in high external electrical fields, the errors may be very highly correlated.

Consider the error correlation matrix (5.6) corresponding to a given affine estimation scheme. The matrix Φ depends generally on the error model, but since we assume here that the employed noise model corresponds to $\Gamma_{\mathrm{n}1}$, Φ does not actually depend on ν. Then, the only term associated with the actual additive error in is $\Phi \Gamma_{\mathrm{n}} \Phi^{\mathrm{T}}$, which in this case takes the form

$$(1 - \nu) \mathrm{Tr}\left(\Phi \Gamma_{\mathrm{n}0} \Phi^{\mathrm{T}}\right) + \nu \mathrm{Tr}\left(\Phi \Gamma_{\mathrm{n}1} \Phi^{\mathrm{T}}\right),$$

which is linear with respect to ν. With a fixed ν, the effect of the error term E_1 is maximal if v is chosen so that it maximizes

$$\mathrm{Tr}\left(\Phi \Gamma_{\mathrm{n}1} \Phi^{\mathrm{T}}\right) = \sigma^2 \|\Phi v\|^2,$$

i.e., if v is the eigenvector of the estimation operator Φ corresponding to the largest eigenvalue. In the case of a Tikhonov estimator, if $\{(u_k, v_k, \lambda_k)\}$ is the singular system of the matrix A, choosing $v = u_m$, we obtain

$$\mathrm{Tr}\left(\Phi \Gamma_{\mathrm{n}1} \Phi^{\mathrm{T}}\right) = \sigma^2 \left(\frac{\lambda_m}{\lambda_m^2 + \alpha}\right)^2$$

where λ_m is the singular value that maximizes the term on the left-hand side.

172   5 Classical Methods Revisited

Examples of the observation and the respective estimates with v chosen as above and $\nu = 0.5$ is shown in Figure 5.17. The structure of the vector v shows clearly in the estimate .

We are interested also in the the dependence of the estimation error on ν. In Figure 5.18 we have plotted the error estimates with two choices of v, one as explained above, the other choice being a unit offset $v = (1/m)\mathbf{1}$. Since the computation of the worst case offset error is based on the approximate model that is not used for data generation, the effect of noise misspecification could in reality be even stronger. In this example, the increase in the error is only about 28% for unit offset error while for the approximate worst case error it is about 130%.

If we compute the CM or MAP estimates and model the additive errors properly, that is, if we use $\Phi = \Phi_\nu$, we note the fact that rank one errors can actually be easily suppressed. The CM estimate errors are computed only upto $\nu = 0.99$. As can be seen, the estimate error decreases monotonically with increasing ν.

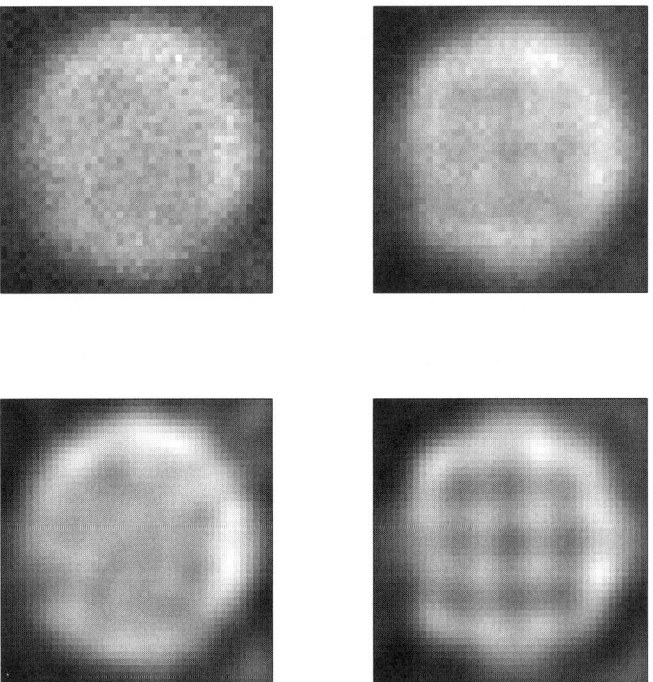

Figure 5.17. Example on the misspecification: Top left: An observation with pure white noise error ($\nu = 0$). Top right: An observation in which half of the error variance is due to white noise and half due to the worst case error structure. Bottom: Reconstructions of the images above assuming the white noise model.

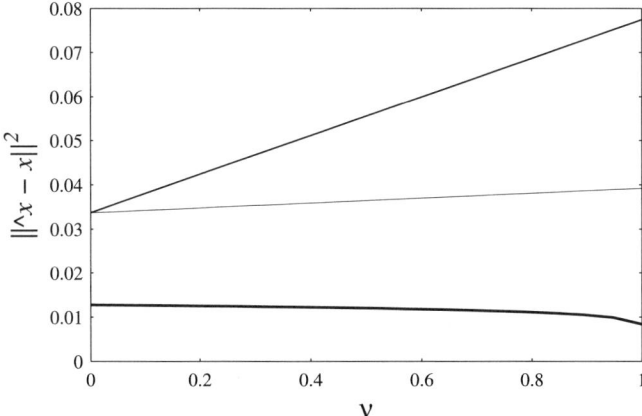

Figure 5.18. The effect of misspecifying the observation error model: the worst case (medium curve) and the offset (weak curve) errors. The parameter ν indicates the balance between the assumed Gaussian white noise model and the mixed model. The bold curve denotes the CM estimate error when the estimator is based on the actual error model.

5.6.3 Additive Cauchy Errors

Here we assume that an additive Gaussian noise model is assumed while the actual errors are Cauchy distributed. In this case we study only a single realization instead of an ensemble. We draw the true image from the smooth distribution π_{smooth}, and the observation model is the medium blur model. Instead of Gaussian additive noise, we have Cauchy-distributed additive noise. The true image and the observation are shown in the upper row of Fig. 5.19. The particular observation includes about 10 pixels in which the error is significantly bigger than in the other, with a single pixel clearly dominating.

We choose the Gaussian noise model so that it matches in some sense with the Cauchy-distributed noise. Since the variance of a Cauchy-distributed variable is infinite, we cannot scale the variances equal. Instead, we write an approximation for the Cauchy distribution using Taylor approximation,

$$\left(\frac{\alpha}{\pi}\right)^m \prod_{j=1}^{m} \frac{1}{1+\alpha^2 e_j^2} = \prod_{j=1}^{m} \exp\left(\log\left(\frac{\alpha}{\pi}\frac{1}{1+\alpha^2 e_j^2}\right)\right)$$

$$\approx \prod_{j=1}^{m} \exp\left(\log\frac{\alpha}{\pi} - 1 - \alpha^2 e_j^2\right).$$

In the light of this approximation, a candidate for the Gaussian variance would be $\sigma^2 = 1/2\alpha^2$. To compensate for the fact that the Cauchy distribution has slowly decreasing tails, we select the Gaussian variance ten times larger than the value above, $\sigma^2 = 5/\alpha^2$.

We approach the estimation problem first from a practical point of view. Since the Cauchy distribution exhibits the property of having infrequent very large draws, we run the observation first through a 3×3 median filter in order to get rid of the large impulse noise. After this operation, we compute an approximate CM estimate using Gaussian error model with noise variance as above. The result of this approximation is also shown in Figure 5.19. As can be seen, this approximate estimate is reasonable and might serve well for most purposes. It is also easy to guess how the estimate that is based on Gaussian approximation would behave without the median filter preprocessing.

If a more accurate estimate is required, we have to return to the actual noise model. To this end we compute both the MAP and CM estimates with the correct likelihood. Again, no inverse crime is committed here, but we assume that we know the Cauchy parameter. We write the posterior in the form

$$\pi(x \mid y) \propto \exp\left\{-\frac{1}{2}\left(2\sum_k \log(1 + \alpha^2(y_k - A_k x)^2) + \|L_{\text{pr}}(x - x_0)\|^2\right)\right\}$$

$$= \exp\left\{-\frac{1}{2}\left(Q(x)^2 + \|L_{\text{pr}}(x - x_0)\|^2\right)\right\} = \exp\left\{-\frac{1}{2}F(x)\right\}.$$

For the computation of the MAP we need to minimize the exponent term of the above. However, we note that since the Hessian of Cauchy part $Q(x)^2$ is generally not positive definite, second-order minimization schemes do not work. We define two different search directions: One is the steepest descent direction $-\nabla F$; another is found by approximating the term $Q(x)$ with a first-order term $Q(x) \approx Q(\bar{x}) + \nabla Q(\bar{x})(x - \bar{x})$ and minimizing the resulting semidefinite quadratic approximation for $F(x)$. Both search directions alone lead to a scheme which stagnates, while altering between these two the minimization proceeds. At each step we check which of the two directions yields a better point in a coarse line search minimization. After this we conduct a more refined line search in the better direction to obtain a new iterate. The convergence with this simple approach is assured, but extremely slow, requiring approximately 200,000 iterations starting from the approximate CM estimate.

For sampling we use an adapted random walk Metropolis–Hastings scheme. The proposal distribution that is based in the second-order approximation of $F(x)$ is not feasible due to indefiniteness of the approximate covariace matrix. We could, in principle, employ mixed-order approximation, i.e., linearize Q as above, but it turns out that in this case the scaling parameter controlling the step size has to be tuned extremely small to result in any of the draws being accepted ever. Instead, we use an updating mask as follows. Let h_z be a truncated Gaussian with center at pixel z. Define the following two random variables: Let $e \sim \mathcal{N}(0, I)$ and z a discrete random variable that is uniform on $(1, 2, \ldots, n)$ and independent of e. Here, n is the number of pixels in the image. Then, we use random walk MH algorithm with the following proposal

$$q^{(\ell+1)} = x^{(\ell)} + c\,\text{vec}(h_z) \odot \left(L_{\text{pr}}^{-1}e + x_0 - \frac{1}{2}\Gamma_{\text{pr}}\nabla Q\right),$$

where $\text{vec}(\cdot)$ denotes the columnwise stacked vector of a matrix and $\odot$ is the pointwise product. The Gaussian h_z is thus used to localize the updating of the pixel values to the neighborhood of the pixel z. The important fact is that the proposal preserves the smoothness and possible jumps in the prior model. If the masking function h_z would be a level set image, smoothness of the image on the boundary of the support of h_z would be lost and practically none of the proposals would be accepted.

The MH algorithm was also started at the approximate CM. The scale parameter c was continuously adapted so that the acceptance ratio was approximately 0.3. The burn-in took about 300,000 rounds. The mean computed from 100,000 draws after the burn-in is shown in Figure 5.19. The difference in the estimation error between the approximate and proper CM estimates is only about 15%. The differences in the details, however, are clear. The MAP esimate is not shown since it cannot visually be distinguished from the CM estimate obtained with the Metropolis–Hastings algorithm.

In Figure 5.20 we show 30,000 draws for two pixels. The periods when the pixels are not in the support of h_z are clearly seen. It is also evident that a more feasible proposal would be needed in order to ensure better mixing. This example shows, among other issues, that both optimization and sampling in large-dimensional spaces can be more art than science, and that there are no systematic approaches to solve all computational problems.

5.7 Discretization and Prior Models

In this section we elaborate a bit further the problem of prior modelling and demonstrate by an example that the dependence on discretization is a more delicate issue than what one might think.

Consider the space of square integrable functions on the unit interval, denoted by $L^2([0,1])$. We construct a nested sequence of closed discretization subspaces $V^j \subset L^2([0,1])$, as follows. Let

$$\varphi(t) = \begin{cases} 1, & \text{if } 0 \leq t < 1, \\ 0, & \text{if } t < 0 \text{ or } t \geq 1. \end{cases}$$

We define the spaces V^j, $0 \leq j < \infty$ as

$$V^j = \text{span}\{\varphi_k^j | 1 \leq k \leq 2^j\},$$

where

$$\varphi_k^j(t) = 2^{j/2}\varphi(2^j t - k - 1).$$

176   5 Classical Methods Revisited

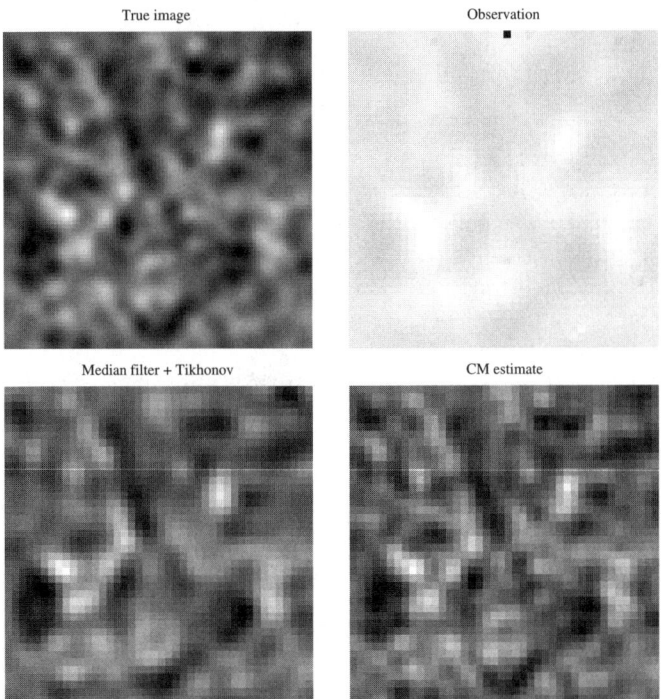

Figure 5.19. Additive Cauchy errors. Rowwise from top left: The true image, the observation, the approximate CM obtained with median filter and CM with Gaussian error approximations, and the CM estimate obtained with Metropolis–Hasting run.

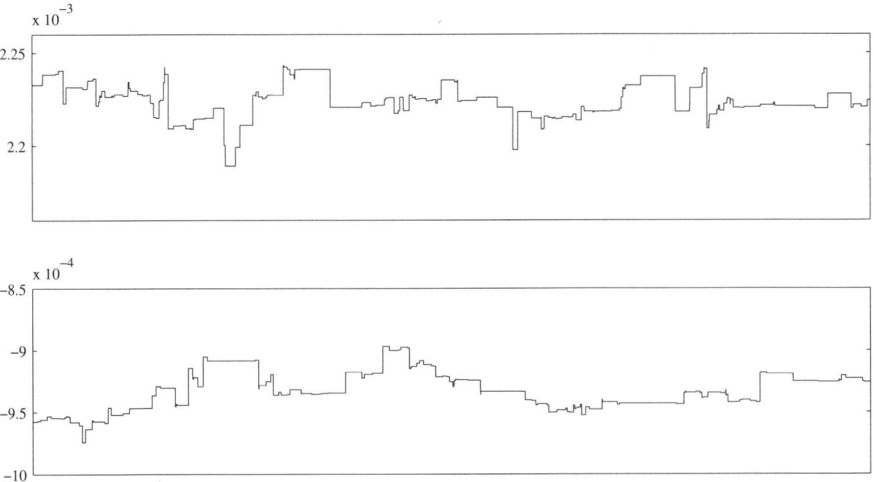

Figure 5.20. Additive Cauchy errors, 30,000 draws for two pixels.

5.7 Discretization and Prior Models

It is easy to see that
$$\langle \varphi_k^j, \varphi_\ell^j \rangle = \delta_{k,\ell},$$
i.e., the functions φ_j^k form an orthonormal basis of the subspace V^j. These spaces also have the properties

1. $V^j \subset V^{j+1}$ for all j, $0 \leq j < \infty$,
2. $f(t) \in V^j$ if and only if $f(2t) \in V^{j+1}$,
3. $\overline{\bigcup V^j} = L^2([0,1])$.

Such sequences of subspaces are called *multiresolution subspaces* of $L^2([0,1])$. The choice of φ above is not the only possible one. In fact, multiresolution subspaces form the basis of wavelet analysis, a topic that is not pursued further here.

The multiresolution spaces are very convenient for computations. To see some of their properties, consider the orthogonal projection

$$P^{j,j-1} : V^j \to V^{j-1}.$$

To find a matrix representation for this projection, consider the embedding $V^{j-1} \subset V^j$ and write

$$\varphi_k^{j-1}(t) = \sum_{\ell=1}^{2^j} c_{k\ell} \varphi_\ell^j(t).$$

The coefficients c_ℓ can be easily calculated by orthogonality, and we have

$$c_{k,\ell} = \langle \varphi_k^{j-1}, \varphi_\ell^j \rangle = h_{\ell-2k},$$

where the coefficients h_k are given as

$$h_k = \begin{cases} 1/\sqrt{2}, & \text{if } k = 0, 1, \\ 0, & \text{otherwise.} \end{cases}$$

Consequently, the projection matrix is readily available. Indeed, if

$$f(t) = \sum_{k=1}^{2^j} x_k^j \varphi_k^j(t),$$

we have

$$P^{j,j-1} f(t) = \sum_{k=1}^{2^{j-1}} \langle f, \varphi_k^{j-1} \rangle \varphi_k^{j-1}(t) = \sum_{k=1}^{2^{j-1}} x_k^{j-1} \varphi_k^{j-1}(t),$$

where the coefficients x_k^{j-1} are

$$x_k^{j-1} = \sum_{\ell=1}^{2^j} h_{\ell-2k} x_\ell^j.$$

In matrix form, if
$$x^j = [x_1^j, \ldots, x_{2^j}^j]^T \in \mathbb{R}^{2^j}, \quad j \geq 0,$$
we have
$$x^{j-1} = H^{j-1,j} x^j,$$
where the coefficient matrix can be expressed in terms of a Kronecker product. Denoting by I_{j-1} the unit matrix of size $2^{j-1} \times 2^{j-1}$ and letting
$$e_1 = \frac{1}{\sqrt{2}}[1,1],$$
it follows that
$$H^{j-1,j} = I_{j-1} \otimes e_1 = \frac{1}{\sqrt{2}}\begin{bmatrix} 1 & 1 & 0 & 0 & \ldots & 0 & 0 \\ 0 & 0 & 1 & 1 & \ldots & 0 & 0 \\ \vdots & & & & & & \vdots \\ 0 & 0 & 0 & 0 & \ldots & 1 & 1 \end{bmatrix} \in \mathbb{R}^{2^{j-1} \times 2^j}.$$

Similarly, the matrix representing the projection $P^{j,j-k} : V^j \to V^{j-k}$ is
$$H^{j-k,j} = I_{j-k} \otimes e_k,$$
where
$$e_k = \left(\frac{1}{\sqrt{2}}\right)^k [1, 1, \ldots, 1] \in \mathbb{R}^{1 \times 2^k}.$$
It is straightforward to check that
$$H^{j-p,j-p-q} H^{j,j-p} = H^{j,j-p-q},$$
i.e., the projections can be performed sequentially.

Consider the following inverse problem: The goal is to estimate a function $f \in L^2([0,1])$ from the noisy data
$$y_\ell = \int_0^1 K(t_\ell, t) f(t) dt + e_\ell, \quad 1 \leq \ell \leq L, \tag{5.14}$$
where e_ℓ is additive noise. Since it is not within the scope of this book to discuss probability densities in infinite-dimensional spaces, assume that there is a discretization level N in the multiresolution basis that can be thought of representing the true signal, i.e.,
$$f(t) = \sum_{k=1}^{2^N} x_k^N \varphi_k^N(t).$$
In other words, we assume that the above discretization approximates the continuous model within working precision. With this assumption, the equation (5.14) can be written as a matrix equation

5.7 Discretization and Prior Models

$$y = A^N x^N + e, \tag{5.15}$$

where the elements of the matrix $A^N \in \mathbb{R}^{L \times 2^N}$ are

$$A^N_{\ell k} = \int_0^1 K(t_\ell, t) \varphi^N_k(t) dt.$$

Assume that the probability distribution of the noise e is known. Then the corresponding stochastic model is written as

$$Y = A^N X^N + E, \tag{5.16}$$

where the probability densites of $X^N \in \mathbb{R}^{2^N}$ and $E \in \mathbb{R}^L$ are presumably known. Let us assume that they are Gaussian, zero mean and independent, with covariance matrices Γ^N_{pr} and Γ_n, respectively.

Assume that we want to solve the inverse problem using a coarser discretization (e.g., for memory limitations), and write the *reduced model*

$$Y = A^j \widetilde{X} + \widetilde{E}, \tag{5.17}$$

where $\widetilde{X} \in \mathbb{R}^{2^j}$, $j < N$. Two questions arise naturally:

1. How much error does the discretization coarsening of passing from equation (5.15) to (5.17) generate?
2. Assuming that we have prior information of X^N, how should we write this information in V^j?

It seems natural to define the low-dimensional model for the unknown as

$$\widetilde{X} = X^j = H^{j,N} X^N,$$

i.e., the *model reduction operator* is the averaging matrix $H^{j,N} \in \mathbb{R}^{2^j \times 2^N}$. This choice immediately defines the prior probability density of the reduced model: X^j is Gaussian, zero mean and with covariance

$$\Gamma^j = H^{j,N} \Gamma^N H^{N,j}, \quad H^{N,j} = (H^{j,N})^{\mathrm{T}}. \tag{5.18}$$

The reduced model matrix A^j has also a natural representation. Letting

$$A^j_{\ell k} = \int_0^1 K(t_\ell, t) \varphi^j_k(t) dt,$$

we have

$$A^j = A^N H^{N,j}, \quad H^{N,j} = (H^{j,N})^{\mathrm{T}}. \tag{5.19}$$

Although the definition (5.18) looks quite innocent, this is *not* the way the reduced models are usually defined in the literature. Indeed, assume that the prior density of X^N is originally a Gaussian smoothness prior of the form

$$\pi_{\mathrm{pr}}(x^N) \propto \exp\left(-\alpha \|L^N x^N\|^2\right),$$

where L^N is discrete approximation of a differential operator. For the sake of definiteness, we choose it to be a three-point finite difference approximation of the second-order derivative with Dirichlet boundary values,

$$L^N = 2^{2N} \begin{bmatrix} -2 & 1 & 0 & \cdots & & 0 \\ 1 & -2 & 1 & & & \\ 0 & 1 & -2 & & & \vdots \\ \vdots & & & \ddots & & 1 \\ 0 & \cdots & & & 1 & -2 \end{bmatrix}.$$

On the discretization level V^j, the corresponding matrix is naturally

$$L^j = 2^{2j} \begin{bmatrix} -2 & 1 & 0 & \cdots & & 0 \\ 1 & -2 & 1 & & & \\ 0 & 1 & -2 & & & \vdots \\ \vdots & & & \ddots & & 1 \\ 0 & \cdots & & & 1 & -2 \end{bmatrix} = H^{j,N} L^N H^{N,j},$$

as one can easily verify. Without further analysis, the natural candidate for the prior density of the reduced model vector $\widehat{X}^j$ would be

$$\pi_{\mathrm{pr}}(x^j) \propto \exp\left(-\alpha \|L^j x^j\|^2\right).$$

The question is now whether this natural construction reproduces the projected covariance Γ^j, i.e., whether the random variable $\widehat{X}^j$ is distributed as $X^j = H^{j,N} X^N$. The answer in our example *is negative*. Indeed, the covariance of $\widehat{X}^j$ is

$$\widehat{\Gamma}^j = \left((L^j)^{\mathrm{T}} L^j\right)^{-1} \neq H^{j,N} \left((L^N)^{\mathrm{T}} L^N\right)^{-1} H^{j,N} = \Gamma^j.$$

We may write the prior density of X^j formally as a smoothness prior, by defining the matrix $\tilde{L}^j \in \mathbb{R}^{2^j \times 2^j}$ to be

$$\tilde{L}^j = \left[(\Gamma^j)^{-1}\right]^{1/2}, \tag{5.20}$$

so that the prior density becomes

$$\pi_{\mathrm{pr}}(x^j) \propto \exp\left(-\alpha (x^j)^{\mathrm{T}} (\Gamma^j)^{-1} x^j\right) = \exp\left(-\alpha \|\tilde{L}^j x^j\|^2\right).$$

The matrix $\tilde{L}^j$ is no longer a tridiagonal discrete approximation of a second-order differential operator.

This example shows that discretization invariant stochastic models are not trivial to construct, and the procedure of constructing the priors affects the results.

The analysis above was based on nested meshes, which are not always available. Assume that we we have two meshes, called mesh A and mesh B in $\mathbb{R}^n$, $n = 1, 2$ or 3, and suppose that mesh A is too fine to be used in calculations. Furthermore, assume that we have constructed a feasible prior covariance matrix in the coarse mesh B. Each row of this matrix represents a discretization of covariance function over the meshed domain. If these functions are not strongly oscillating, we may interpolate them from the finer mesh, thus obtaining an interpolated covariance matrix for the fine mesh A. Let r_ℓ^C denote a mesh point in mesh $C \in \{A, B\}$, and let x_ℓ^C denote the corresponding discretized value of the unknown at that point. We have then

$$\operatorname{cov}(x_\ell^A - x_k^A) \approx \operatorname{cov}(x_\ell^B - x_k^B)$$

at nearby mesh points. This property is not necessarily valid if instead of interpolating the covariance matrix, we interpolate, for example, the inverse of its Cholesky factor.

Generally, it can be said that prior models usually have larger variances than the likelihood models. For this reason the accuracy of the statistical modelling of the likelihood is often a more decisive topic than the modelling of the prior. The next chapter discusses the modelling errors of the likelihood that are induced by the discretization.

5.8 Statistical Model Reduction, Approximation Errors and Inverse Crimes

In the previous section, we considered stochastic model reduction, i.e., lowering of the dimensionality of the problem from the one that is believed to represent accurately enough the true measurement process. It is an inverse crime to believe that the reduced model represents exactly the process. The statistical model reduction is an attempt to counteract the crime by including the modelling errors into the analysis.

Let us consider a linear Gaussian model

$$Y = AX + E, \quad X \sim \mathcal{N}(x_0, \Gamma_{\mathrm{pr}}), \quad E \sim \mathcal{N}(0, \Gamma_{\mathrm{n}}), \tag{5.21}$$

and assume that this model is obtained by careful modelling of the underlying reality. If the equation is obtained by discretization, we assume that $X \in \mathbb{R}^N$ represents a discretized variable in a mesh so fine that the predictions of the model coincide with observations within working precision. We also assume that we have access to a reliable prior model.

As in the previous section, suppose that, for example, for computational reasons, we need to work with a reduced model of lower dimensionality. Let us assume that the model reduction corresponds to a linear operator,

$$P : \mathbb{R}^N \to \mathbb{R}^n, \quad \widetilde{X} = PX.$$

Typically, P is a projection to coarse mesh basis functions. Let $\widetilde{A} \in \mathbb{R}^{m \times n}$ denote the discretized model corresponding to the reduced model. We write

$$Y = AX + E = \widetilde{A}\widetilde{X} + (A - \widetilde{A}P)X + E$$
$$= \widetilde{A}\widetilde{X} + \widetilde{E}. \tag{5.22}$$

An inverse crime is committed by neglecting the second term on the right, i.e., assuming that the coarsely discretized model plus independent error are sufficient to describe the data, or $\widetilde{E} \approx E$.

It is a straightforward matter to calculate the mean and covariance of $\widetilde{E}$. Assuming for simplicity that X and E are uncorrelated, we have

$$\widetilde{E} \sim \mathcal{N}(\widetilde{e}_0, \widetilde{\Gamma}_n), \tag{5.23}$$

where

$$\widetilde{e}_0 = \mathrm{E}\{\widetilde{E}\} = (A - \widetilde{A}P)x_0, \tag{5.24}$$
$$\widetilde{\Gamma}_n = \mathrm{cov}(\widetilde{E}) = (A - \widetilde{A}P)\Gamma_{\mathrm{pr}}(A - \widetilde{A}P)^{\mathrm{T}} + \Gamma_n. \tag{5.25}$$

We refer to the noise model (5.23) as the *enhanced noise model*.

If either of the covariance terms in (5.25) is essentially smaller than the other, we'll refer to either *noise-dominating* or *approximation error dominating* case. If we have

$$\mathrm{Tr}\left(\Gamma_n\right) \gg \mathrm{Tr}\left((A - \widetilde{A}P)\bar{\Gamma}_{\mathrm{pr}}(A - \widetilde{A}P)^{\mathrm{T}}\right),$$

we could expect that we have a noise-dominated case and that the approximation errors do not play a significant role. However, as we shall see, this might not be the case since the covariance structure due to the approximation error may be nontrivial, i.e., not even close to white noise. This means also that the trivial ad hoc remedy of masking the modelling error with white noise, that is, replacing Γ_n by $\kappa\sigma^2 I$, $\kappa > 1$ large enough, may not be a good idea since we might have to set κ so large as to render all observations virtually much more erroneous that they actually are. This is particularly bad in severely ill-posed problems where the signal-to-noise ratio is a crucial issue.

If, on the other hand, we have

$$\mathrm{diag}\,\Gamma_n \gg \mathrm{diag}\,(A - \widetilde{A}P)\bar{\Gamma}_{\mathrm{pr}}(A - \widetilde{A}P)^{\mathrm{T}},$$

the approximation error is not a crucial issue and the inverse crime reduces effectively to an inverse misdemeanor.

With affine estimators the surplus estimation error due to coarse discretization is of the form

$$\mathcal{B}^+ = \Phi(\Gamma_n)(A - \widetilde{A}P)\bar{\Gamma}_{\mathrm{pr}}(A - \widetilde{A}P)^{\mathrm{T}}\Phi(\Gamma_n)^{\mathrm{T}},$$

5.8 Statistical Model Reduction, Approximation Errors and Inverse Crimes

where the matrix $\Phi = \Phi(\Gamma_n)$ depends on the employed error model. It is easy to see that $\text{Tr}\,\mathcal{B}^+$ increases as $\Gamma_n \to 0$.

The enhanced noise model (5.23) is useful, for example, for improving the performance of Tikhonov regularization. From the statistical point of view, however, there is more to it. We shall derive the posterior density and the conditional mean estimate when the error model is taken into account.

Consider the joint probability distribution of the observation Y and the reduced variable $\widetilde{X}$. We have from the accurate model (5.21),

$$\mathrm{E}\{Y\} = Ax_0, \quad \mathrm{cov}(Y) = A\Gamma_{\mathrm{pr}}A^{\mathrm{T}} + \Gamma_{\mathrm{n}}.$$

Similarly,

$$\mathrm{E}\{\widetilde{X}\} = Px_0, \quad \mathrm{cov}(\widetilde{X}) = P\Gamma_{\mathrm{pr}}P^{\mathrm{T}}.$$

The cross-covariance of the variables is

$$\mathrm{E}\{(\widetilde{X} - Px_0)(Y - Ax_0)\} = \mathrm{E}\{P(X - x_0)(A(X - x_0) + E)\} = P\Gamma_{\mathrm{pr}}A^{\mathrm{T}}.$$

Hence, the joint covariance is, not surprisingly,

$$\mathrm{cov}\left(\begin{bmatrix} \widetilde{X} \\ Y \end{bmatrix}\right) = \begin{bmatrix} P\Gamma_{\mathrm{pr}}P^{\mathrm{T}} & P\Gamma_{\mathrm{pr}}A^{\mathrm{T}} \\ A\Gamma_{\mathrm{pr}}P^{\mathrm{T}} & A\Gamma_{\mathrm{pr}}A^{\mathrm{T}} + \Gamma_{\mathrm{n}} \end{bmatrix}.$$

From the joint probability density we can immediately deduce the posterior covariance of $\widetilde{X}$. By Theorem 3.7, we have

$$\pi(\widetilde{x} \mid y) \propto \exp\left(-\frac{1}{2}(\widetilde{x} - \widetilde{x}_0)^{\mathrm{T}} \Gamma_{\mathrm{post}}^{-1}(\widetilde{x} - \widetilde{x}_0)\right),$$

where

$$\widetilde{x} = Px_0 + P\Gamma_{\mathrm{pr}}A^{\mathrm{T}}(A\Gamma_{\mathrm{pr}}A^{\mathrm{T}} + \Gamma_{\mathrm{noise}})^{-1}(y - Ax_0), \tag{5.26}$$

and

$$\Gamma_{\mathrm{post}} = P\Gamma_{\mathrm{pr}}P^{\mathrm{T}} - P\Gamma_{\mathrm{pr}}A^{\mathrm{T}}(A\Gamma_{\mathrm{pr}}A^{\mathrm{T}} + \Gamma_{\mathrm{noise}})^{-1}A\Gamma_{\mathrm{pr}}P^{\mathrm{T}}.$$

The derived formulas are not formally very different from the ones corresponding to the (5.21). There is, however, a crucial point here. The computation of the CM (or MAP) estimate (5.26) requires the solution of a system of size $m \times n$, yet it uses the information of the accurate model. This may have a significant effect on the performance as we shall see in examples.

A natural way to approximate the above enhanced error model can be obtained as follows. Writing $Y = \widetilde{A}\widetilde{X} + \widetilde{E}$, where $\widetilde{E} = E + \epsilon$, we compute the covariance matrix $\widetilde{\Gamma}_n = \mathrm{cov}\,(E + \epsilon)$ and further its Cholesky factor, $L_{E+\epsilon} = \mathrm{chol}\,(\widetilde{\Gamma}_n^{-1})$. Then, we can modify the Tikhonov problem by considering an augmented error problem of the form

$$\min\left\{\|L_{E+\epsilon}(y - Ax)\|^2 + \|L_{\mathrm{pr}}(x - x_0)\|^2\right\}. \tag{5.27}$$

It turns out that in many cases this yields a good alternative to using the proper form (5.26). It should be pointed out that in this approximation, we implicitly assume that the pair $(X, \widetilde{E})$ is mutually independent.

5.8.1 An Example: Full Angle Tomography and CGNE

As an example, consider using CGNR to compute an estimate for an anisotropic smoothness target from full angle tomography observations. The example demonstrates thus the inverse crime problem in a more traditional setting. We emphasize that the results shown below are not peculiar to the CGNR algorithm especially, but, in fact, the same behavior can be expected with all algorithms when inverse crimes are committed.

The actual data is generated in a 84×84 mesh from the structured distribution and the employed observation model A is based on the 42×42 mesh. The proper observation data is computed as explained earlier and the inverse crime (IC) data is generated as $Y = \widetilde{A}PX + E$. We set $E \sim \mathcal{N}(0, \sigma^2 I)$, with different noise variance levels σ^2. Let κ denote the ratio of the noise variance and the modelling error variance,

$$\kappa = \frac{\text{noise variance}}{\text{modelling error variance}}$$
$$= \frac{M\sigma^2}{\text{Tr}\left((A - \widetilde{A}P)\bar{\Gamma}_{\text{pr}}(A - \widetilde{A}P)^{\text{T}}\right)}.$$

We let σ^2 vary such that $0.1 \leq \kappa \leq 10$. Hence, the lower bound $\kappa = 0.1$ corresponds to a modelling error-dominated case while the upper bound $\kappa = 10$ is the noise-dominated case. The CGNR iteration is terminated with the Morozov criterion with discrepancy equal to $\tau\sqrt{n}\sigma$, where n is the number of pixels and $\tau = 1.1$. The same computations are performed with both the proper and IC data. In addition to the CGNR estimates we also compute the conditional mean estimates for the proper data using both the standard and the enhanced error models.

Examples of the CM with standard error model, CGNR estimates with IC data and CGNR estimates with proper data, are shown in Figure 5.21. The middle column shows the results based on IC data. The drastical level of overoptimism rendered by using inverse crime data can clearly be seen once the additive noise level falls below the approximation error level.

Ensemble behavior of the estimates in all four cases can be seen from Figure 5.22. In all these cases, except for the enhanced error CM estimate, the error model is based on the additive noise only. The apparent errors for CGNR estimates with IC data seem to fall nicely and monotonically while it actually blows up shortly after the modelled additive error level drops below the approximation error level. Here we use the approximate enhanced error model (5.27) which shows in this example only as a slight raise of the estimate error level at very low additive error levels. It should be noted that the performance of the uncorrected CM estimates also decreases with decreasing noise levels.

It was mentioned earlier in this chapter that it is, of course, possible to make the ad hoc adjustment to the additive noise level by using $\Gamma_{E+\epsilon} =$

5.8 Statistical Model Reduction, Approximation Errors and Inverse Crimes   185

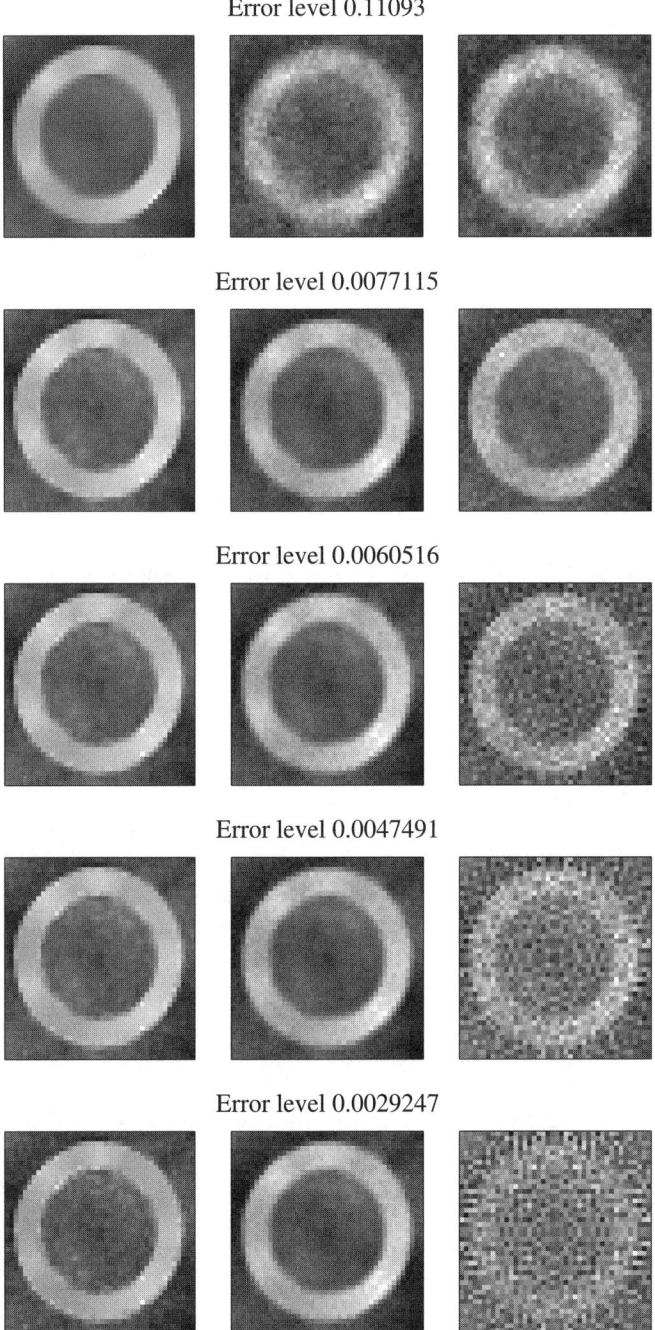

Figure 5.21. An example of the effect of committing an inverse crime. Left column: CM estimates with proper data. Middle column: CGNE estimates with improper (inverse crime) data. Right column: CGNE estimates with proper data.

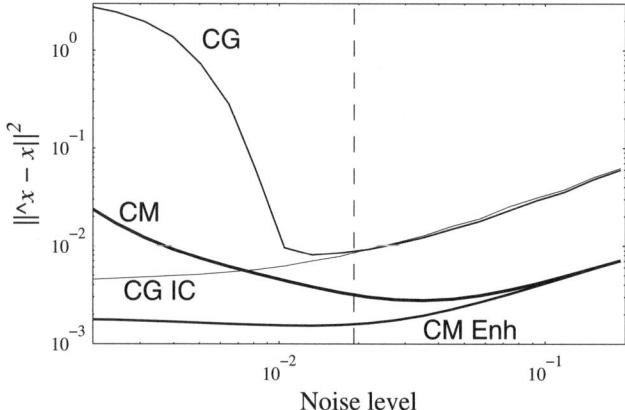

Figure 5.22. The estimation errors for CGNE with proper and IC data as well as CM with proper data and CM estimate with proper data but enhanced observation error model. The vertical dashed line denotes the modelling error level

$c\Gamma_E$ with $c > 1$. The most popular covariance structure for Γ_E is $\Gamma_E = \sigma^2 I$ and the overall statistics are assumed to be Gaussian. The tails of the Gaussian distribution fall rapidly and it would be imperative to choose c large enough in order to mask the approximation error. The main diagonal and the first off-diagonal entries of the covariance of the approximation error for measurements that are related to the first projection angle are shown in Figure 5.23. It is clear that while several measurements are relatively free from the approximation error, for other measurements the situation is much worse. Also, although the off-diagonal entries are not overly large, they are nevertheless nonvanishing. If the adjustment parameter c is to be set big enough to mask the diag Γ_ϵ, it has to correspond to max diag Γ_ϵ. Then, we are effectively arguing that all measurements are correspondingly untrustworthy and we lose much information that is available in the measurements.

5.9 Notes and Comments

Throughout this book, we have sought to avoid inverse crimes by generating the data using a finer meshing in the forward model than in the one used for solving the inverse problem. It is tempting to use, instead of fine meshing, a *continuous model* as a reference, that is, to refer directly, say, to the integral equation representations as a true model. We have avoided this approach in this book for the simple reason that the theory of random variables in infinite-dimensional - and possibly nonseparable - spaces is rather involved and would distract the reader to some extent from the core ideas of the book.

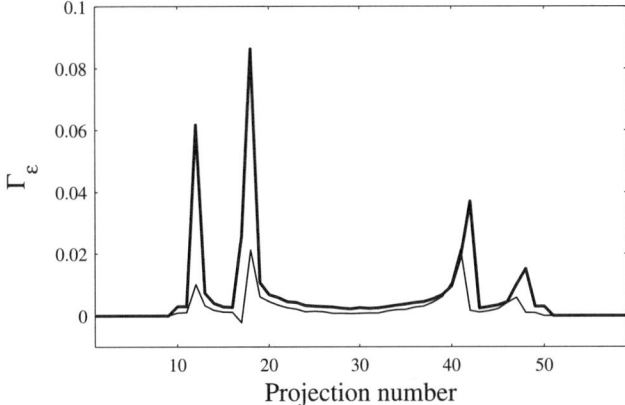

Figure 5.23. The covariance due to the approximation error, the first projection angle: Autocovariances (diagonals, bold line) and cross-covariances of neighboring projections (first off-diagonal, weak line).

Basic textbooks on statistical estimation and decision theory include [88, 120]. The standard estimation theory assumes that the true statistics is both known and used in the construction of the estimators.

The word estimation is used differently in different contexts such as in error estimation in numerical analysis. In connection with statistical estimation theory the focus is on the distribution of the estimate and its error. With well-posed problems the unknown can be treated as a deterministic unknown, but with ill-posed problems we need to treat the unknown as a random variable with some prior distribution. Thus, if *statistical* error analysis is to be carried out in the case of ill-posed problems, the error distribution depends on the prior model. It is then of course necessary that the feasibility of the prior model - implicit or explicit - can be verified.

An important issue related to the proper prior modelling is the possibility to define *discretization invariant* prior models, that is, models that lead to consistent posterior estimates regardless of the discretization level. Research towards this direction has been made for example in [80] and [81].

As explained in the beginning of this chapter, the aim has not been to compare the performance of the classical methods with the optimal estimator in which the true statistics - although only modelled - has been assumed to be known and has also been employed in the construction of the estimator. Rather, we again stress that the message is the relevance of *carrying out prior modelling* in more difficult inverse problems. We also demonstrated that if the prior model is structurally feasible, the increase of the estimate errors due, for example to scaling errors, may be tolerable. Furthermore, as discussed in Chapters 3 and 7, some parameters of the prior model can be handled with the hyperprior approach.

Apart from prior models, the almost canonical assumption regarding measurement noise, has been that the noise is Gaussian, additive, zero mean and has covariance of the form $\Gamma_n = \lambda I$. All these assumptions may fail. Indeed, it could be said that a measurement system that fulfills the above assumptions, can be desribed as an idealization. Physical measurement systems can be very complicated and the actual measurement error model can be very nontrivial. For example, the measurement noise model for weak light sources may be a combination of a Poisson process due to small number of photons, Gaussian noise due to electrical conversions and Cauchy-type noise that is due to electrical contact problems. These three sources may be mutually independent or dependent - possibly multiplicative.

In statistical literature, such noise methods that take into account infrequent large measurement errors are referred to as *robust methods*; see [61]. These are more or less ad hoc iterative methods that change weights for each measurement according to how well the observation model predicted the measurement at the previous iteration. Thus, if the difference between a measurement and its prediction is considerably larger than for other measurements, it will be assigned a smaller weight - or larger variance - for the next iteration.

Regarding the correlatedness of measurement noise, it turns out that in many cases correlated noise leads to severely deteriorated estimates if an uncorrelated noise model is employed. On the other hand, if the correlation structure is known and employed, the effect of the noise on the estimates can be much smaller than in the case of uncorrelated noise.

6
Model Problems

In this chapter, we discuss in detail some inverse problems that serve as model problems when we illustrate the statistical inversion methods, particularly in Chapter 7. We start with linear inverse problems that arise in biomedical applications. First we discuss the inverse problem of *X-ray tomography*; then we proceed to electromagnetic source problems arising in *electric and magnetic encephalography* (EEG/MEG) and in *cardiography* (ECG/MCG).

Two nonlinear inverse problems studied here have applications not only in biomedical imaging but also in industrial process monitoring. The first one is *electrical impedance tomography (EIT)*. This inverse problem has been extensively studied in recent years both from the theoretical and practical point of view (see "Notes and Comments" at the end of the chapter). The second one is a problem formally related to the EIT problem, the near-infrared *optical absorbtion and scattering tomography (OAST)* problem.

6.1 X-ray Tomography

X-ray imaging is based on acquiring several projection images of a body from different directions. The word "tomography" from the two Greek words $\tau o\mu o\sigma$ and $\gamma\rho\alpha\varphi o\sigma$, which mean cut and writing, respectively, refers traditionally to methods where one seeks to produce a sectional image of the interior of a body based on the observed attenuation of the rays in a given plane of intersection. In the modern use, the word tomography is no longer restricted to sectional imaging. Rather, in modern X-ray imaging the goal is to produce three-dimensional reconstructions of the target. Such procedures are sometimes referred to as tomosynthesis. In its current use, the word refers to any noninvasive imaging of an inaccessible region.

It has been known, that since the classical work of Johann Radon on integral geometry in 1917 that if projection images from all around a two-dimensional slice of the body are available, the inner structure of the body along this slice can be reconstructed. This fact was rediscovered by Cormack

6 Model Problems

and Hounsfield in the sixties, leading to commercial computerized tomography (CT) imaging technologies.

In practice, there are often geometric and radiation dosage restrictions that prohibits collecting complete projection data. The projection data may therefore be restricted to a limited view angle, leading to a *limited angle tomography* problem. The restrictions on radiation dose often imply that only few illumination directions can be used. In such a case, the problem is referred to as a *sparse projection data* problem. Together, these type of problems can be called *X-ray tomography with few radiographs*. In these problems, the classical results of Radon cannot be applied.

6.1.1 Radon Transform

In this subsection we review the basic mathematical model of X-ray imaging as well as some of the classical results concerning tomography based on full projection data.

Let $\Omega \subset \mathbb{R}^n$, $n = 2, 3$, be a bounded domain representing the studied object ($n = 3$) or a cross-sectional slice of it ($n = 2$). Assume that a pointlike X-ray source is placed on one side of the object. The radiation passes through the object and is detected on the other side by an X-ray film or a digital sensor; see Figure 6.1.

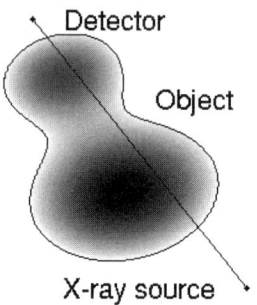

Figure 6.1. X-ray measurement setting.

The most commonly used model assumes that the scattering of the X-rays by the traversed material is insignificant, i.e., only absorption occurs, and that the rays are not deflected by the material interaction. If we further assume that the *mass absorption coefficient* is proportional to the density of the material, the attenuation of the intensity I along a line segment $d\ell$ is given by

$$dI = -I\mu d\ell,$$

where $\mu = \mu(p) \geq 0$, $p \in \Omega$ is the mass absorption of the material. We assume that μ is compactly supported in $\overline{\Omega}$ and bounded. If I_0 is the intensity of

6.1 X-ray Tomography

the transmitted X-ray sent along the line ℓ, the received intensity I after the X-ray has traversed the body is obtained from the equation

$$\log\left(\frac{I}{I_0}\right) = \int_{I_0}^{I} \frac{dI}{I} = -\int_{\ell} \mu(p)d\ell(p).$$

The inverse problem of X-ray tomography can thus be stated as a problem of integral geometry: *Estimate the function $\mu : \Omega \to \mathbb{R}_+$ from the values of its integrals along a set of straight lines passing through Ω.*

The nature of the X-ray tomography problem depends on how many lines of integration are available. In the ideal case, we have data along all possible lines passing through the object. The classical results are based on the availability of this complete data. In the following, we consider briefly the *Radon transform* of a function defined in $\mathbb{R}^n$ and its inversion. In $\mathbb{R}^2$, the Radon transform coincides with the complete X-ray tomography data as described above. In $\mathbb{R}^3$, the Radon transform is related to the X-ray data through an integration procedure which will be explained later.

Let μ be a piecewise continuous function with a compact support in a domain $\Omega \subset \mathbb{R}^n$. Consider a hyperplane L of dimension $n-1$. If $\omega \in \mathbb{S}^{n-1}$ is a unit vector normal to the plane, we may represent L as

$$L = L(\omega, s) = \{p \in \mathbb{R}^n \mid \langle \omega, p \rangle = s\},$$

where s is the signed distance of L from the origin. The Radon transform of the function μ at (ω, s) is the integral along the hyperplane $L(\omega, s)$,

$$\mathcal{R}\mu(\omega, s) = \int_{L(\omega,s)} \mu(p) dS(p),$$

where dS is the Lebesgue surface measure on the plane $L(\omega, s)$.

The following theorem shows that the Radon transform determines the function μ uniquely. We use the notation

$$\mathcal{R}_\omega \mu : \mathbb{R} \to \mathbb{R}, \quad s \mapsto \mathcal{R}\mu(\omega, s),$$

where $\omega \in \mathbb{S}^{n-1}$ is fixed.

Theorem 6.1. *For a piecewise continuous compactly supported μ, we have the identity*

$$\widehat{\mathcal{R}_\omega \mu}(\tau) = \widehat{\mu}(\tau\omega),$$

where the hats denote Fourier transforms. In particular, the knowledge of the Radon transform determines the function μ uniquely.

Proof: For arbitrary $p \in \mathbb{R}^n$ we write

$$p = \langle \omega, p \rangle \omega + (p - \langle \omega, p \rangle \omega) = \langle \omega, p \rangle \omega + p_\perp ,$$

where $\langle \omega, p_\perp \rangle = 0$. Using this decomposition, the Radon transform of μ can be represented as

$$\mathcal{R}\mu(\omega, s) = \int_{\mathbb{R}^{n-1}} \mu(s\omega + p_\perp) dp_\perp,$$

and its one-dimensional Fourier transform with respect to s is

$$\widehat{\mathcal{R}_\omega \mu}(\tau) = \int_{-\infty}^{\infty} \int_{\mathbb{R}^{n-1}} e^{-i\tau s} \mu(s\omega + p_\perp) dp_\perp ds = \int_{\mathbb{R}^n} e^{-i\langle s\omega, p\rangle} \mu(p) dp = \widehat{\mu}(\tau\omega),$$

proving the claim. Hence, since the Radon transform of μ determines its Fourier transform, the function itself can be recovered by the formula

$$\mu(p) = \frac{1}{2} \left(\frac{1}{2\pi} \right)^n \int_0^\infty \int_{\mathbb{S}^{n-1}} e^{i\tau \langle p, \omega \rangle} \widehat{\mathcal{R}_\omega \mu}(\tau) \tau^{n-1} dS(\omega) d\tau. \qquad (6.1)$$

Here we used the simple fact that $\mathcal{R}\mu(-\omega, -s) = \mathcal{R}\mu(\omega, s)$. □

We remark that it is possible to design an inversion algorithm based on the formula (6.1). The main reason why this approach is not often used in applications is that the interpolation of the complex phase from a rectangular grid to a spherical grid is numerically challenging and, when carelessly done, it produces large errors in the reconstruction.

As mentioned before, the Radon transform in $\mathbb{R}^3$ does not give the X-ray data along single rays but instead along planes of intersection. Let $L(\omega, s)$ be such a plane, and let θ_1 and θ_2 be two orthogonal unit vectors that span this plane. Then the Radon transform can be written as

$$\mathcal{R}\mu(\omega, s) = \int_{-\infty}^{\infty} \int_{-\infty}^{\infty} \mu(s\omega + t_1 \theta_1 + t_2 \theta_2) dt_1 dt_2.$$

From this formula we see that integrals over lines determine also integrals over planes, i.e., the complete X-ray data determines the Radon transform, hence the function μ by the previous theorem.

There are several other analytic methods which lead from complete Radon transform data to the mass absorption function. The most classical one is the inverse Radon transform formula, dating back to the original works of Radon himself. Another classical method is known as *filtered backprojection*. We do not treat these methods here in more detail. For references to the literature, see "Notes and Comments" at the end of this chapter.

6.1.2 Discrete Model

As mentioned before, the analytic inversion methods in tomography rely on complete projection data. Since in many application such data are not available (see Figure 6.2), in practice we need to discretize the observation model directly and approximately solve the problem computationally.

6.1 X-ray Tomography

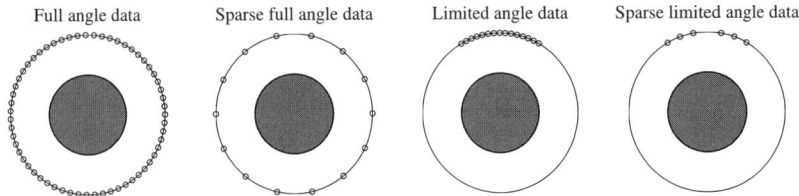

Figure 6.2. A schematic picture of various tomography problems with few projection data. The shaded disc represents the object; the circles indicate the illumination source locations.

To illustrate the latter approach, consider the two-dimensional case. Similar techniques can be used in three dimensions as well. Assume that we have a nonnegative function μ in $\mathbb{R}^2$, whose compact support is contained in the unit rectangle, $[0,1] \times [0,1]$. We divide the image area in $N \times N$ pixels, and approximate the density by a piecewise constant function,

$$\mu(p) = \mu_{ij}, \quad \text{when } p \in P_{ij}, \ 1 \le i,j \le N.$$

Here, P_{ij} denotes the pixel, $(j-1)/N < p_1 < j/N$, $(i-1)/N < 1-p_2 < i/N$, i.e., the image is given by the matrix (μ_{ij}).

To describe the X-ray data in vector form, we stack the matrix entries μ_{ij} in a vector x of length N^2 so that

$$x((j-1)N + i) = \mu_{ij}, \quad 1 \le i,j \le N.$$

Let $L_1, \ldots, L_M$ denote lines corresponding to X-rays passing through the object from souce to detectors. For the mth line, L_m, we have

$$y_m = \int_{L_m} \mu(p)dp = \sum_{i,j=1}^{N} |L_m \cap P_{ij}| \mu_{ij} = \sum_{k=1}^{N^2} A_{mk} x_k,$$

where $|L_m \cap P_{ij}|$ denotes the length of the intersection of the line L_m and the pixel $P_{i,j}$, and we denoted

$$A_{mk} = |L_m \cap P_{ij}|, \quad k = (j-1)N + i.$$

The discretized model for the X-ray imaging can then be written in the form

$$Ax = y.$$

Hence, the X-ray inverse problem can be recast as a linear matrix problem typically of very large size but also with a very sparse model matrix.

Later, in numerical examples, we shall consider the limited angle tomography, where the angles of illumination are restricted to an angle strictly less than π. In these cases, the matrix A has a large null space and the inverse problem becomes severely underdetermined.

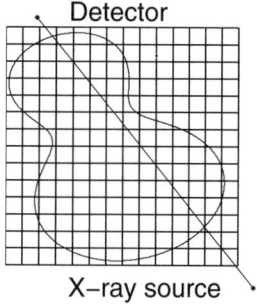

Figure 6.3. Discretized representation of the X-ray data.

6.2 Inverse Source Problems

In this section, we discuss the electromagnetic inverse source problems that arise in biomedical applications. In electromagnetic inverse source problems, the objective is to estimate the electromagnetic source in an inaccessible region by observing the fields outside that region. In the applications discussed here, the frequencies are relatively low, implying that the radiation is negligible, and one can use the static or quasi-static approximations of Maxwell's equations to describe the physical setting.

There are two groups of problems that differ from each other in the data acquisition modality. In the first one, the electric voltage potentials are observed, while in the second group the magnetic fields are measured. The first group includes the *electrocardiography* (ECG) for monitoring the cardiac functions and the *electroencephalography* (EEG) used for studying cerebral activities. The corresponding modalities using magnetic measurements are *magnetocardiography* (MCG) and *magnetoencephalography* (MEG). Both measurement modalities have advantages and drawbacks. They provide complementary information of the same phenomena.

6.2.1 Quasi-static Maxwell's Equations

Let $\Omega \subset \mathbb{R}^3$ be a bounded domain corresponding, e.g., to the human body or to a portion of it. Assume that Ω is simply connected, that its complement is connected and that its boundary $\partial \Omega$ is smooth. The assumption that the complement is connected implies that $\partial \Omega$ consists of one surface.

Consider the electromagnetic fields in Ω, induced by given sources described later. Let $x \in \Omega$ and denote by $t \in \mathbb{R}$ the time. Let $\mathcal{E}(x,t)$ and $\mathcal{H}(x,t)$ be the electric and magnetic fields, and $\mathcal{D}(x,t)$ and $\mathcal{B}(x,t)$ the electric displacement and magnetic flux density, respectively. The *Maxwell–Faraday* and *Maxwell–Ampère* laws state that

$$\nabla \times \mathcal{E} = -\frac{\partial \mathcal{B}}{\partial t}, \quad \nabla \times \mathcal{H} = \frac{\partial \mathcal{D}}{\partial t} + \mathcal{J},$$

where $\mathcal{J} = \mathcal{J}(x,t)$ is the current density of free charges. We say that the medium is *linear* and *isotropic*, if the fields are related to each other through the *constitutive relations*

$$\mathcal{D}(x,t) = \int_{-\infty}^{t} \epsilon(x,t-s)\mathcal{E}(x,s)ds, \quad \mathcal{B}(x,t) = \int_{-\infty}^{t} \mu(x,t-s)\mathcal{H}(x,s)ds,$$

where the strictly positive scalar functions ϵ and μ are the electric permittivity and magnetic permeability of the medium, respectively. In anisotropic media, $\epsilon(x,s)$ and $\mu(x,s)$ are positive definite symmetric matrices in $\mathbb{R}^{3\times 3}$. Here, only the isotropic case is considered.

The current density $\mathcal{J}$ consists of two parts, the source term and the Ohmic term,

$$\mathcal{J} = \mathcal{J}_0 + \mathcal{J}_\Omega,$$

where the Ohmic part is due to the electric field,

$$\mathcal{J}_\Omega(x,t) = \int_{-\infty}^{t} \sigma(x,t-s)\mathcal{E}(x,s)ds.$$

Here, σ denotes the electric conductivity. It is a nonnegative function. In anisotropic material, its value is a positive semidefinite matrix in $\mathbb{R}^{3\times 3}$.

If the fields are time-harmonic, we may write, using the complex notation,

$$\mathcal{E}(x,t) = \mathrm{Re}\big(E(x)e^{-i\omega t}\big), \quad \mathcal{B}(x,t) = \mathrm{Re}\big(B(x)e^{-i\omega t}\big), \quad \mathcal{J}_0 = \mathrm{Re}\big(J_0(x)e^{-i\omega t}\big),$$

where $\omega > 0$ is the harmonic angular frequency and $\mathrm{Re}(z)$ denotes the real part of z.

A substitution into Maxwell's equations shows that the original Maxwell's equations are satisfied if the complex amplitudes satisfy the frequency domain equations,

$$\nabla \times E(x) = i\omega \widehat{\mu}(x,\omega) H(x), \tag{6.2}$$

$$\nabla \times H(x) = (-i\omega \widehat{\epsilon}(x,\omega) + \widehat{\sigma}(x,\omega))E(x) + J_0(x), \tag{6.3}$$

where the material parameters appear Fourier transformed with respect to time,

$$\widehat{\mu}(x,\omega) = \int_0^\infty e^{i\omega t} \mu(x,t) dt,$$

and $\widehat{\epsilon}$ and $\widehat{\sigma}$ are defined similarly.

In the quasi-static approximation, which holds for relatively low frequencies and nonferromagnetic materials, the product $\omega \widehat{\mu}$ is negligible, yielding

$$\nabla \times E \approx 0.$$

In particular, since the domain Ω is simply connected, we may express E in terms of a scalar potential $u = u(x)$ as

$$E = -\nabla u.$$

By substituting this into equation (6.3) and applying the divergence on both sides, we obtain
$$\nabla \cdot \gamma \nabla u = \nabla \cdot J_0, \tag{6.4}$$
where γ denotes the *admittivity*,
$$\gamma(x, \omega) = \hat{\sigma}(x, \omega) - \mathrm{i}\omega\hat{\epsilon}(x, \omega). \tag{6.5}$$

The inverse of γ, denoted by $\rho = 1/\gamma$, is the *impedivity* of the body. Observe that when the fields are static, i.e., $\omega \to 0+$, the admittivity is real and coincides with the static conductivity. In this case, the impedivity is just the *resistivity* of the body. In the rest of this chapter, we assume static fields, i.e., $\gamma(x) \in \mathbb{R}$.

For the discussion to ensue, we also fix an admissible class of admittivities and current densities as follows.

Definition 6.2. *An admittivity distribution $\gamma : \Omega \to \mathbb{R}_+$ and the corresponding source current density $J_0 : \Omega \to \mathbb{R}^n$ are in the admissible class, denoted as*
$$\gamma \in \mathcal{A}(\Omega), \quad J_0 \in \mathcal{J}(\Omega),$$
if the following conditions are satisfied:

1. *For some $N \geq 1$, there is a family $\{\Omega_j\}_{j=1}^N$ of open disjoint sets, $\Omega_j \subset \Omega$, having piecewise smooth boundaries and for which*
$$\overline{\Omega} = \bigcup_{j=1}^N \overline{\Omega}_j.$$
Furthermore, we require that $\gamma|_{\Omega_j}, J_0|_{\Omega_j} \in C(\overline{\Omega}_j)$, $1 \leq j \leq N$, i.e., γ and J_0 restricted to each subset Ω_j allow a continuous extension up to the boundary of the subset.
2. *For some constants c and C,*
$$0 < c \leq \gamma(x) \leq C < \infty.$$
3. *The normal components of the current density J_0 are continuous through the interfaces of the subsets Ω_j.*

In medical applications, the subsets Ω_j in the forward problem may represent organs. In the inverse problem, the set of admissible admittivities provides a natural discretization basis.

Due to possible discontinuities of the admittivity, equation (6.4) has to be considered in the weak sense as we do in the sequel.

6.2.2 Electric Inverse Source Problems

Consider first the forward problem of finding the voltage potential u from the equation (6.4) corresponding to a given J_0 in Ω. To this end, we need to specify the boundary condition. In biomedical applications it is natural to assume that no electric currents exit the body through the boundary, i.e.,

$$n \cdot J|_{\partial\Omega} = n \cdot (J_0 - \gamma \nabla u)|_{\partial\Omega} = 0.$$

In particular, if the source current J_0 has a vanishing normal component at the boundary,

$$n \cdot J_0|_{\partial\Omega} = 0, \tag{6.6}$$

as we assume in the sequel, we have the Neumann boundary condition

$$n \cdot \gamma \nabla u|_{\partial\Omega} = 0. \tag{6.7}$$

To fix the ground voltage level, we may further require, e.g., that

$$\int_{\partial\Omega} u \, dS = 0. \tag{6.8}$$

Under the assumption that γ and J_0 are admissible in the sense of Definition 6.2, one may now show that the forward problem (6.4), (6.7) and (6.8) has a unique solution in the weak sense. Indeed, let $H^1(\Omega)$ denote the L^2-based Sobolev space. Define further

$$H^1_N(\Omega) = \{u \in H^1(\Omega) \mid n \cdot \gamma \nabla u|_{\partial\Omega} = 0\},$$

and introduce the homogenous Sobolev seminorm

$$\|u\|_{\dot{H}^1(\Omega)} = \left(\int_\Omega |\nabla u|^2 dx\right)^{1/2}. \tag{6.9}$$

Observe that $\|u\|_{\dot{H}^1(\Omega)} = 0$ if and only if $u = \text{constant}$. Therefore, we introduce the quotient space

$$\dot{H}^1_N(\Omega) = H^1_N(\Omega)/\mathbb{R},$$

i.e., we identify elements differing by a constant. In this quotient space, (6.9) defines a norm.

Let v be a test function, and assume for now that u satisfies the equation (6.4) and the Neumann boundary condition (6.7). By Green's formula, we have

$$\int_\Omega v \nabla \cdot \gamma \nabla u \, dx = -\int_\Omega \nabla v \cdot \gamma \nabla u \, dx.$$

Similarly, the right side of the equation (6.4) together with the boundary condition (6.6) gives

$$\int_\Omega v \nabla \cdot J_0 \, dx = -\int_\Omega \nabla v \cdot J_0 \, dx.$$

The uniqueness of the weak solution of the forward problem is proved in the following theorem.

Theorem 6.3. *Let $\gamma \in \mathcal{A}(\Omega)$ and $J_0 \in \mathcal{J}(\Omega)$. Then there is a unique $u \in \dot{H}_N^1(\Omega)$ satisfying the weak form equation*

$$\mathcal{B}_\gamma(v, u) = \int_\Omega \nabla v \cdot \gamma \nabla u \, dx = \int_\Omega \nabla v \cdot J_0 \, dx \qquad (6.10)$$

for all $v \in \dot{H}_N^1(\Omega)$.

Proof. From the definition of admissible class of admittivities, it follows that

$$c\|u\|_{\dot{H}^1(\Omega)}^2 \leq \int_\Omega \gamma |\nabla u|^2 dx = \mathcal{B}_\gamma(u, u) \leq C\|u\|_{\dot{H}^1(\Omega)}^2,$$

i.e., the quadratic form $\mathcal{B}_\gamma$ defines an equivalent norm in $\dot{H}_N^1(\Omega)$. Furthermore, since

$$\left| \int_\Omega \nabla v \cdot J_0 \, dx \right| \leq C\|v\|_{\dot{H}^1(\Omega)},$$

we see that the right side of (6.10) defines a continuous functional in $\dot{H}_N^1(\Omega)$. The claim follows now by the Riesz representation theorem; see Appendix A. □

Observe that the theorem above it follows that the solution is unique only up to a constant. This ambiguity is removed by the ground condition (6.8).

The *electric inverse source problem* can now be stated as follows.

Problem 6.4. Assume that the impedivity γ of a body Ω is known. Given the observations[1] $\{u(p_1), \ldots, u(p_L)\}$ of the voltage potential at given points $p_\ell \in \partial\Omega$, estimate the current density J_0 in Ω.

This inverse problem is essentially what EEG and ECG need to solve. In electroencephalography, the source is the electric brain activity, while in electrocardiography one seeks to estimate the electric functions of the heart. The pointwise voltage measurements are idealizations of voltage recordings through contact electrodes attached to the surface of the body.

We remark here that the above formulation of the EEG/ECG problem is not the only one considered in literature. Often, rather than seeking the electric sources, one seeks to solve the electric voltage potential distribution on an intrinsic surface, cardiac or cerebral, from the corresponding voltage distributions on the body's outer surface.

6.2.3 Magnetic Inverse Source Problems

The magnetic forward problem is based on the same assumptions as the electric one, i.e., that the normal components of the source current and the Ohmic

---

[1] Strictly speaking, to define pointwise boundary values of the solutions, we should prove additional regularity of the solutions. For brevity, this discussion is omitted.

current vanish at the boundary $\partial\Omega$. Thus, let J_0 be the source current density in Ω with the boundary condition (6.6), and let u denote the corresponding voltage potential satisfying the Neumann boundary condition (6.7). Assuming that $\widehat{\mu}(x,\omega) = \mu_0 =$ constant, equation (6.3) gives, within the applicability of the quasi-static approximation,

$$\nabla \times B = \mu_0 \nabla \times H = \mu_0 J, \quad J = -\gamma \nabla u + J_0.$$

Furthermore, equation (6.2) implies that

$$\nabla \cdot B = \nabla \cdot (\mu_0 H) = 0.$$

Hence, B satisfies the equation

$$-\Delta B = ((\nabla \times)^2 - \nabla \nabla \cdot) B = \mu_0 \nabla \times J.$$

To solve this vector-valued Poisson equation, we need to specify the asymptotic behavior of the magnetic field. It is required that at infinity, the asymptotic behavior of the field is

$$|B(x)| = \mathcal{O}\left(\frac{1}{|x|}\right).$$

By using the fact that

$$\Phi(x-y) = \frac{1}{4\pi} \frac{1}{|x-y|}, \quad x \neq y,$$

is the fundamental solution of the operator $-\Delta$ with the required asymptotics, we may represent the magnetic field as

$$B(x) = \Phi * (\nabla \times J)(x), \quad x \in \mathbb{R}^3 \setminus \overline{\Omega},$$

where the convolution must be interpreted in the sense of distributions, since $\nabla \times J$ is, in general, a distribution supported in $\overline{\Omega}$. By the definition of distribution derivatives, we obtain a classical integral representation

$$B(x) = \frac{\mu_0}{4\pi} \int_\Omega J(y) \times \frac{x-y}{|x-y|^3} dy, \quad x \in \mathbb{R}^3 \setminus \overline{\Omega}. \tag{6.11}$$

This representation is known as the *Biot–Savart law*. Further, by substituting $J = J_0 - \gamma \nabla u$, we have

$$B(x) = B_0(x) - \frac{\mu_0}{4\pi} \sum_{j=1}^N \int_{\Omega_j} \gamma(y) \nabla u(y) \times \frac{x-y}{|x-y|^3} dy, \tag{6.12}$$

where the B_0 is the magnetic field of the primary current,

$$B_0(x) = \frac{\mu_0}{4\pi} \int_\Omega J_0(y) \times \frac{x-y}{|x-y|^3} dy. \tag{6.13}$$

Consider the restriction of the integrals in (6.12) to the subset Ω_j. Let $n_j(x)$ denote the exterior unit normal of $\partial\Omega_j$. Integrating by parts yields

$$\int_{\Omega_j} \gamma(y)\nabla u(y) \times \frac{x-y}{|x-y|^3} dy = \int_{\partial\Omega_j} \gamma(y)u(y)n_j(y) \times \frac{x-y}{|x-y|^3} dS$$
$$- \int_{\Omega_j} u(y)\nabla\gamma(y) \times \frac{x-y}{|x-y|^3} dy,$$

since

$$\nabla \times \frac{x-y}{|x-y|^3} = \nabla \times \nabla \frac{1}{|x-y|} = 0.$$

Let $\{\Gamma_k \mid 1 \leq k \leq K\}$ be the set of oriented interfaces between the subdomains Ω_j, and denote by $[\gamma]_k$ the jump of the admittivity across the interface Γ_k. Hence, if Γ_k separates the subdomains Ω_i and Ω_j, and n_i is the orientation of Γ_k, we have

$$[\gamma]_k(y) = \gamma\big|_{\Omega_i}(y) - \gamma\big|_{\Omega_j}(y), \quad y \in \Gamma_k.$$

At the outer surface $\partial\Omega$, we set $\gamma = 0$ outside Ω. With these notations, and by observing that, at the interface Γ_k, the normal vectors n_i and n_j have opposite directions, formula (6.11) yields

$$B(x) = B_0(x) + \frac{\mu_0}{4\pi}\bigg(\sum_{j=1}^{N}\int_{\Omega_j} u(y)\nabla\gamma(y) \times \frac{x-y}{|x-y|^3} dy \quad (6.14)$$
$$- \sum_{k=1}^{K}\int_{\Gamma_k} [\gamma]_k(y)u(y)n_j(y) \times \frac{x-y}{|x-y|^3} dS\bigg).$$

In the special case when $\nabla\gamma = 0$ in each subset Ω_j, i.e., γ is piecewise constant, formula (6.14) is often referred to as the *Geselowitz formula*.

The magnetic inverse source problem can be stated in the following form.

Problem 6.5. Assume that γ in the body Ω is known. Let $v_1, \ldots, v_L$ be a given set of unit vectors. Given the observations $\{v_1 \cdot B(p_1), \ldots, v_L \cdot B(p_L)\}$, where $p_\ell \notin \Omega$, estimate the source current J_0 in Ω.

Similarly as in the electric inverse source problem, the desired current density J_0 represents either the activity of the brain (MEG) or the heart (MCG). The unit vectors v_j define the orientations of the magnetometers, i.e., the axes of the coils used to register the magnetic fields.

Under certain geometric assumptions, the model for the magnetic field can be considerably simplified. One case is when we have radially symmetric geometry. Assume that Ω is a ball, the surfaces Γ_k are concentric spheres, and γ depends only on the distance from the center of the ball. Setting the origin at the center of the ball, we have $\gamma(y) = g(|y|)$, and $\nabla\gamma(y) = g'(|y|)r(y)$, where

$r(y) = y/|y|$ is the radial unit vector. Also, $n_j(y) = r(y)$. Assume that the magnetometers measure the radial component of the magnetic field outside Ω, i.e., $v_\ell = r(p_\ell)$. Since

$$x \cdot (y \times (x - y)) = 0,$$

we observe that

$$r(x) \cdot B(x) = r(x) \cdot B_0(x),$$

in other words, the radial component of the magnetic field is insensitive to the radial conductivity of the body.

A similar result holds when the body Ω is a half-space, the conductivity depends only on the depth and the observed component of the magnetic field is vertical.

Consider the field B_0 outside the body Ω. Assume that with respect to an arbitrarily fixed origin, the current density J_0 in Ω is radial, i.e., for some scalar function j,

$$J(y) = r(y)j(y), \quad r(y) = \frac{y}{|y|}.$$

Since $x \cdot (y \times (x - y)) = 0$, we immediately have from formula (6.13) that the radial component of B_0 vanishes, i.e.,

$$r(x) \cdot B_0(x) = \frac{\mu_0}{4\pi} \int_\Omega j(y) \frac{x \cdot y \times (x-y)}{|x||y||x-y|^3} dy = 0.$$

In particular, for a spherically symmetric body Ω, radial components of the source currents are completely invisible for observations of the radial component.

So far, we have assumed that the source current J_0 is in the admissible class $\mathcal{J}(\Omega)$. This assumption is largely dictated by the solvability of the equation (6.10) defining the electric potential. However, as it was shown above, in particular geometries the Ohmic currents, and therefore also the electric potential, play no role. In those cases, it is common to represent the source current J_0 by using *current dipoles* that do not fit into the admissible class $\mathcal{J}(\Omega)$ defined here.

A current dipole at $x = x_0 \in \Omega$ with dipole moment vector q is defined as a localized singular source,

$$J_0(x) = q\,\delta(x - x_0).$$

Assume that the souce current consists of N current dipoles, i.e.,

$$J_0(x) = \sum_{n=1}^N q_n\,\delta(x - x_n).$$

If there is no magnetic field due to the Ohmic part of the current, we obtain a discrete representation of the magnetic field as

$$B(x) = \frac{\mu_0}{4\pi} \int_\Omega \frac{(x-y) \times J_0(y)}{|x-y|^3} dy = \frac{\mu_0}{4\pi} \sum_{n=1}^{N} \frac{(x-x_n) \times q_n}{|x-x_n|^3}. \qquad (6.15)$$

This representation will be used in some of the examples discussed later in this book.

Observe that in both the electric and magnetic inverse source problems, it was assumed that the impedivity γ is known. In practice, this assumption is seldom valid. In the special cases when the Ohmic part can be ignored, this is of course no problem. However, in proper modelling, the estimation of the impedivity should be included as a part of the problem. Such modelling makes the inverse problems highly nonlinear and considerably more complex than the problems considered above.

6.3 Impedance Tomography

The *electrical impedance tomography* (EIT) problem can be described physically in the following way. We have a body with unknown electric resistivity, or more generally, impedivity distribution. A set of contact electrodes is attached on the surface of the body and, through these electrodes, prescribed electric currents are injected into the body. The corresponding voltages needed to maintain these currents are measured on these electrodes. This process may be repeated with various currents. The aim is to estimate the resistivity distribution in the body from this set of data.

The leading idea for solving this problem is *Ohm's law*, which states that in a simple resistor, the resistance is the voltage divided by the current. Here, however, instead of a single resistor, we have a possibly complicated resistivity distribution, and no straightforward method of getting it from the described measurement exists.

The mathematical description of the measurement setting is as follows: Let $\Omega \subset \mathbb{R}^n$, $n = 2, 3$, be a bounded domain with a connected complement corresponding to the body. We assume that Ω is simply connected and it has a smooth boundary $\partial\Omega$. The assumption that the complement is connected implies that $\partial\Omega$ consists of one curve in $n = 2$, or one surface in $n = 3$.

Consider time-harmonic electromagnetic fields in Ω with low frequencies. The fields are described by the time-harmonic Maxwell's equations (6.2)–(6.3). In EIT, we assume that no internal current sources are present, i.e., $J_0 = 0$.

As in the discussion of the inverse source problems, we consider the quasi-static approximation, allowing us to represent the electric field in terms of a scalar potential,

$$E(x) = -\nabla u(x).$$

By substituting this to the equation (6.3) and applying the divergence on both sides, we obtain

$$\nabla \cdot \gamma \nabla u = 0. \qquad (6.16)$$

Since we restrict here the discussion to this static case, $\omega \to 0+$, we drop the frequency dependency from the notations, i.e., we write the resistivity simply as

$$\rho = \rho(x) = \frac{1}{\gamma(x)},$$

and $\gamma = \gamma(x) = \hat{\sigma}(x, 0+)$ is simply the conductivity distribution of the body.

Let us consider the boundary conditions. Assume that L contact electrodes are attached to the body's surface. Mathematically, the electrodes are modelled as strictly disjoint surface patches $e_\ell \subset \partial\Omega$, $1 \leq \ell \leq L$, with $\bar{e}_\ell \cap \bar{e}_k = \emptyset$ for $\ell \neq k$. In dimension two, the e_ℓ's are just disjoint line segments. In the three-dimensional case, we assume that the boundaries of the electrodes are piecewise smooth, regular closed curves.

Let I_ℓ be the electric current injected through the electrode e_ℓ. We call the vector $I = [I_1, \ldots, I_L]^T \in \mathbb{R}^L$ a *current pattern* if it satisfies the charge conservation condition

$$\sum_{\ell=1}^{L} I_\ell = 0. \tag{6.17}$$

To describe the boundary condition corresponding to the current injection, observe that we know only the total current injected through each electrode, while the actual distribution of current along the electrode is not known. We assume also that no current flows in or out the body between the electrodes. These observations lead to the boundary conditions

$$\int_{e_\ell} \gamma \frac{\partial u}{\partial n} dS = I_\ell, \quad 1 \leq \ell \leq L, \tag{6.18}$$

$$\gamma \frac{\partial u}{\partial n}\bigg|_{\partial\Omega \setminus \cup e_\ell} = 0. \tag{6.19}$$

Consider next the boundary conditions related to the voltages. Let U_ℓ denote the voltage on the ℓth electrode, the ground voltage being chosen so that

$$\sum_{\ell=1}^{L} U_\ell = 0. \tag{6.20}$$

The vector $U = [U_1, \ldots, U_L]^T \in \mathbb{R}^L$ is called a *voltage pattern*. If the electrodes were perfectly conducting, the tangential electric field should vanish along the electrodes and the proper boundary condition would be

$$u\big|_{e_\ell} = U_\ell, \quad 1 \leq \ell \leq L. \tag{6.21}$$

In practice, however, the electrodes are not perfect conductors. Rather, there is usually a thin contact impedance layer between the electrode and the body. Assuming that this impedance, denoted by z_ℓ, is constant along the electrode e_ℓ, a more accurate boundary condition than (6.21) is

204   6 Model Problems

$$\left(u + z_\ell \gamma \frac{\partial u}{\partial n}\right)\bigg|_{e_\ell} = U_\ell, \quad 1 \leq \ell \leq L. \tag{6.22}$$

We can now state the *EIT forward problem* as follows.

Problem 6.6. Given the admittivity distribution γ in Ω, the contact impedances $z = [z_1, \ldots, z_L]^\mathrm{T}$ and a current pattern $I = [I_1, \ldots, I_L]^\mathrm{T}$ satisfying the conservation of charge condition (6.17), find the voltage potential u *and* the electrode voltage vector $U = [U_1, \ldots, U_L]^\mathrm{T}$ satisfying the equation (6.4) and the boundary conditions (6.18), (6.19) and (6.22).

Before discussing the solvability of this problem or the assumptions concerning γ, let us discuss and formulate the EIT inverse problem.

Due to the linearity of Maxwell's equations, the dependence of the electrode voltages on the current is linear. We may define the *resistance matrix* of the complete model as follows: If the admittivity γ and the contact impedances z_ℓ, are given, the resistance matrix $R(\gamma, z)$ is the $L \times L$ complex matrix with the property

$$U = R(\gamma, z)I,$$

where I is any current pattern that satisfies (6.17) and U is the corresponding voltage pattern satisfying condition (6.20). Let $\{I^{(1)}, \ldots, I^{(K)}\}$ be a set of linearly independent current patterns. Observe that, due to the charge conservation condition (6.17), $K \leq L - 1$. If $K = L - 1$, the set is maximal, and it is called a *frame*. The corresponding voltage vectors are $U^{(k)} = R(\gamma, z)I^{(k)}$, $1 \leq k \leq K$. The *EIT inverse problem* can be stated as follows:

Problem 6.7. Given a set $\{I^{(1)}, \ldots, I^{(K)}\}$ of linearly independent current patterns, $K \leq L - 1$, and the observations of the corresponding voltage patterns $\{U^{(1)}, \ldots, U^{(K)}\}$, estimate the admittivity distribution γ in Ω and the contact impedances z_ℓ.

After these considerations of the underlying ideas, we can turn to a more rigorous treatment of the problem. First, we fix the assumptions concerning the admittivity. As stated before, we assume that both the admittivity and the contact impedances are real-valued and strictly positive. This assumption means that we consider the static measurement. An extension of the discussion to ensue can be extended to complex valued parameters in a straightforward manner. We also assume that the admittivity is in the admissible class of Definition 6.2, i.e., $\gamma \in \mathcal{A}(\Omega)$.

Consider the forward problem of EIT. Due to possible discontinuities of $\gamma \in \mathcal{A}$, equation (6.16) must be interpreted in the weak sense as follows. Let v be a test function and $V = [V_1, \ldots, V_L]^\mathrm{T} \in \mathbb{R}^L$. If u satisfies equation (6.16), Green's formula yields

$$0 = \int_\Omega v \nabla \cdot \gamma \nabla u \, dx = \int_{\partial \Omega} v \gamma \frac{\partial u}{\partial n} dS - \int_\Omega \gamma \nabla v \cdot \nabla u \, dx.$$

On the other hand, if u satisfies the boundary conditions (6.18) and (6.19), we have

$$\int_{\partial\Omega} v\gamma\frac{\partial u}{\partial n}dS = \sum_{\ell=1}^{L}\int_{e_\ell} v\gamma\frac{\partial u}{\partial n}dS$$

$$= \sum_{\ell=1}^{L}\int_{e_\ell}(v-V_\ell)\gamma\frac{\partial u}{\partial n}dS + \sum_{\ell=1}^{L}V_\ell I_\ell.$$

Further, from the boundary condition (6.22) we see that since $z_\ell > 0$,

$$\gamma\frac{\partial u}{\partial n}\bigg|_{e_\ell} = \frac{1}{z_\ell}(U_\ell - u)\big|_{e_\ell}.$$

By substituting this equality into the previous ones we obtain

$$\int_\Omega \gamma\nabla v\cdot\nabla u\,dx + \sum_{\ell=1}^{L}\frac{1}{z_\ell}\int_{e_\ell}(v-V_\ell)(u-U_\ell)dS = \sum_{\ell=1}^{L}V_\ell I_\ell.$$

This is the variational form for solving the forward problem. To give a rigorous formulation, we introduce the notations

$$\mathbb{H} = H^1(\Omega) \oplus \mathbb{R}^L, \tag{6.23}$$

where $H^1(\Omega)$ is the L^2-based Sobolev space and $\mathbb{R}^L$ is equipped with the Euclidian norm. Further, let

$$\dot{\mathbb{H}} = \mathbb{H}/\mathbb{R}, \tag{6.24}$$

equipped with the quotient norm,

$$\|(u,U)\|_{\dot{\mathbb{H}}} = \inf_{c\in\mathbb{R}}\|(u-c,U-c)\|_{\mathbb{H}}. \tag{6.25}$$

Thus, $(u,U)\in\mathbb{H}$ and $(v,V)\in\mathbb{H}$ are in the same equivalence class in $\dot{\mathbb{H}}$ if

$$u - v = U_1 - V_1 = \cdots = U_L - V_L = \text{constant}. \tag{6.26}$$

With these notations, the following proposition fixes the notion of the weak solution of the electrode model.

Theorem 6.8. *Let $\gamma \in \mathcal{A}(\Omega)$. The problem (6.4), (6.18), (6.19) and (6.22) has a unique weak solution $(u,U) \in \dot{\mathbb{H}}$ in the following sense: There is a unique $(u,U) \in \dot{\mathbb{H}}$ satisfying the equation*

$$\mathcal{B}_{\gamma,z}((u,U),(v,V)) = \sum_{\ell=1}^{L} I_\ell V_\ell \tag{6.27}$$

for all $(v,V) \in \dot{\mathbb{H}}$, where the quadratic form $\mathcal{B}_{\gamma,z}$ is given as

$$\mathcal{B}_{\gamma,z}((u,U),(v,V)) = \int_\Omega \gamma\nabla u \cdot \nabla v\, dx + \sum_{\ell=1}^L \frac{1}{z_\ell}\int_{e_\ell}(u-U_\ell)(v-V_\ell)dS. \quad (6.28)$$

Furthermore, the quadratic form is coercive in $\dot{\mathbb{H}}$, i.e., we have the inequalities

$$\alpha_0 \|(u,U)\|_{\dot{\mathbb{H}}}^2 \leq \mathcal{B}_{\gamma,z}((u,U),(u,U)) \leq \alpha_1 \|(u,U)\|_{\dot{\mathbb{H}}}^2 \quad (6.29)$$

for some constants $0 < \alpha_0 \leq \alpha_1 < \infty$.

Proof. We show that, in fact, the quadratic form $\mathcal{B}_{\gamma,z}$ defines a norm of $\dot{\mathbb{H}}$ that is equivalent to the quotient norm (6.25), i.e., there are constants $0 < a \leq A < \infty$ such that for all $(u,U) \in \dot{\mathbb{H}}$,

$$a\mathcal{B}_{\gamma,z}(u,U) \leq \|(u,U)\|_{\dot{\mathbb{H}}}^2 \leq A\mathcal{B}_{\gamma,z}(u,U). \quad (6.30)$$

Then the claim of the theorem follows from the Riesz representation theorem (see Appendix A).

Consider the first inequality. Let $(u,U) \in \dot{\mathbb{H}}$, and let $c \in \mathbb{R}$ be a constant such that

$$\|u - c\|_{H^1}^2 + \|U - c\|^2 < \|(u,U)\|_{\dot{\mathbb{H}}}^2 + \varepsilon,$$

where $\varepsilon > 0$ is arbitrary. Then

$$\mathcal{B}_{\gamma,z}(u,U) \leq C\left(\int_\Omega |\nabla(u-c)|^2 dx + \sum_{\ell=1}^L \int_{e_\ell} |(u-c) - (U_\ell - c)|^2 dS\right)$$

$$\leq C\left(\|u-c\|_{H^1}^2 + 2\int_{\partial\Omega} |u-c|^2 dS + 2\sum_{\ell=1}^L |e_\ell| |U_\ell - c|^2\right),$$

where $|e_\ell|$ denotes the size of the electrode. Since

$$\int_{\partial\Omega} |u-c|^2 dS \leq \|u-c\|_{H^{1/2}(\partial\Omega)}^2 \leq C\|u-c\|_{H^1(\Omega)}^2, \quad (6.31)$$

by the trace theorem of Sobolev spaces (see Appendix 1), we obtain the estimate

$$\mathcal{B}_{\gamma,z}(u,U) \leq C(\|u-c\|_{H^1}^2 + \|U-c\|^2) < C(\|(u,U)\|_{\dot{\mathbb{H}}}^2 + \varepsilon)$$

for all $\varepsilon > 0$. Choosing $a = 1/C$ the first inequality of (6.30) follows.

To prove the second part, assume that the inequality does not hold. Then we could pick a sequence $(u^{(n)}, U^{(n)}) \in \dot{\mathbb{H}}$, $1 \leq n < \infty$, such that

$$\|(u^{(n)}, U^{(n)})\|_{\dot{\mathbb{H}}}^2 = 1, \quad \mathcal{B}_{\gamma,z}(u^{(n)}, U^{(n)}) < \frac{1}{n}.$$

Let $(c^{(n)})$ denote a sequence of constants such that the sequence $(w^{(n)}, W^{(n)}) = (u^{(n)} - c^{(n)}, U^{(n)} - c^{(n)})$ satisfies

6.3 Impedance Tomography

$$1 \leq \|w^{(n)}\|_{H^1}^2 + \|W^{(n)}\|^2 < 1 + \frac{1}{n}. \tag{6.32}$$

By the compact embedding theorem of Sobolev spaces (see Appendix A), we may choose a subsequence of $(w^{(n)})$ that converges in $L^2(\Omega)$. Let us denote this subsequence for simplicity also by $(w^{(n)})$, and $w^{(n)} \to w$ in $L^2(\Omega)$. Moreover, by assumption

$$\|\nabla w^{(n)}\|_{L^2}^2 = \|\nabla u^{(n)}\|_{L^2}^2 \leq C\mathcal{B}_{\gamma,z}(u^{(n)}, U^{(n)}) < \frac{C}{n},$$

so $(w^{(n)})$ is a Cauchy sequence in $H^1(\Omega)$, implying that $w \in H^1(\Omega)$ and $\nabla w = \lim \nabla w^{(n)} = 0$, i.e., w is constant, $w = c_0$.

Consider now the integrals over the electrodes. For each ℓ, we have

$$\frac{C}{n} > = \int_{e_\ell} |(w^{(n)} - c_0) - (W^{(n)} - c_0)|^2 dS$$

$$= \int_{e_\ell} |w^{(n)} - c_0|^2 dS - 2|W^{(n)} - c_0| \int_{e_\ell} |w^{(n)} - c_0| dS + |e_\ell||W^{(n)} - c_0|^2$$

$$\geq -2|W^{(n)} - c_0| \int_{e_\ell} |w^{(n)} - c_0| dS + |e_\ell||W^{(n)} - c_0|^2.$$

This estimate, together with the inequality similar to (6.31), gives

$$|W^{(n)} - c_0|^2 \leq C \left(\frac{1}{n} + |W^{(n)} - c_0| \|w^{(n)} - c_0\|_{H^1(\Omega)} \right).$$

Since by (6.32), the sequence $(W^{(n)})$ is bounded, we see that in fact $W^{(n)} \to c_0$ as $n \to \infty$. But this leads to a contradiction since by assumption

$$\|w^{(n)} - c_0\|_{H^1}^2 + \|W^{(n)} - c_0\|^2 \geq \|(u^{(n)}, U^{(n)})\|_{\dot{\mathbb{H}}}^2 = 1.$$

The theorem is thus proved. □

A common approach to the EIT problem is to use local linearization. Assume that $\gamma, \gamma_0 \in \mathcal{A}(\Omega)$ and that the contact impedances are fixed. Let (u, U) and (u_0, U_0) denote the corresponding solutions of the forward problem with a given current pattern I, i.e.,

$$\mathcal{B}_{\gamma,z}((u, U), (v, V)) = \mathcal{B}_{\gamma_0,z}((u_0, U_0), (v, V)) = \sum_{\ell=1}^{L} I_\ell V_\ell$$

for all $(v, V) \in \dot{\mathbb{H}}$. Let us denote

$$(u, U) - (u_0, U_0) = (w, W), \quad \delta\gamma = \gamma - \gamma_0.$$

By substracting, and using the linearization approximation $\delta\gamma w \approx 0$, we arrive at the approximate identity used for computing (w, W),

$$\mathcal{B}_{\gamma_0,z}((w,W),(v,V)) = -\int_\Omega \delta\gamma \nabla u_0 \cdot \nabla v\, dx. \tag{6.33}$$

Hence, the linearized approximation of the perturbed solution can be computed by means of the same quadratic form as the unpertubed solution u_0 but with a different right-hand side.

The actual numerical implementation of the EIT forward solver by using finite elements is not discussed here. References concerning the implementation and linearization are given in "Notes and Comments" at the end of the chapter.

6.4 Optical Tomography

In optical tomography, measurements of scattered and transmitted near-infrared light on the surface of the body are used to estimate the internal optical properties of the body. More precisely, the measurement setting is as follows. A target with unknown internal optical properties is illuminated by one or several near-infrared sources attached on the surface of the object. The scattered field transmitted through the object is recorded at various locations on the same surface. From these data, the scattering and absoption properties of the material needs to be estimated.

When the target material is strongly scattering, as is typically the case in several medical and industrial applications, the signal propagation in the medium is diffusive, so in this case we can say that optical tomography is a *diffuse tomographic imaging method* similar to the electrical impedance tomography.

6.4.1 The Radiation Transfer Equation

The propagation of near-infrared light in the medium is governed by Maxwell's equations. However, the wavelength in the near-infrared region is so small compared to characteristic distances in the medium that the use of exact wave propagation models is of little use. Furthermore, since the scattering of light is typically strong, the radiation is completely incoherent. Therefore, the light propagation in optical tomography is modelled by using the *radiation transfer equation* (RTE), also known as the *Boltzmann equation*. This will be our starting point also here.

We begin by recalling the basic notations and concepts. Let $\Omega \subset \mathbb{R}^n$, $n = 2$ or $n = 3$ be a bounded body with a smooth boundary and connected complement. We consider radiation in the body Ω. The first concept that we introduce is the *radiation flux density*. Let $\theta \in \mathbb{S}^{n-1}$ be a direction vector. The radiation flux density at $x \in \Omega$ at the time $t \in \mathbb{R}$ to the infinitesimal solid angle ds in direction θ is written as

$$dJ(x,t,\theta) = I(x,t,\theta)\theta ds(\theta).$$

Here, the scalar amplitude $I(x, t, \theta)$ is called the *radiance*. In the framework of the transport theory, this function satisfies the radiation transfer equation. Assuming that the propagation speed of the light is constant in the medium, the RTE can be written as

$$\frac{1}{c} I_t(x, t, \theta) + \theta \cdot \nabla I(x, t, \theta) + (\mu_a(x) + \mu_s(x)) I(x, t, \theta) \qquad (6.34)$$

$$- \mu_s(x) \int_{\mathbb{S}^{n-1}} f(x, \theta, \omega) I(x, t, \omega) ds(\omega) = q(x, t, \theta),$$

where the right side q is the source term and c is the presumably constant speed of light. The scalar functions $\mu_a \geq 0$ and $\mu_s \geq 0$ are the *scattering* and *absorption coefficients*, respectively. They represent the probability density per unit volume of an absorption or a scattering event. The kernel $f \geq 0$ is the *scattering phase function*. We can think of $f(x, \theta, \omega) ds(\omega)$ as the probability that radiation propagating in the direction θ scatters into the angle $ds(\omega)$. The phase function satisfies the *reciprocity condition*,

$$f(x, \theta, \omega) = f(x, -\omega, -\theta).$$

Furthermore, the probability of scattering into the full sphere is unity regardless of the incoming direction, i.e.,

$$\int_{\mathbb{S}^{n-1}} f(x, \theta, \omega) ds(\omega) = \int_{\mathbb{S}^{n-1}} f(x, \theta, \omega) ds(\theta) = 1. \qquad (6.35)$$

The unity of the latter integral is a consequence of the reciprocity property.

In *isotropic* media, the scattering depends only on the relative angle between the incoming and the scattering direction, i.e.,

$$f(x, \theta, \omega) = h(x, \theta \cdot \omega), \qquad (6.36)$$

for some function $h : \Omega \times \mathbb{R} \to \mathbb{R}_+$. If this is not the case, the material is *anisotropic*.

Given the radiation flux density, the flux through an infinitesimal oriented surface patch αdS, $\alpha \in \mathbb{S}^{n-1}$, is obtained by integrating over all radiation directions the flux density,

$$d\Phi(x, t) = \left(\int_{\mathbb{S}^{n-1}} dJ(x, t, \theta)) \right) \cdot \alpha dS$$

$$= \left(\int_{\mathbb{S}^{n-1}} I(x, t, \theta) \theta ds(\theta)) \right) \cdot \alpha dS \qquad (6.37)$$

$$= J(x, t) \cdot \alpha dS,$$

where the vector field J is the *energy current density*. For later use, we define also the scalar functions

$$\varphi(x,t) = \int_{\mathbb{S}^{n-1}} I(x,t,\theta) ds(\theta), \tag{6.38}$$

called the *energy fluence*.

To define the boundary conditions, let $\nu = \nu(x)$ denote the exterior unit normal vector of the outer surface at $x \in \partial\Omega$. The *photon flux density* at $x \in \partial\Omega$ into the body is given by the integral

$$\Phi_-(x,t) = -\int_{\{\theta \cdot \nu(x) < 0\}} I(x,t,\theta)\theta \cdot \nu(x) dS(\theta), \tag{6.39}$$

where the minus sign is added to make the quantity positive. Similarly, the flux out of the body is given by

$$\Phi_+(x,t) = \int_{\{\theta \cdot \nu(x) > 0\}} I(x,t,\theta)\theta \cdot \nu(x) dS(\theta). \tag{6.40}$$

Let us denote by $R = R(x)$ the *reflection coefficient* of the surface at $\partial\Omega$, where $0 \leq R \leq 1$. The reflection coefficient describes the ratio of the outflowing photons that are reflected back into the body by the mismatch of the refractive index between the body and the surrounding material. If no photons are injected into the body through a surface patch $dS(x)$, the inward flow is purely due to the reflected photons, i.e., we have the boundary condition

$$\Phi_-(x,t) = R(x)\Phi_+(x,t), \tag{6.41}$$

and the photon flux actually leaving the material is

$$\Phi_{\text{out}}(x,t) = (1 - R(x))\Phi_+(x,t). \tag{6.42}$$

To describe the source model, assume that, on the surface, a set of optical fibers has been attached to illuminate the material, while another set is used to record the scattered light. We assume that the source and the receivers are on for a period $T > 0$ of time. Ideally, the optical fibers are pointlike. Let $x_\ell \in \partial\Omega$, $1 \leq \ell \leq L$ be surface locations representing feed fibers, and $y_j \in \partial\Omega$, $1 \leq j \leq J$ the recording fibers. At the feed fiber locations, the total photon flux entering the body consists of the feed flux denoted by Φ_{in}, which we assume to be known, plus the reflected photons. Therefore, instead of the equation (6.41), we have

$$\Phi_-(x_\ell, t) = R(x_\ell)\Phi_+(x_\ell, t) + \Phi_{\text{in}}(x_\ell, t), \quad 1 \leq \ell \leq L. \tag{6.43}$$

In optical tomography experiments, we control the feed function Φ_{in}. We can now state the inverse problem of optical tomography based on the radiation transfer equation.

Problem 6.9. Assume that $I(x,t,\omega)$ satisfies the radiation transfer equation (6.34) in Ω with $q(x,t) = 0$ and with the boundary condition (6.41) at $x \neq$

x_ℓ and (6.43) at given locations x_ℓ. The problem is to estimate the optical properties $\mu_a(x)$, $\mu_s(x)$ and $f(x,\theta,\omega)$ from the knowledge of all possible feed functions $\Phi_{\text{in}}(x_\ell, t)$, $1 \leq \ell \leq L$, $0 \leq t \leq T$ and the corresponding photon fluxes, $\Phi_{\text{out}}(y_j, t)$, $0 \leq j \leq J$, $0 \leq t \leq T$.

Before continuing our discussion, a few remarks are in order. First, the solvability of the radiation transfer equation is not a simple matter in general. In this book, the question is not discussed in detail, and we refer to literature (see "Notes and Comments" at the end of the section).

In the above formulation of the problem, the source was modeled by a boundary condition (6.43). Often, in practice, the photon feed through an optical fiber at x_ℓ is modelled by an equivalent source term q_ℓ inside the body. The source term appears then as a nonvanishing right side in the equation (6.34), while the boundary condition is everywhere on $\partial\Omega$ given by $\Phi_{\text{in}} = 0$.

Finally, we assumed the reflection coefficient R to be known. If this is not the case or if it is poorly known, the estimation of R should be included as a part of the problem, similarly as the estimation of the contact impedances in impedance tomography should be included in the inverse problem.

6.4.2 Diffusion Approximation

The radiation transfer equation being an integro-differential equation, it leads easily to numerical problems of prohibitive size unless simplifications are made. The common simplification, that is justified at least in the case of strongly scattering media, is the *diffusion approximation*. The idea is that in strongly scattering media, the radiance $I(x,t,\theta)$ varies only moderately when considered as a function of the propagation direction θ. A standard way to accomplish the approximation is to expand the function in terms of spherical harmonics and truncate the series. Here, we derive the diffusion approximation in a slightly different, but equivalent, way.

We begin with a simple lemma.

Lemma 6.10. *Let us denote* $\theta = (\theta_1, \ldots, \theta_n) \in \mathbb{S}^{n-1}$, *and*

$$H_1 = \text{sp}\{1, \theta_j,\ 1 \leq j \leq n\} \subset L^2(\mathbb{S}^{n-1}).$$

The orthogonal projection $P : L^2(\mathbb{S}^{n-1}) \to H_1$ *is given as*

$$Pg = \alpha + a \cdot \theta,$$

where $\alpha \in \mathbb{R}$, $a \in \mathbb{R}^n$ *are given by the formulas*

$$\alpha = \frac{1}{|\mathbb{S}^{n-1}|} \int_{\mathbb{S}^{n-1}} g(\theta) ds,$$

$$a = \frac{n}{|\mathbb{S}^{n-1}|} \int_{\mathbb{S}^{n-1}} \theta\, g(\theta) ds.$$

Proof. Consider the functional $F : \mathbb{R} \times \mathbb{R}^n \to \mathbb{R}$,
$$F(\alpha, a) = \|g - \alpha - a \cdot \theta\|_2^2.$$

Clearly, this function attains its minimum when $\alpha + a \cdot \theta = Pg$. By symmetry, we have
$$\int_{\mathbb{S}^{n-1}} \theta \, ds = 0,$$
and
$$\int_{\mathbb{S}^{n-1}} \theta_i \theta_j \, dS(\theta) = \frac{|\mathbb{S}^{n-1}|}{n} \delta_{i,j}. \qquad (6.44)$$

The last equality follows from the fact that by symmetry,
$$\int_{\mathbb{S}^{n-1}} \theta_1^2 \, dS(\theta) = \cdots = \int_{\mathbb{S}^{n-1}} \theta_n^2 \, dS(\theta),$$
so
$$\int_{\mathbb{S}^{n-1}} \theta_j^2 \, dS(\theta) = \frac{1}{n} \sum_{j=1}^n \int_{\mathbb{S}^{n-1}} \theta_j^2 \, dS(\theta) = \frac{|\mathbb{S}^{n-1}|}{n}.$$

Hence,
$$F(\alpha, a) = \|g\|_2^2 - 2\left(\alpha \int_{\mathbb{S}^{n-1}} g(\theta) \, ds + a \cdot \int_{\mathbb{S}^{n-1}} \theta \, g(\theta) \, ds\right)$$
$$+ |\mathbb{S}^{n-1}|\left(\alpha^2 + \frac{\|a\|^2}{n}\right).$$

By setting
$$\nabla_{\alpha, a} F = 0,$$
it follows that α and a must take on the values stated in the theorem. $\square$

We turn now to the radiative transfer equation, which we write compactly as
$$\mathcal{B}I = q,$$
where $\mathcal{B}$ denotes the integro-differential operator on the left side of the equation (6.34).

Definition 6.11. *The first-order polynomial approximation, or P_1 approximation for short, of the radiative transfer equation (6.34) is*
$$P\mathcal{B}PI = Pq.$$

Remark As mentioned before, the P_1 approximation is formulated in terms of spherical harmonics. Since the first-order polynomials are always harmonic in $\mathbb{R}^n$, $n \geq 2$, there is no need here to introduce spherical harmonics. However, for higher-order approximations (referred to as P_k approximations), the spherical harmonics are indispensable.

From the definitions of the previous section and Lemma 6.10, we may immediately write for the radiance I the equation

$$(PI)(x,t) = \frac{1}{|\mathbb{S}^{n-1}|}(\varphi(x,t) + nJ(x,t) \cdot \theta),$$

so the P_1 approximation becomes

$$P\mathcal{B}(\varphi + nJ \cdot \theta) = Pq.$$

The following lemma gives the explicit form of the operator $P\mathcal{B}P$.

Lemma 6.12. *The explicit form of the P_1 approximation is given by*

$$P\mathcal{B}(\varphi + nJ \cdot \theta) = \left(\frac{1}{c}\varphi_t + \mu_a\varphi + \nabla \cdot J\right)$$
$$+ n\left(\frac{1}{c}J_t + \frac{1}{n}\nabla\varphi + (\mu_a + \mu_s(1-B))J\right) \cdot \theta,$$

where $B = B(x) \in \mathbb{R}^{n \times n}$ is a symmetric positive definite matrix with entries

$$B_{i,j}(x) = \frac{n}{|\mathbb{S}^{n-1}|}\int_{\mathbb{S}^{n-1}}\int_{\mathbb{S}^{n-1}} \theta_i\omega_j f(x,\theta,\omega)ds(\theta)ds(\omega).$$

Furthermore, the spectrum of B lies in the interval $[0,1]$.

Proof. Clearly, we only need to consider the gradient and integral parts of the radiative transfer equation. We have

$$P\theta \cdot \nabla\varphi = \theta \cdot \nabla\varphi$$

and

$$P(\theta \cdot \nabla(J \cdot \theta)) = \frac{1}{|\mathbb{S}^{n-1}|}\left(\int_{\mathbb{S}^{n-1}} \omega \cdot \nabla(J \cdot \omega)ds(\omega)\right.$$
$$\left. + n\theta \cdot \int_{\mathbb{S}^{n-1}} \left(\omega\left(\omega \cdot \nabla(J \cdot \omega)\right)\right)ds(\omega)\right).$$

By symmetry arguments, the latter integral vanishes while the former one gives

$$\int_{\mathbb{S}^{n-1}} \omega \cdot \nabla(J \cdot \omega)ds = \frac{|\mathbb{S}^{n-1}|}{n}\nabla \cdot J,$$

so

$$P\theta \cdot \nabla(\varphi + nJ \cdot \theta) = \theta \cdot \nabla\varphi + \nabla \cdot J.$$

Consider next the integral term of $\mathcal{B}$. By (6.35), we have

$$\int_{\mathbb{S}^{n-1}}\int_{\mathbb{S}^{n-1}} f(\theta,\omega)\,\varphi ds(\theta)ds(\omega) = |\mathbb{S}^{n-1}|\varphi,$$

and similarly,

$$\int_{\mathbb{S}^{n-1}}\int_{\mathbb{S}^{n-1}} f(\theta,\omega) J\cdot\omega ds(\theta)ds(\omega) = \int_{\mathbb{S}^{n-1}} J\cdot\omega ds = 0.$$

Furthermore,

$$\int_{\mathbb{S}^{n-1}}\int_{\mathbb{S}^{n-1}} f(\theta,\omega)\varphi\,\theta\,d(\theta)ds(\omega) = \varphi\int_{\mathbb{S}^{n-1}}\theta ds = 0,$$

and, finally,

$$\int_{\mathbb{S}^{n-1}}\int_{\mathbb{S}^{n-1}} \theta f(\theta,\omega) J\cdot\omega ds(\theta)ds(\omega) = |\mathbb{S}^{n-1}| BJ,$$

with $B \in \mathbb{R}^{n\times n}$ as in the claim. Hence,

$$P\int_{\mathbb{S}^{n-1}} f(\theta,\omega)(\varphi + nJ\cdot\omega)dS(\omega) = \varphi + n\theta\cdot BJ.$$

By combining the results above, we obtain the desired form for the P_1 approximation.

Finally, we show that the spectrum of B lies in the interval $[0,1]$. Observe that B is positive definite because of the positivity of the scattering phase, and it is symmetric due to the reciprocity property of f. Therefore, its spectrum is in the positive real axis. To see that the eigenvalues are bounded by unity, assume that $v \in \mathbb{R}^n$. We have

$$0 \le v^T B v = \frac{n}{|\mathbb{S}^{n-1}|}\int_{\mathbb{S}^{n-1}}\int_{\mathbb{S}^{n-1}} (v\cdot\theta)\,f(x,\theta,\omega)\,(\omega\cdot v)ds(\theta)ds(\omega).$$

Using the fact that the scattering phase f is nonnegative, we obtain by Schwarz inequality the estimate

$$v^T B v \le \frac{n}{|\mathbb{S}^{n-1}|}\left(\int_{\mathbb{S}^{n-1}}\int_{\mathbb{S}^{n-1}} (v\cdot\theta)^2 f(x,\theta,\omega)ds(\theta)ds(\omega)\right)^{1/2}$$
$$\left(\int_{\mathbb{S}^{n-1}}\int_{\mathbb{S}^{n-1}} (v\cdot\omega)^2 f(x,\theta,\omega)ds(\theta)ds(\omega)\right)^{1/2}$$
$$= \frac{n}{|\mathbb{S}^{n-1}|}\left(\int_{\mathbb{S}^{n-1}} (v\cdot\theta)^2 ds(\theta)\right)^{1/2}\left(\int_{\mathbb{S}^{n-1}} (v\cdot\omega)^2 ds(\omega)\right)^{1/2}.$$

The last equality follows from the fact that the integral of the phase function equals unity; see (6.35). By the same argument used to derive the identity (6.44), we deduce that

$$\int_{\mathbb{S}^{n-1}} (v\cdot\theta)^2 ds(\theta) = \int_{\mathbb{S}^{n-1}} (v\cdot\omega)^2 ds(\omega) = \frac{|\mathbb{S}^{n-1}|}{n}\|v\|^2.$$

Therefore, we have the upper bound

$$0 \leq v^T B v \leq \|v\|^2,$$

hence, the spectrum of B must lie in the interval $[0, 1]$. □

We refer to the matrix B as the *anisotropy matrix*.

From the above result, we see that the P_1 approximation leads to a coupled system of equations, one involving the part independent of θ, the other one the multiplier of θ. In fact, φ and J should satisfy the system

$$\frac{1}{c}\varphi_t = -\nabla \cdot J - \mu_a \varphi + q_0, \tag{6.45}$$

$$\frac{1}{c}J_t = -\frac{1}{n}\nabla\varphi - (\mu_a + (1-B)\mu_s)J + q_1, \tag{6.46}$$

where the source terms q_0 and q_1 are

$$q_0(x,t) = \int_{\mathbb{S}^{n-1}} q(x,t,\theta)ds, \quad q_1(x,t) = \int_{\mathbb{S}^{n-1}} \theta\, q(x,t,\theta)ds.$$

We could use this coupled system to describe the light propagation. A further simplification is attained by assuming that

$$\frac{1}{c}J_t \approx 0, \tag{6.47}$$

or alternatively, that J_t is proportional to J. Assuming that (6.47) holds, equation (6.46) reduces to an equivalent formulation of *Fick's law*,

$$J = -K\nabla\varphi + n\kappa q_1, \quad K = \frac{1}{n}\Big((\mu_a + (1-B)\mu_s)\Big)^{-1}. \tag{6.48}$$

Observe that the *diffusion matrix* $K \in \mathbb{R}^{n \times n}$ is well defined if $\mu_a > 0$ or the spectrum of B does not contain the unity. A substitution to equation (6.45) leads to an equation of parabolic type, a *diffusion equation*

$$\frac{1}{c}\varphi_t = \nabla \cdot K\nabla\varphi - \mu_a \varphi + Q, \tag{6.49}$$

where

$$Q = q_0 - n\nabla \cdot K q_1. \tag{6.50}$$

Before discussing the boundary conditions in the diffusion approximation, we remark that the anisotropy in the radiative transfer equation is related to the anisotropic diffusion. Indeed, if the scattering phase function f depends only on the angle between the incoming and scattering direction as in formula (6.36), we have

$$B_{i,j}(x) = \frac{n}{|\mathbb{S}^{n-1}|} \int_{\mathbb{S}^{n-1}} \int_{\mathbb{S}^{n-1}} \theta_i \omega_j h(x, \theta \cdot \omega) ds(\theta) ds(\omega) = 0, \text{ if } i \neq j.$$

This claim follows from a symmetry argument: By denoting with ϑ the angle between the ω and θ and writing

$$\theta = \omega \cos \vartheta + \theta_\perp, \quad \omega \cdot \theta_\perp = 0,$$

we observe that

$$\int_{\mathbb{S}^{n-1}} \theta h(x, \theta \cdot \omega) ds(\theta) = \omega \int_{\mathbb{S}^{n-1}} \cos \vartheta h(x, \cos \vartheta) ds(\theta),$$

since integrals of $\omega_\perp$ over $\mathbb{S}^{n-2}$ cancel. Furthermore, by passing to spherical coordinates of $\mathbb{R}^n$, we have

$$\omega \int_{\mathbb{S}^{n-1}} \cos \vartheta h(x, \cos \vartheta) ds(\theta) = \omega |\mathbb{S}^{n-2}| \int_{-1}^{1} t(1-t^2)^{(n-3)/2} h(x,t) dt.$$

Hence, we find that

$$B_{j,j}(x) = \frac{n}{|\mathbb{S}^{n-1}|} \left(\int_{\mathbb{S}^{n-1}} \omega_j^2 ds(\omega) \right) |\mathbb{S}^{n-2}| \int_{-1}^{1} t(1-t^2)^{(n-3)/2} h(x,t) dt$$

$$= |\mathbb{S}^{n-2}| \int_{-1}^{1} t(1-t^2)^{(n-3)/2} h(x,t) dt = b(x).$$

This means that, in isotropic medium, the anisotropy matrix B and thus the diffusion matrix K is also isotropic,

$$K(x) = k(x)I, \quad k(x) = \frac{1}{n(\mu_a(x) + (1-b(x))\mu_s(x))}.$$

Conversely, given a positive definite symmetric matrix valued function $K(x)$, it is possible to construct a scattering phase function such that K is its diffusion approximation. Indeed, by defining a scattering phase as

$$f(x,\theta,\omega) = \theta^T B(x) \omega, \quad B(x) = I - \frac{1}{\mu_s(x)} \left(\frac{1}{n} K(x)^{-1} - \mu_a(x) \right),$$

it is straightforward to see that the corresponding diffusion matrix is K.

To fix the appropriate boundary values for the forward problem, we start with the following lemma.

Lemma 6.13. *Within the P_1 approximation, the total photon flux inwards $(-)$ and outwards $(+)$ at a point $x \in \partial\Omega$ is given by*

$$\Phi_\pm(x) = \gamma \varphi \pm \frac{1}{2} \nu \cdot J, \tag{6.51}$$

where $\nu = \nu(x)$ is the exterior unit normal of $\partial\Omega$ and the dimension-dependent constant γ is

$$\gamma = \gamma_n = \frac{\Gamma(n/2)}{\sqrt{\pi}(n-1)\Gamma((n-1)/2)}.$$

In particular, in dimensions $n = 2$ and $n = 3$ we have

$$\gamma_2 = \frac{1}{\pi}, \quad \gamma_3 = \frac{1}{4}.$$

Proof. The total flux inwards at $x \in \partial\Omega$ within the P_1 approximation is

$$\Phi_-(x) = -\frac{1}{|\mathbb{S}^{n-1}|} \int_{\{\nu(x)\cdot\theta<0\}} (\varphi(x,t) + nJ(x,t)\cdot\theta)\theta\cdot\nu(x)ds(\theta).$$

By passing to spherical coordinates, we obtain

$$-\int_{\{\nu\cdot\theta<0\}} \theta\cdot\nu ds = |\mathbb{S}^{n-2}| \int_0^1 t(1-t^2)^{(n-3)/2} dt = \frac{|\mathbb{S}^{n-2}|}{n-1}.$$

To integrate the second term, write first

$$J = (J\cdot\nu)\nu + J_\perp, \quad \nu\cdot J_\perp = 0.$$

Then,

$$\int_{\{\nu\cdot\theta<0\}} (J\cdot\theta)(\theta\cdot\nu)ds = (J\cdot\nu)\int_{\{\nu\cdot\theta<0\}} (\nu\cdot\theta)^2 ds + \int_{\{\nu\cdot\theta<0\}} (\theta\cdot J_\perp)(\theta\cdot\nu)ds.$$

The latter integral vanishes due to the antisymmetry of the integrand in the plane perpendicular to ν. To evaluate the former integral, we can write, by symmetry,

$$\int_{\{\nu\cdot\theta<0\}} (\theta\cdot\nu)^2 ds = \frac{1}{2}\int_{\mathbb{S}^{n-1}} (\nu\cdot\theta)^2 ds = \frac{|\mathbb{S}^{n-1}|}{2n}.$$

Taking into account the formula

$$|\mathbb{S}^{n-1}| = \frac{2\pi^{n/2}}{\Gamma(n/2)}$$

we have

$$\Phi_-(x) = \frac{1}{|\mathbb{S}^{n-1}|} \left(\frac{|\mathbb{S}^{n-2}|}{n-1}\varphi - n(J\cdot\nu)\frac{|\mathbb{S}^{n-1}|}{2n} \right)$$

$$= \frac{\Gamma(n/2)}{\sqrt{\pi}(n-1)\Gamma((n-1)/2)}\varphi - \frac{1}{2}\nu\cdot J$$

as claimed. The outward flux is obtained by switching the direction of the normal vector. □

218   6 Model Problems

We may now specify the boundary data for the inverse boundary value problem. At $x \in \partial\Omega$, the total flux inwards must be equal to the known input flux $\Phi_{\text{in}}(x)$ from the outside source plus the reflected photon flux due to the refractive index mismatch. If $R = R(x)$, $0 \le R \le 1$, is the reflection coefficient at $x \in \partial\Omega$. By assuming that the boundary condition describing the reflection is local, we may write

$$\Phi_-(x) = R(x)\Phi_+(x) + \Phi_{\text{in}}(x).$$

The measured data (in the ideal case) is the total outgoing flux outside the body's surface,

$$\Phi_{\text{out}}(x) = (1 - R(x))\Phi_+(x).$$

Now we may define the boundary data for the inverse problem for strongly scattering media.

Definition 6.14. *Let $\Omega \subset \mathbb{R}^n$ be a bounded domain with a smooth connected boundary. Assuming that there are no internal sources and that the approximation (6.48) holds, the complete boundary data consists of the pairs of the presumably known input flux*

$$\Phi_{\text{in}} = (1 - R)\gamma\varphi + \frac{1}{2}(1 + R)\nu \cdot K\nabla\varphi, \tag{6.52}$$

and the corresponding measured output flux

$$\Phi_{\text{out}} = (1 - R)\gamma\varphi - (1 + R)\nu \cdot K\nabla\varphi. \tag{6.53}$$

The forward problem for the diffusion approximation is to find φ satisfying the equation (6.49) with $Q = 0$ and the Robin boundary condition (6.52). We shall not prove the solvability of this boundary value problem in detail. The proof goes along the lines of the proofs presented previously for the quasistatic Maxwell problems in the preceding sections.

In practice, the illumination of the target is done using optical fibers attached to the body's surface. Such sources are strongly anisotropic and the diffusion approximation directly under the sources is quite poor. Therefore, the boundary sources are often modelled by equivalent point sources inside the body. This means that the boundary value problem corresponding to the forward problem becomes

$$\frac{1}{c}\varphi_t = \nabla \cdot K\nabla\varphi - \mu_a\varphi + Q,$$

with the homogenous Robin boundary condition

$$(1 - R)\gamma\varphi + \frac{1}{2}(1 + R)\nu \cdot K\nabla\varphi = 0,$$

and the source Q consists of point sources under the optical feed fibers,

$$Q(x) = \sum_{\ell=1}^{L} q_\ell \delta(x - x_\ell).$$

The corresponding boundary data for the inverse problem consists of the outflow at given receiver locations $y_k \in \partial\Omega$,

$$\Phi_{\text{out}}(y_k) = \left((1-R)\gamma\varphi - (1+R)\nu \cdot K\nabla\varphi\right)\Big|_{y=y_k}.$$

6.4.3 Time-harmonic Measurement

In some applications of optical tomography, time-harmonic light modulation is used. Here we consider the case where the input field is modulated with a fixed harmonic frequency, i.e.,

$$I(x, t, \hat{\theta}) = e^{-i\omega t}\hat{I}(x, \theta).$$

Denoting the wave number by $k = \omega/c$, the P_1 approximation yields the pair of equations

$$ik\varphi = \nabla \cdot J + \mu_a \varphi + q_0, \qquad (6.54)$$

$$ikJ = \frac{1}{n}\nabla\varphi + \left(\mu_a + (1-B)\mu_s\right)J + q_1, \qquad (6.55)$$

where, for simplicity, we have denoted the time-harmonic amplitudes of φ, J and the source terms by the same symbols. In the time-harmonic approximation, we do not need to asume that the time derivative of J is insignificant. Instead, we may solve equation (6.55) for J and substitute into (6.54) to get the diffusion equation

$$\nabla \cdot D\nabla\varphi + (\mu_a - ik)\varphi = Q,$$

where

$$D = \frac{1}{n}\left(ik + \mu_a + (1-B)\mu_s\right)^{-1} \qquad (6.56)$$

and

$$Q = q_0 - n\nabla \cdot Dq_1.$$

Observe that the diffusion matrix D differs from the matrix K of equation (6.48) by the factor ik.

6.5 Notes and Comments

The inverse problems presented in this chapter have been studied extensively in the literature, not only from the computational point of view.

The literature concerning X-ray tomography and its different variants is vast. We mention the textbooks [96] and [106] concerning this topic, and the article [118]. The reference to the original article of Johann Radon mentioned in the text is [104].

From the literature concerning Maxwell's equations, we refer to the classical monograph [93]. Standard references on the mathematical model for MEG are [42] and [110]. For a review of the MEG inverse problem, see [52]. A more recent review is [9].

Electrical impedance tomography is a widely studied inverse problem. Since the classic article of Calderón [18], the problem has usually been formulated as follows. Let $u \in H^1(\Omega)$ satisfy the Dirichlet boundary value problem

$$\nabla \cdot \sigma \nabla = 0, \text{ in } \Omega, \quad u\big|_{\partial\Omega} = f \text{ on } \partial\Omega.$$

Define the *Dirichlet-to-Neumann map*,

$$\Lambda_\sigma : H^{1/2}(\partial\Omega) \to H^{-1/2}(\partial\Omega), \quad f \mapsto \frac{\partial u}{\partial n}\bigg|_{\partial\Omega}.$$

Can one determine σ from the knowledge of Λ_σ? There are a wealth of results concerning this question, the main difference being, what is assumed about the conductivity. A major breakthrough in this problem was made in the articles [123], [60], [105] and [94], and [95] where the uniqueness was proved with various smoothness requirements on σ. The articles are based on the use of complex geometrical optics. Recently, the two-dimensional version of the original question of Calderón with bounded conductivities was solved in [7].

The electrode model of the EIT problem goes back to the article [23], and the solvability of the forward problem was shown in [119]. In the case of complex admittivities and contact impedances, one has to assume that $\operatorname{Re} \gamma > 0$ and $\operatorname{Re} z_\ell > 0$ and employ the Lax–Milgram theorem instead of the Riesz representation theorem.

The finite-dimensional data is naturally inadequate for determining the conductivity except for simple cases.

The linearized formula (6.33) was derived in Section 6.3 without carefully analyzing the error term. A more careful analysis, showing that the mapping $(\gamma, z) \mapsto (u, U)$ is indeed Frèchet differentiable can be found in the article [64]. The reference contains also a description of the finite-element calculation of the solutions and the Jacobians.

The optical tomography problem is discussed in the review article [6]. The nonuniqueness of the optical tomography problem in anisotropic material has been demonstrated in [5].

The inverse problem 6.9 is often referred to as the *inverse problem of transport theory*, and in the literature it is stated in terms of the *albedo operator*,

$$A : I\big|_{\Gamma_+} \mapsto I\big|_{\Gamma_-},$$

where $\Gamma_\pm\{(x,t,\omega) \mid x \in \partial\Omega,\ t > 0,\ \pm\nu(x)\cdot\theta > 0\}$. For discussion of this problem, see, for example, [121] and references there.

The solvability of the radiation tranfer equation is dicussed, for example, in [27]. The radiation transfer equation (6.34) is based on the assumption of constant propagation speed and straight rays, which is tantamount to assuming a constant refractive index. These restrictions can be removed by replacing the advection term $\theta \cdot \nabla$ by a vector field corresponding to a geodesic flow on the underlying Riemannian manifold, see, for example, [100].

The asumption that the medium is strongly scattering is not always a good one. For instance, the human head contains weakly scattering regions that are filled with cerebrospinal fluid (CSF). These nonscattering regions must be treated separately without relying on diffusion approximation. The article [62] contains an analysis of the forward problem. The inverse problem with nonscattering regions has been discussed in [28].

The source modelling in optical tomography is problematic, since the surface sources are not diffuse and the diffusion approximation thus fails. The sources can be handled by modelling the light propagation near the sources by the RTE; see [100] and [138].

7
Case Studies

This chapter is devoted to extensive case studies in which the statistical inversion theory is applied to a number of real-world applications or prototypes of them. The applications are based on the model problems that in the classical form are described in Chapter 2. Here, the basic models are specified further and augmented by various additional features.

7.1 Image Deblurring and Recovery of Anomalies

In the preceding chapters, it was emphasized that the prior probability density should reflect our belief of what the unknown parameter typically is like. The construction of an informative prior can be based on our understanding of the underlying physics, or it may be derived from previous observations of the unknown quantity. From a practical point of view, however, such priors may be problematic. Often in applications such as in medical imaging, one is not interested in imaging of typical features. Instead, the whole procedure may aim at finding deviations from the normal such as tumors in normal healthy tissue or fractures in normal bone. A prior density based on a large sample of normal tissues may give a negligible prior probabilty to the interesting but rare anomalies that, statistically speaking, are outliers. Consequently, the Bayesian approach biases too much towards the normal, to the extent that anomalies may be completely ignored. In this section, we demonstrate one possible way of circumventing this type of biasing problem, while also showing the importance of treating properly approximation errors due to modelling.

7.1.1 The Model Problem

The inverse problem that we consider is the two-dimensional deconvolution or deblurring problem. The continuous observation model is

$$g(t) = \int_D K(t-s)f(s)ds + e(t), \quad D \subset \mathbb{R}^2, \tag{7.1}$$

that is, we assume the additive noise model. For the sake of definiteness, we suppose that D is the unit square, $D = [0,1] \times [0,1]$, and that the true image f vanishes outside D.

Assume that the blurred image $g(t)$ is given on a relatively coarse grid. We assume that the grid points form a regular rectangular mesh. Let the grid points be $(x_i, y_j) = (i/m, j/m) \in D$, $0 \le i, j \le m$. We number the data points $t_p \in D$, $1 \le p \le M = (m+1)^2$ by reading the rectangular image column by column, that is,

$$t_p = t_{(i-1)(m+1)+m-j+1} = \left(\frac{i}{m}, \frac{j}{m}\right), \quad 0 \le i, j \le m.$$

Following the principle of this book, we pass to the pixelized random model of the above equation. The reconstruction of the original image can be done in any pixel mesh we wish. Choosing a fine discretization increases the computational task and renders the problem more underdetermined. On the other hand, the use of a coarse grid may generate approximation errors. In this example, we choose to reconstruct f at the same points t_p in which the data is given. However, we use the ideas of the previous chapter to reduce the effects of approximation errors.

We start by discretizing the problem in a refined mesh. Let n be a multiple of m, the number of the discretization intervals of the data. We divide the image D into $(n+1) \times (n+1) = N$ pixels whose center points are $(x_i, y_j) = (i/n, j/n)$, $0 \le i, j \le n$. Similarly to the renumbering of the image points of the blurred image, we set

$$s_q = t_{(i-1)(n+1)+n-j+1} = \left(\frac{i}{n}, \frac{j}{n}\right), \quad 0 \le i, j \le n. \tag{7.2}$$

We write the discretized observation model as

$$y = \mathcal{A}x + e, \tag{7.3}$$

Here, $y \in \mathbb{R}^M$, $x \in \mathbb{R}^N$, $N > M$, with

$$y_p = g(t_p), \quad x_q = f(s_q),$$

and $\mathcal{A} \in \mathbb{R}^{M \times N}$ is given as

$$\mathcal{A}_{pq} = \frac{1}{n^2} K(t_p - s_q), \quad 1 \le p \le M, \ 1 \le q \le N.$$

We assume that the refined mesh is so fine that equation (7.3) can be considered as an accurate discrete model, and we write the corresponding stochastic model,

$$Y = \mathcal{A}X + E,$$

where $X \in \mathbb{R}^N$, $Y, E \in \mathbb{R}^M$ are random vectors.

7.1 Image Deblurring and Recovery of Anomalies

Assume that the prior information of X in the *typical case* is coded in the prior probability density
$$\pi_{\mathrm{pr}}(x) = \pi_1(x),$$
but that we have reasons to believe that X may in fact contain features that deviate considerably from the typical case. We also assume that we know, in terms of statistics, what kind of deviations are possible. In this example, we assume that the deviation from the normal can be expressed by means of an additive model. Hence, we suppose that the true image can be decomposed as
$$X = U + V, \quad U \sim \pi_1(u), \ V \sim \pi_2(v),$$
where U represents the typical part of X, distributed according to the statistics defined by the density π_1, while V is the *anomalous part*, a deviation from the typical, with probability density $\pi_2(v)$. The observation model is then
$$Y = \mathcal{A}U + \mathcal{A}V + E. \tag{7.4}$$
Assuming that the typical and anomalous parts are independent, this model leads to the joint prior density
$$\pi(u, v) = \pi_1(u)\pi_2(v).$$
Before discussing the problem further, we clarify this with an example. Let the convolution kernel K be a Gaussian kernel,
$$K(s) = \exp\left(-\kappa|s|^2\right), \tag{7.5}$$
where $\kappa = 80$, that is, the half-width of the kernel is $2\log 2/\sqrt{2\kappa} \approx 0.2$.

Assume that the true image consists of a smooth object that represents the typical case, plus few singular pointlike objects, representing the anomalous part. In Figure 7.1 (left), we have plotted a smooth object consisting of three Gaussian humps. As an anomaly, we add a Dirac delta into the image. The location of the anomaly is marked by a cross in the figure. We compute the blurred image in a relatively coarse mesh. The blurred image is shown in Figure 7.1 (right). The pixel size in the blurred image is 0.05×0.05, that is, $m = 21$ and $M = (21)^2 = 441$. The size of the original image is $N = (4m+1)^2 = 6561$. It is quite difficult to tell apart the regular object and the anomalous one from the blurred data.

7.1.2 Reduced and Approximation Error Models

We could estimate the variables U and V from the model (7.4). However, when n is large, the problem becomes computationally heavy, in particular when MCMC schemes are used. Therefore, we reduce the model. We aim to estimate the image in the same grid in which the data was given. The most straightforward way is simply to choose $n = m$ and discretize. It turns out

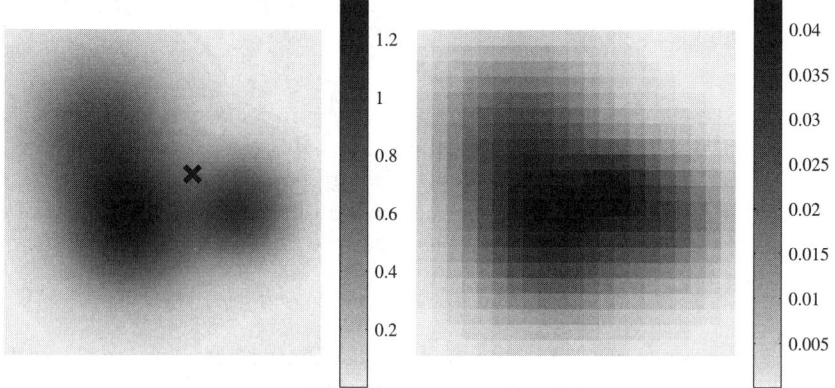

Figure 7.1. The true image (left). The location of the pointlike anomalous part is marked with a cross. The blurred image (right) is given in a coarser grid.

that such discretization of the model equation (7.1) introduces remarkable modelling errors that have to be taken into account in the computations. In the next subsection, the effect of modelling errors is demonstrated numerically.

We define a *reduced model* as follows: Assume that $n = km$, that is, the fine discretization intervals are obtained by dividing the coarse intervals in k parts. Hence, each coarse pixel consists of k^2 finer pixels. Let $P : \mathbb{R}^N \to \mathbb{R}^M$ denote the model reduction operator that integrates the finely pixelized image into a coarser one. More precisely, denoting by D_p^N the pixels of the fine discretization, $1 \leq p \leq N$, and by D_q^M those of the coarse discretization, $1 \leq q \leq M$, we set

$$P_{pq} = \frac{1}{k^2}, \text{ if } D_q^N \subset D_p^M, \text{ and } P_{pq} = 0 \text{ otherwise.}$$

We define
$$\widetilde{U} = PU, \quad \widetilde{V} = PV.$$

Furthermore, we define a coarse discretized model matrix $A \in \mathbb{R}^{M \times M}$ by

$$A_{pq} = \frac{1}{m^2} K(t_p - s_q), \quad 1 \leq p, q \leq M.$$

The reduced model is now

$$Y = A\widetilde{U} + A\widetilde{V} + \epsilon,$$

where the noise ϵ contains both the additive measurement error and the discretization error,

$$\epsilon = E + (\mathcal{A} - AP)U + (\mathcal{A} - AP)V.$$

In the following, we shall ignore the modelling error due to the anomalous part, that is, we assume that

$$(\mathcal{A} - AP)V \approx 0. \tag{7.6}$$

It is not obvious at this point why this approximation is legitimate. In fact, it is done only to simplify the problem. We discuss it in more detail in the following subsection and take it for granted here.

Let us now specify the prior densities of U, V and the additive error E. First, we assume that E is Gaussian zero mean white noise with variance σ^2, and independent of U and V, that is,

$$\pi_{\text{noise}}(e) \propto \exp\left(-\frac{1}{2\sigma^2}\|e\|^2\right).$$

The covariance matrix of E is denoted by $\Gamma_e = \sigma^2 I$.

The regular part U is also assumed to be Gaussian. We believe a priori that the regular part of the image is smooth, and this belief is expressed in terms of the second-order smoothness prior,

$$\pi_1(u) \propto \exp\left(-\frac{\alpha}{2}\|Lu\|_2^2\right), \tag{7.7}$$

where L is the five-point finite-difference approximation of the Dirichlet Laplacian, that is,

$$(Lu)_q = N\left(\sum_{q' \in \mathcal{N}(q)} u_{q'} - 4u_q\right). \tag{7.8}$$

Here, $\mathcal{N}(q)$ denotes the set of indices of the neighboring pixels of q. The parameter α is assumed to be known. For later reference, we denote the covariance matrix of U by

$$\Gamma_u = \frac{1}{\alpha} L^{-1} (L^{\text{T}})^{-1}.$$

To define the statistics of the anomaly $\widetilde{V}$, assume that we can expect impulse-type anomalies, that is, we expect that the image may contain few pixels that deviate strongly from the background and the pixels are uncorrelated with one another. Furthermore, we assume that the anomalies are positive. To model such anomalies, we choose to use the ℓ^1-prior, that is, we set

$$\pi_2(\widetilde{v}) \propto \pi_+(\widetilde{v}) \exp(-\beta\|\widetilde{v}\|_1), \quad \|\widetilde{v}\|_1 = \sum_{q=1}^m |\widetilde{v}_q|.$$

As usual, π_+ is the positivity constraint, that is, $\pi_+(\widetilde{v}) = 1$ if all components of $\widetilde{v}$ are positive, otherwise $\pi_+(\widetilde{v}) = 0$.

To compose the appropriate conditional densities, consider first the joint probability density of U and Y when $\widetilde{V} = \widetilde{v}$ is given. Denoting by $\mathrm{E}_{|\widetilde{v}}$

the conditional expectation conditioned on $\widetilde{V} = \widetilde{v}$, and noticing, in view of assumption (7.6),
$$Y = \mathcal{A}U + \mathcal{A}\widetilde{V} + E,$$
we have
$$\mathrm{E}_{|\widetilde{v}}\{Y\} = \mathcal{A}\widetilde{v}.$$

Similarly, the conditional joint covariance of $\widetilde{U}$ and Y is

$$\mathrm{E}_{|\widetilde{v}}\left\{\begin{bmatrix} \widetilde{U} \\ Y - \mathcal{A}\widetilde{v} \end{bmatrix} \begin{bmatrix} \widetilde{U}^{\mathrm{T}} & (Y - \mathcal{A}\widetilde{v})^{\mathrm{T}} \end{bmatrix}\right\} = \begin{bmatrix} P\Gamma_u P^{\mathrm{T}} & P\Gamma_u \mathcal{A}^{\mathrm{T}} \\ \mathcal{A}\Gamma_u P^{\mathrm{T}} & \mathcal{A}\Gamma_u \mathcal{A}^{\mathrm{T}} + \Gamma_e \end{bmatrix} = \Gamma. \quad (7.9)$$

Since $\widetilde{U}$ and Y, with a fixed $\widetilde{V} = \widetilde{v}$, are jointly Gaussian, we may immediately write the conditional density,

$$\pi(\widetilde{u}, y \mid \widetilde{v}) \propto \exp\left(-\frac{1}{2} \begin{bmatrix} \widetilde{u}^{\mathrm{T}} & (y - \mathcal{A}\widetilde{v})^{\mathrm{T}} \end{bmatrix} \Gamma^{-1} \begin{bmatrix} \widetilde{u} \\ y - \mathcal{A}\widetilde{v} \end{bmatrix}\right). \quad (7.10)$$

This gives the joint probability density,

$$\pi(\widetilde{u}, \widetilde{v}, y) = \pi(\widetilde{u}, y \mid \widetilde{v})\pi_{\mathrm{pr}}(\widetilde{v}) \quad (7.11)$$

$$\propto \pi_+(\widetilde{v}) \exp\left(-\frac{1}{2} \begin{bmatrix} \widetilde{u}^{\mathrm{T}} & (y - \mathcal{A}\widetilde{v})^{\mathrm{T}} \end{bmatrix} \Gamma^{-1} \begin{bmatrix} \widetilde{u} \\ y - \mathcal{A}\widetilde{v} \end{bmatrix} - \gamma\|\widetilde{v}\|_1\right).$$

From formulas (7.10) and (7.11), we can now write the conditional probability densities $\pi(\widetilde{u} \mid y, \widetilde{v})$ and $\pi(\widetilde{v} \mid y, \widetilde{u})$ that are needed in the sequel.

Let us denote by Γ_{ij}, $1 \le i, j \le 2$, the blocks of Γ in (7.9). From Theorem 4.5, it follows that

$$\pi(\widetilde{u} \mid y, \widetilde{v}) \propto \exp\left(-\frac{1}{2}(\widetilde{u} - \widetilde{u}_0)^{\mathrm{T}} \widetilde{\Gamma}_{22}^{-1}(\widetilde{u} - \widetilde{u}_0)\right), \quad (7.12)$$

where
$$\widetilde{u}_0 = \Gamma_{12}\Gamma_{22}^{-1}(y - \mathcal{A}\widetilde{v}) = P\Gamma_u \mathcal{A}^{\mathrm{T}}(\mathcal{A}\Gamma_u\mathcal{A}^{\mathrm{T}} + \Gamma_e)^{-1}(y - \mathcal{A}\widetilde{v}),$$

and $\widetilde{\Gamma}_{22}$ is the Schur complement of Γ_{22},
$$\widetilde{\Gamma}_{22} = \Gamma_{11} - \Gamma_{12}\Gamma_{22}^{-1}\Gamma_{21}.$$

The density of $\widetilde{v}$ conditioned on y and $\widetilde{u}$ follows from Bayes' formula and formula (7.11). Let us denote by R_{ij} the blocks of Γ^{-1}, that is,

$$\Gamma^{-1} = \begin{bmatrix} \widetilde{\Gamma}_{22}^{-1} & -\widetilde{\Gamma}_{22}^{-1}\Gamma_{12}\Gamma_{22}^{-1} \\ -\widetilde{\Gamma}_{11}^{-1}\Gamma_{21}\Gamma_{11}^{-1} & \widetilde{\Gamma}_{11}^{-1} \end{bmatrix} = \begin{bmatrix} R_{11} & R_{12} \\ R_{21} & R_{22} \end{bmatrix}.$$

We have $\widetilde{v}_k \ge 0$ so that we can we can write $\|\widetilde{v}\| = \mathbf{1}^{\mathrm{T}}\widetilde{v}$, where $\mathbf{1} = [1, \ldots, 1]^{\mathrm{T}}$. Then, Bayes' formula asserts that

$$\pi(\widetilde v\mid \widetilde u,y)\propto \pi_+(\widetilde v)\exp\left(-\frac12(\widetilde v^{\mathrm T}A^{\mathrm T}R_{22}A\widetilde v-2\widetilde v^{\mathrm T}(A^{\mathrm T}R_{21}\widetilde u+A^{\mathrm T}R_{22}y))-\gamma\|\widetilde v\|_1\right)$$

$$=\pi_+(\widetilde v)\exp\left(-\frac12(\widetilde v^{\mathrm T}B\widetilde v-2\widetilde v^{\mathrm T}q)\right), \tag{7.13}$$

where
$$B=A^{\mathrm T}R_{22}A,\quad q=A^{\mathrm T}R_{21}\widetilde u+A^{\mathrm T}R_{22}y-\gamma\mathbf 1\,.$$

7.1.3 Sampling the Posterior Distribution

We have now all the necessary tools to build an MCMC algorithm for estimating $\widetilde u$ and $\widetilde v$. For notational simplicity, we denote $u=\widetilde u$, $v=\widetilde v$ in the algorithm. The basic structure of the algorithm is a block-form Gibbs sampler described as follows:

Initialize $(u,v)=(u^0,v^0)$, set $k=0$.
do until satisfactory sample size
    $(u,v)=(u^k,v^k)$
    draw u^{k+1} from $\pi(u\mid v,y)$
    for $p=1:M$
        draw $v_p^{k+1}\geq 0$ from $\pi(v_p\mid u,v_{-p},y)$
    end
    $k\leftarrow k+1$
end

Above, the notation v_{-p} denotes the vector v with the element v_p deleted.

Before discussing specific examples, let us explain how the drawing is done in practice. Consider first the updating of $\widetilde u$, when $\widetilde v$ is fixed, with the density given by (7.12). We start by precomputing the Cholesky factorization of $\widetilde\Gamma_{22}$,

$$\widetilde\Gamma_{22}=U_2^{\mathrm T}U_2,$$

where the matrix U_2 is upper triangular. This can be done before starting the MCMC run. When the updated $\widetilde v$ is given, we first compute

$$\widetilde u_0=\Gamma_{12}\Gamma_{22}^{-1}(y-A\widetilde v), \tag{7.14}$$

and then update $\widetilde u$ as

$$\widetilde u=\widetilde u_0+U_2^{\mathrm T}\xi,\quad \xi\sim\mathcal N(0,I).$$

Clearly, $\widetilde u$ has the right conditional density.

The componentwise updating of $\widetilde v$, given $\widetilde u$ and y, which is more involved, requires the conditional densities $\pi(\widetilde v_p\mid \widetilde v_{-p},\widetilde u,y)$. For convenience, let us consider the case $p=1$, which is the general case up to reordering. We partition the arrays B and q using the notations

230   7 Case Studies

$$B = \begin{bmatrix} b^2 & \beta^\mathrm{T} \\ \beta & B' \end{bmatrix}, \quad q = \begin{bmatrix} q_1 \\ q' \end{bmatrix},$$

where $b^2 \in \mathbb{R}$, $\beta \in \mathbb{R}^{M-1}$ and $B' \in \mathbb{R}^{(M-1)\times(M-1)}$, and correspondingly, $q_1 \in \mathbb{R}$, $q' \in \mathbb{R}^{M-1}$. From (7.13), we obtain the one-dimensional conditional density of $\tilde{v}_1$,

$$\pi(\tilde{v}_1 \mid \tilde{v}_{-1}, \tilde{u}, y) \propto \pi_+(\tilde{v}_1)\exp\left(-\frac{1}{2}(b^2\tilde{v}_1^2 - 2(q_1 - \beta^\mathrm{T}\tilde{v}_{-1})\tilde{v}_1)\right)$$

$$\propto \pi_+(\tilde{v}_1)\exp\left(-\frac{1}{2}(b\tilde{v}_1 - c)^2\right), \qquad (7.15)$$

where

$$c = \frac{q_1 - \beta^\mathrm{T}\tilde{v}_{-1}}{b}.$$

Before starting the MCMC iteration, we precompute the matrices B, $A^\mathrm{T} R_{22}$ as well as

$$q_0 = A^\mathrm{T} R_{22} y - \gamma \mathbf{1}.$$

Inside each loop, when $\tilde{u}$ has been updated, we compute

$$q = q_0 + A^\mathrm{T} R_{21} \tilde{u}.$$

To update the component $\tilde{v}_1$, we compute the corresponding numbers b and c and draw $\tilde{v}_1$ from the density (7.15). Letting

$$t_0 = -\frac{c}{\sqrt{2}},$$

we draw a random number $t \in \mathbb{R}$ from the density

$$\pi(t) \propto \pi_+(t - t_0)e^{-t^2}, \qquad (7.16)$$

and set

$$\tilde{v}_1 = \frac{1}{b}(\sqrt{2}t + c).$$

Hence, we have reduced the updating to the simple task of drawing from the density (7.16). Using the notation

$$\mathrm{Erf}(t) = \frac{2}{\sqrt{\pi}} \int_0^t e^{-s^2}\,ds,$$

we see that the distribution function of t is

$$\Phi(t) = \int_{t_0}^t \pi(s)ds = \frac{1}{1 - \mathrm{Erf}(t_0)}(\mathrm{Erf}(t) - \mathrm{Erf}(t_0)), \quad t \geq t_0.$$

It follows that the variable t can be drawn as

7.1 Image Deblurring and Recovery of Anomalies

$$t = \Phi^{-1}(\xi) = \text{Erf}^{-1}(\text{Erf}(t_0) + (1 - \text{Erf}(t_0))\xi), \quad \xi \sim \mathcal{U}(0,1). \quad (7.17)$$

A word of caution is in order here. If $-c$, hence t_0, is large, the formula (7.17) becomes useless, the reason being that if $t \geq 5$, $\text{Erf}(t) = 1$ within the working precision of 16 digits and formula (7.17) produces roundoff noise. Hence, for large values of t_0, one should write, for example,

$$\Phi(t) = \int_{t_0}^{\infty} \pi(s)ds - \int_{t}^{\infty} \pi(s)ds = 1 - \frac{\text{Erfc}(t)}{\text{Erfc}(t_0)},$$

where Erfc is the complementary error function,

$$\text{Erfc}(t) = \frac{2}{\sqrt{\pi}} \int_{t}^{\infty} e^{-s^2} ds.$$

Further, by defining the scaled complementary error function

$$\text{Erfcx}(t) = e^{t^2} \frac{2}{\sqrt{\pi}} \int_{t}^{\infty} e^{-s^2} ds,$$

we have

$$\Phi(t) = 1 - e^{-(t^2 - t_0^2)} \frac{\text{Erfcx}(t)}{\text{Erfcx}(t_0)},$$

which allows a stable numerical evaluation. We remark that the inverse of Φ needs to be evaluated numerically.

Consider now a numerical example. We let the regular part of the true image consist of Gaussian functions, and the singular part of Dirac delta functions, that is,

$$f(s) = \sum_{\ell=1}^{L} c_\ell \exp\left(-\alpha_\ell |s - s_\ell|^2\right) + \sum_{k=1}^{K} d_k \delta(s - s_k).$$

The convolution of the true image with the Gaussian kernel $\mathcal{K}$ given in (7.5) can be evaluated analytically, yielding

$$\mathcal{K}(t) * f(t) = \sum_{\ell=1}^{L} \frac{\pi c_\ell}{\alpha_\ell + \kappa} \exp\left(-\gamma_\ell |t - s_\ell|^2\right) + \sum_{k=1}^{K} d_k \exp\left(-\kappa(t - s_k)^2\right),$$

where

$$\gamma_k = \frac{\kappa \alpha_k}{\alpha_k + \kappa}.$$

In our example, we choose the true image to be that in Figure 7.1. We add Gaussian noise with standard deviation 1% of the maximum of the noiseless signal to the blurred image.

First, we consider the reconstruction based on the assumption that the true image would consist of a smooth object only, that is, we assume that X is a priori distributed according to π_1. For the sake of completeness, we compute the estimations in two different ways. First, we ignore the modelling error and use the matrix A as a true blurring kernel. In this case the MAP estimate is

$$x_{\text{MAP}} = \begin{bmatrix} \sigma^{-1}A \\ \sqrt{\alpha}L \end{bmatrix}^\dagger \begin{bmatrix} \sigma^{-1}y \\ 0 \end{bmatrix}.$$

A more sophisticated way is to take into account the modelling error and calculate the MAP estimate from the joint probability density of $\widetilde{u}$ and y, assuming that no anomaly exists. This amounts to using formula (7.14) with $\widetilde{v} = 0$. The results, shown in Figure 7.2, indicate clearly that the infeasible prior model leads to a loss of the nature of the anomaly in either case.

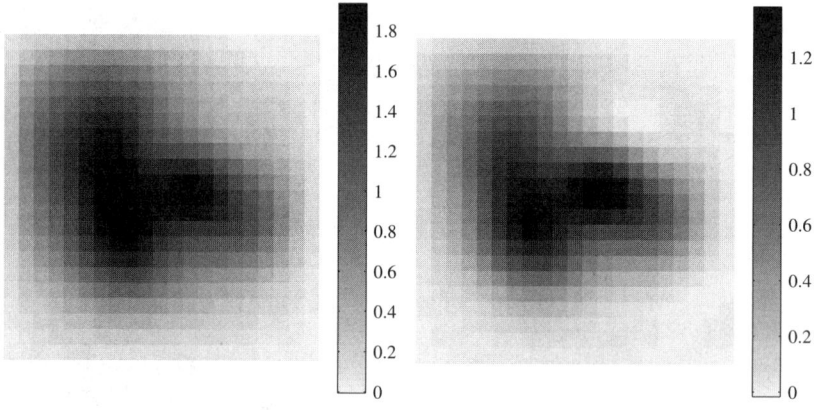

Figure 7.2. The MAP estimates using the smoothness prior. In the first image, the modelling error is ignored; in the second one it is taken into account

Next, we run the proposed MCMC algorithm. In calculating the covariance matrices, the parameter m defining the data and reconstruction grids is $m = 21$. The fine mesh used for discretization error modelling is $n = 4m = 84$. We run the MCMC algorithm producing a sample of size 10,000, starting from the initial state $[\widetilde{u}^0; \widetilde{v}^0] = [0; 0]$. The burn-in period, that is, the number of iterations excluded from the beginning of sampling is chosen to be $n_{\text{burn-in}} = 300$.

In Figure 7.3, we display the autocorrelations of the pixel histories of the regular and anomalous parts. These pixels represent different features of the image. We select three pixels: One pixel is from the flat background; the second one is from the maximum of the regular part; and the third one is the location of the anomaly of the projected image $\widetilde{v}$. We calculate the scaled time series autocovariances (3.42) for both the regular and anomalous pixel values at these pixels. Clearly, the pixel values at the anomaly are strongly correlated and hence the estimated pixel values may be more unreliable than those at the background pixels.

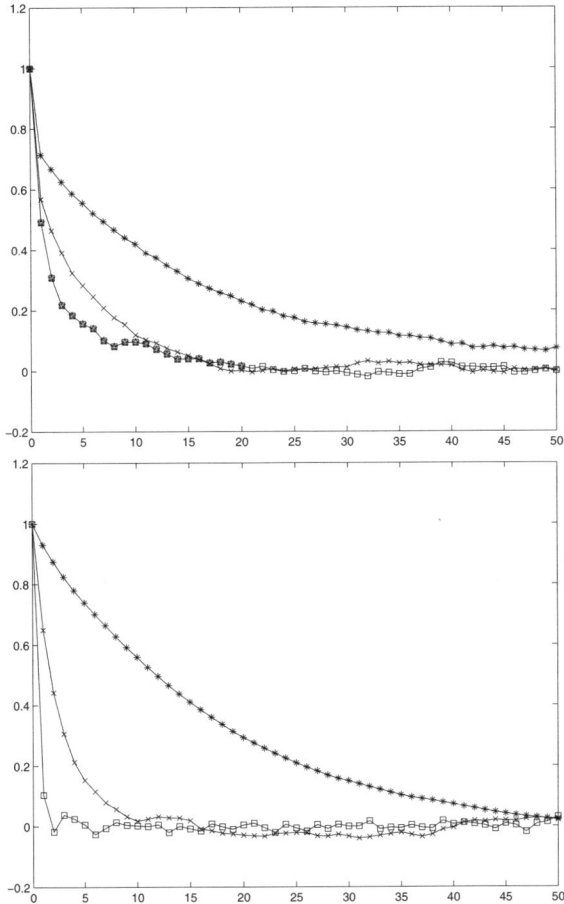

Figure 7.3. Autocovariances of various pixel values: regular part (upper figure), anomalous part (lower figure). The pixel containing the anomaly is marked by an asterisk (∗); the pixel at the maximum of the regular part by a plus (+); and a background pixel by as square (□).

Figure 7.4 shows the conditional mean estimates of the regular and the anomalous parts based on the MCMC sampling, as well as the autocovariances.

The results show clearly that the algorithm is able to localize the anomaly from the smooth background. The time series autocovariances of the various pixel histories shown in Figure 7.3 indicate that the pixel *values* of the anomalous pixel are relatively strongly correlated between draws. One can conclude from this that the estimated pixel value at the anomaly is not very reliable, and this conclusion is reinforced by the estimated autocovariances: The autocovariance of the anomalous pixel is relatively high compared to the background pixel autocovariances. However, one can argue that the *location*

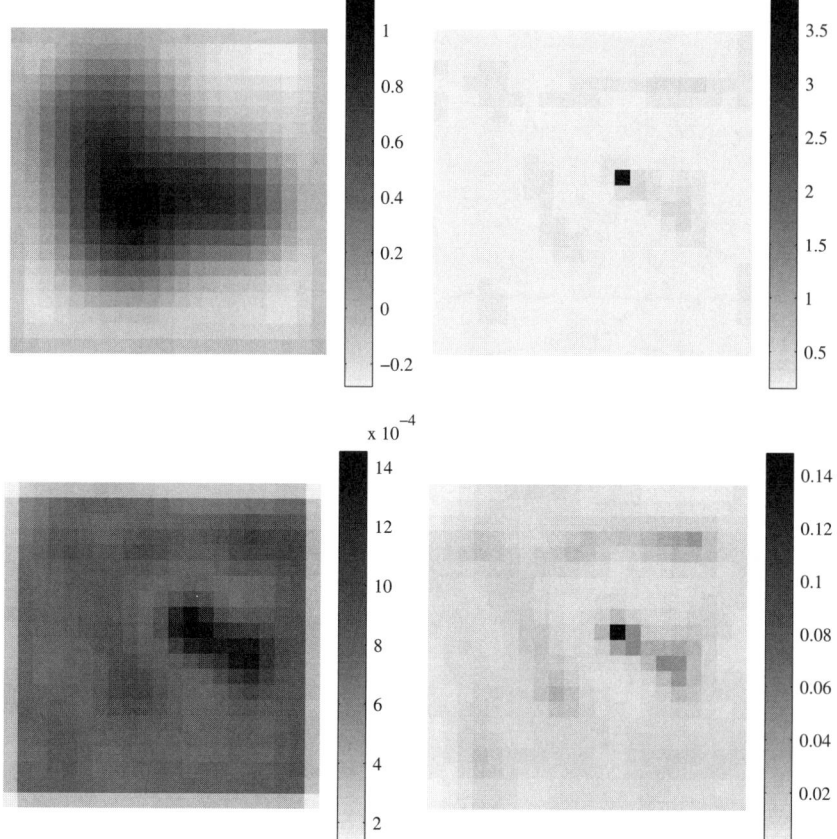

Figure 7.4. The MCMC-based CM estimates for the regular and anomalous parts (top row) and the corresponding autocovariances (bottom row). The visible boundary effect in the regular part is due to the forced Dirichlet boundary condition in the definition of the smoothness prior.

of the anomaly is found reliably. Indeed, since the background pixel histories are less correlated, their estimated values are more trustworthy. Therefore, it can be asserted with great confidence that they are *not* anomalous ones, that is, possible anomalies must be outside of them.

If the actual value of the anomalous pixel is of interest, one could design a fast search algorithm for estimating it.

7.1.4 Effects of Modelling Errors

In the previous subsection, we ignored for computational convenience the discretization error of the anomalous part but modelled the one due to the

7.1 Image Deblurring and Recovery of Anomalies

regular part. Here, we study the effects of the different modelling errors and by numerical evidence justify the aforementioned choice.

Let us consider first the importance of modelling the discretization error of the regular part. Hence, assume that we write the model directly in the coarse mesh,

$$Y = AU + AV + E,$$

where $A \in \mathbb{R}^{M \times M}$, and the additive error E is independent Gaussian measurement noise. Furthermore, assume that the prior model is written also directly in this mesh as (7.7) using a discretized Laplacian (7.8) of the $(m+1) \times (m+1)$ pixel map. This seemingly innocent simplification changes only the conditional covariace matrix Γ in (7.9).

With this simplification, we run the same MCMC algorithm as described before. In Figure 7.5, we have plotted the estimates of the conditional means of the regular and anomalous parts as well as their autocovariances. We used the same analytically computed data with additive noise as in the previous section. A comparison of the results clearly shows a deterioration of the estimate of the regular part. Also, the anomalous part contains visible background noise, compensating for the errors in the regular part. The increased estimation error covariance is clearly visible in the autocovariance plots.

Let us estimate the discretization error due to the anomalous part V. Denote by D_q^N and D_p^M, $1 \leq q \leq N$, $1 \leq p \leq M$, the pixels in the division of D into N or M pixels, respectively. Assume that the true anomaly consists of a single Dirac delta at point $t^* \in D$. Assume that this point is contained in the pixel $D_{q^*}^N \subset D_{p^*}^M$. Then, we have

$$v_q = n^2 \delta_{q,q^*}, \quad \tilde{v}_p = Pv = m^2 \delta_{p,p^*}.$$

The modelling error in this case is

$$\left| [(\mathcal{A} - AP)v]_p \right| = \left| e^{-\kappa |t_p - s_{q^*}|^2} - e^{-\kappa |t_p - t_{p^*}|^2} \right| \quad (7.18)$$

$$\leq \left| 1 - e^{-\kappa(|t_p - s_{q^*}|^2 - |t_p - t_{p^*}|^2)} \right|.$$

From this estimate, we observe that if $s_{q^*} = t_{p^*}$, the modelling error vanishes. By the construction of the nested refining of the pixel map, this condition is satisfied only when the indices (i, j) in (7.2), $0 \leq i, j \leq n$ to the pixel s_{q^*} are multiples of the refining factor k. Geometrically, this means that singularities are allowed to be located in particular subpixels of D_p^M.

It is not hard to see that the upper bound (7.18) is largest when $p = p^*$. Since $D_{q^*}^N \subset D_{p^*}^M$, we have $|t_{p^*} - s_{q^*}| < 1/\sqrt{2}m$, and

$$\left| [(\mathcal{A} - AP)V]_p \right| \leq \left| 1 - e^{-\kappa/(2m^2)} \right|.$$

In general, this error is not negligible. In the present example with a single delta peak, the upper bound is of the order 0.09, or about 9% of the signal.

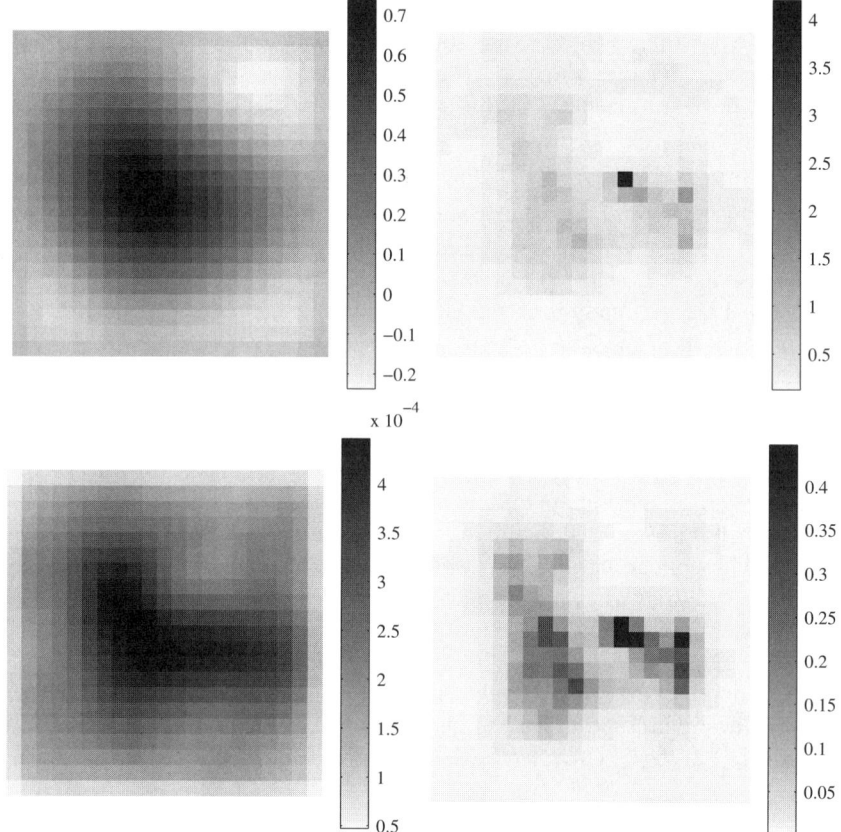

Figure 7.5. The CM estimates for the regular and anomalous parts (top row) and the corresponding autocovariances (bottom row) using the linear model without modelling discretization errors.

To see how severe the approximation error is in practice, we generate data in which the condition $s_{q^*} = t_{p^*}$ is violated. We set the anomalous pixel in the refined mesh as far as possible from the mesh points t_q of the coarse mesh. In Figure 7.6, we have plotted the reconstructed regular and anomalous parts in this case. The difference is visible only locally near the anomalous pixel, thus showing that the discretization error of the anomalous part does not affect significantly the localization properties of the algorithm.

7.2 Limited Angle Tomography: Dental X-ray Imaging

In Chapter 6 we considered the reconstruction of the mass absorbtion coefficient from X-ray attenuation measurements. In that case we assumed that the

7.2 Limited Angle Tomography: Dental X-ray Imaging

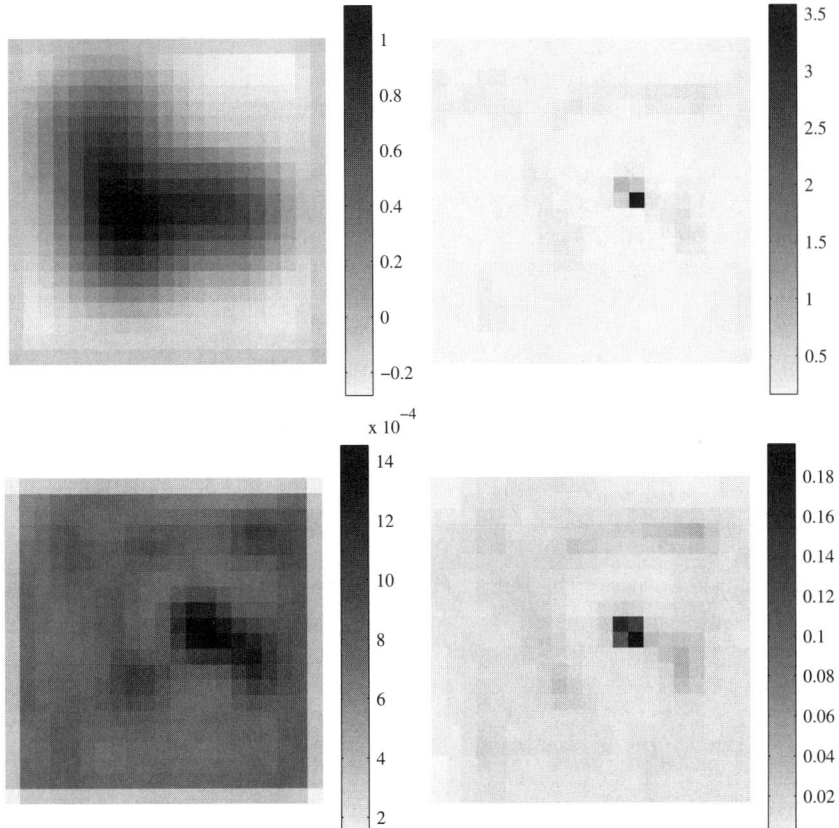

Figure 7.6. The CM estimates and the autocovariances of the regular and anomalous parts when the anomalous signal in the fine mesh does not coincide with the mesh point of the coarse grid.

anatomical structure of the target, a skull-type layer, was known, and a Gaussian structural prior model could be constructed. In this section we consider a similar but more difficult case in which we know that the target consists of a number of complex subdomains within which the mass absorption coefficient is almost constant. This case is relevant in medical X-ray tomography with adequately high photon energies so that muscular and other soft tissues are essentially transparent to the X-rays. Thus the projections contain information essentially from different types of bone.

In medical applications full angle data is usually available, but there are a few instances in which this is not the case. The first one is dental X-ray imaging, in which the detector is placed in the patient's mouth just behind the teeth. Unlike in conventional CT geometry, the detector plane is thus fixed, and the source rotates typically over a 60° angle, or an angle of ±30°

238   7 Case Studies

with respect to the normal of the detector plane. In addition, only a small number of projections can be acquired.

Another application is intersurgical X-ray in which the source and the detector are connected to a portable C-shaped arc which can be rotated around the target over a limited angle dictated by the surgical installation.

Limited angle tomography occurss also naturally in geophysical borehole tomography as well as in atmospheric research applications.

We note that traditionally in medical practice the X-ray projection images would be investigated individually by visual means as conventional X-ray images. Recently tomographic images have been pursued based on conventional projections obtained in a controlled geometry.

The state-of-the-art method in dental imaging, called the tuned aperture computerized tomography method (TACT) [140], is based on adding projections up with a shift as follows.

Assume that we have projection images y_k, $1 \leq k \leq K$ of the same target. Each image is created by shifting a pointlike X-ray source along a line parallel to the detector plane. Assume for simplicity that the source is far enough from the target that it can be thought of as a parallel beam source, that is, for each source position, the beams arrive at the detector plate at a different constant slant angle. viewing the actual three-dimensional parameter distribution as layers parallel to the detector plane, the image y_k can be thought of as consisting of shadow images of these layers, each one shifted by an amount proportional to its distance of the detector plate. The TACT image is then a linear combination of shifted images y_k whose shadow images of one material layer at a time are aligned. Such procedure focuses the projection images onto each material layer. This is in principle a crude version of the parallel-beam unfiltered backprojection method used in CT full angle tomography.

The X-ray measurements are inherently Poisson distributed since the detector current depends linearly, at least ideally, on the absorbed photon count. However, there is also electronic noise, which often can be assumed to be Gaussian. With high transmitted intensities, that is, high equivalent photon counts, the Poisson distribution can be approximated reliably by a Gaussian distribution. Let $y = [y_1; \ldots; y_m]$ denote the vector of the logarithms of the detector currents. If we approximate the likelihood density with a Gaussian model and assume that contribution due to electronic noise is negligible, we readily obtain the approximation

$$\pi(y \mid x) \propto \exp\left(-\frac{1}{2}(y - Ax)^{\mathrm{T}} \Gamma_{\mathrm{n}}^{-1}(y - Ax)\right)$$

for the likelihood, where $\Gamma_{\mathrm{n}} = \mathrm{diag}\,(cy)$ and c is a constant that depends on the detector characteristics and the incident beam intensity. If the electronic noise is not insignificant but independent of the photon count, we have an approximation $\Gamma_{\mathrm{n}} = \mathrm{diag}\,(cy) + \sigma_e^2 I$, where σ_e^2 is the electronic noise variance.

In the prior model, our task is to encode the knowledge that the target consists of unkown subregions with small internal variation of the mass

7.2 Limited Angle Tomography: Dental X-ray Imaging   239

absorption coefficient. If the geometry of the target subregions is adequately simple, a reparametrization is feasible: Instead of a pixel image, one could estimate the boundary surfaces of the subregions. Such models are discussed later in this chapter in connection with optical tomography. In the case of objects like teeth, the regions are probably too complex for straightforward reparametrization. Furthermore, if there were only a few values for the mass absorption coefficient in all subdomains, level set methods could be used. However, the current standard choice is to employ the total variation prior or some other "long-tailed" distribution.

7.2.1 The Layer Estimation

In this example we consider the problem of estimating the three-dimensional mass absorption coefficient of a tooth based on few projection data. The X-ray source is pointlike, and it moves along a circular arc around the detector. The opening angle of the arc is 60°, and 7 projections, corresponding to evenly spaced source locations, are recorded.

Since the computation of the forward operator matrix A in three-dimensional is in many cases infeasible due to the large dimension of the problem, we transform the problem of estimating the mass absorption coefficient x into a succession of two-dimensional problems as follows.

Let X represent the three-dimensional voxelized mass absoption, that is, we divide the volume into voxel layers. The layers are perpendicular to the detector plate and parallel to the source orbit. Let $X^{(k)}$ denote the mass absorption coefficients in the kth layer, where the layers are numbered from the top to the bottom of the voxel box.

To model the prior, we start with writing a two-dimensional TV prior model for the top layer $X^{(1)}$. For the successive slices $X^{(k)}$, we then write a prior model that is a combination of the two-dimensional TV prior within the layer and an ℓ^1 prior that corresponds to a small deviation from the estimate of the previous layer. More precisely,

$$\pi(x^{(1)}) \propto \pi_+(x^{(1)}) \exp\left(-\alpha \text{TV}(x^{(1)})\right),$$
$$\pi(x^{(k)}|x^{(k-1)}) \propto \pi_+(x^{(k)}) \exp\left(-\alpha \text{TV}(x^{(k)}) - \gamma \|x^{(k)} - x^{(k-1)}\|_1\right), \quad k > 1.$$

In our model, the pixel size in each layer is approximately 0.16 mm and the two-dimensional image size of each $X^{(k)}$ is 166 × 166. The number of layers is 600 so that the layer thickness becomes 0.045 mm.

The pixel size of the slices in our example is selected so that the slice thickness corresponds to a row in the detector array of the X-ray sensor Sigma by Instrumentarium Imaging Corporation used to produce the data.

To model the data, consider a single slice with absorption coefficients $X^{(k)}$. If the X-ray source is far away from the target, the rays that pass through the voxels of this layer hit a single row of the detector array. Let us denote by

$Y^{(k)}$ the random vector that models the data corresponding to the slice $X^{(k)}$ and let $A^{(k)}$ denote the matrix corresponding to the two-dimensional forward map from $X^{(k)}$ to $Y^{(k)}$. If the array is small and the distance to the X-ray source is sufficiently large as in our example, we can approximate $A^{(k)}$ by a single matrix A for all layers. Thus the posterior of the kth layer is

$$\pi(x^{(k)}|y^{(k)}, x^{(k-1)}) \propto \pi_+(x^{(k)}) \exp\left(-F(x^{(k)}; y^{(k)}, x^{(k)})\right),$$

where

$$F(x^{(k)}; y^{(k)}, x^{(k-1)}) = \frac{1}{2}\left(y^{(k)} - Ax^{(k)}\right)^{\mathrm{T}} \Gamma_{\mathrm{n}}^{-1}\left(y^{(k)} - Ax^{(k)}\right)$$
$$+ \alpha \mathrm{TV}(x^{(k)}) + \gamma \|x^{(k)} - x^{(k-1)}\|_1.$$

The parameters α and γ could be chosen to be hierarchical random variables as discussed earlier in Chapter 3. However, in a problem such as this, these parameters can also be selected with auxiliary methods such as using the full angle data and exact three-dimensional reconstructions to determine feasible α and γ. The statistics of the measurement error can be estimated based on a large number of repeated measurements. Here, we set $\alpha = \gamma = 1250$ and $\Gamma_{\mathrm{n}} = \mathrm{diag}\,(\sigma_1^2, \ldots, \sigma_N^2) \approx 4 \cdot 10^{-4} I$.

We estimate the vector X by solving a sequence of two-dimensional problems, starting from the top slice and proceeding through the voxel box downward.

7.2.2 MAP Estimates

From the point of view of optimization, both the TV and the ℓ^1–norms are inconvenient as they are nondifferentiable. Thus we approximate the absolute value function in both norms by

$$|t| \approx h_\beta(t) = \beta^{-1} \log \cosh(\beta t)$$

where $\beta > 0$. While larger β yields a better approximation, the second derivative of h_β grows like $1/\beta$ at the origin. Hence, for numerical stability, we eventually aim at gradient-based methods rather than second-order approaches.

The minimization of F with $x^{(k)} \geq 0$ is a typical nonlinear programming problem. In this example, we use an exterior point algorithm which augments the functional F with a penalty term that is modified during the iteration. In this case the exterior point algorithm can be described as follows. Let G_j denote a penalty functional,

$$G_j(z) = \sum_{\ell=1}^{N} \varphi_j(z(\ell)), \quad z \in \mathbb{R}^N,$$

where

$$\varphi_j\bigl(z(\ell)\bigr) = \begin{cases} \zeta_j z^2(\ell), & z(\ell) < 0 \\ 0, & z(\ell) \geq 0. \end{cases}$$

We choose an increasing sequence of parameters ζ_j and seek to compute the sequence $x_j^{(k)}$ of minimizers of functionals

$$\Phi_j(x^{(k)}) = F(x^{(k)}; y^{(k)}, x^{(k-1)}) + G_j(x^{(k)}).$$

When the value of the parameter ζ_j increases, the penalty term forces the iterates stronger to the feasible region. Solving the minimizer $x_j^{(k)}$ of Φ_j with j fixed requires an iteration. We can either solve the above problem as two nested iterations by minimizing Φ_j and then increasing j, or we can increase ζ_j after each inner iteration loop that seeks to minimize Φ_j.

Once our problem has been cast as an unconstrained minimization problem, a possible approach would be to solve it by using a gradient direction method with a line search. Among the possible strategies for selecting the step length automatically without multiple function evaluations per search direction, the Barzilai–Borwein algorithm [11] proceeds as follows. Let d_j be the search direction at the jth iteration. In this case, d_j is the negative gradient of the augmented functional Φ_j. The Barzilai–Borwein step length parameter is

$$\kappa_j = \frac{\|x_j - x_{j-1}\|^2}{(x_j - x_{j-1})^{\mathrm{T}}(d_j - d_{j-1})},$$

where x_j is the estimate after the jth iteration.

We demonstrate the method with real X-ray dental data. A typical TACT image and the corresponding MAP estimate for a layer parallel to the detector plane are shown in Figure 7.7. The TACT image shows clearly the vertical stripes that are due to the raw X-ray image shifts.

The search of the MAP estimate is based on the nested Barzilai–Borwein iterations. With an appropriate choice of increasing values of ζ_j, essential convergence occurs after five outer iterations with six inner iterations, i.e., keeping ζ_j fixed. The same procedure is repeated in each slice.

The difference in the contrast in the two reconstructions is clear. The faintly visible row of small circular artifacts is due to an added marker in the target that is used to align the shifts in the TACT images.

7.2.3 Sampling: Gibbs Sampler

Due to the high dimensionality, the Metropolis–Hastings scheme is very difficult to implement in a way that feasible sampling is attained. Therefore, we resort to the Gibbs sampler which is feasible although possibly very slow. Due to the dimensionality problem, in this case we treat the slices as independent, that is, we set the coupling constant $\gamma = 0$. As opposed to the MAP estimation, we do not have to approximate the absolute value function or employ a penalty function to enforce the positivity constraint. The Gibbs sampler

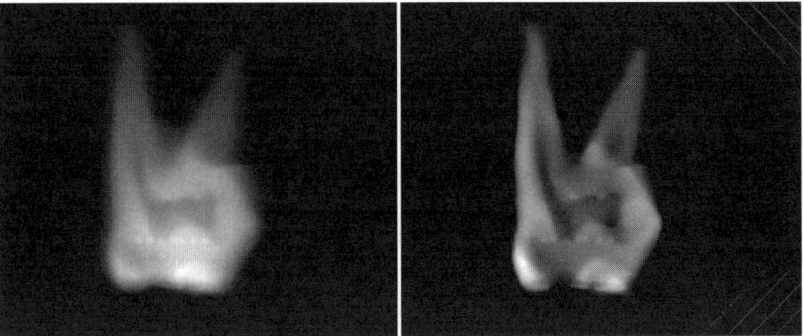

Figure 7.7. A TACT image that is focused on a layer (left). The corresponding MAP estimate (right).

was realized by computing numerical approximations for the full conditional densities at each sample update.

For the initialization, we compute the single-layer MAP estimates and use these as initial values for sampling. With this choice of initial values there is no need for a burn-in sequence removal. We draw 15,000 samples and compute the sample mean as well as the pixelwise marginal distributions. In this case, the resulting CM estimate is visually slightly worse than the MAP estimate since the transversal interlayer information is not employed.

However, as noted earlier, one of the most important features in MCMC sampling is that it allows us to compute different statistics such as the marginal distributions of pixels. These are important especially in the verification of the accuracy of the reconstructions.

The marginal distributions of two pixels are shown in Figure 7.8. The first pixel is from the middle of the tooth bone and near a boundary whose normal is almost perpendicular with the normal of the detector plane. It has been analytically shown that the limited angle reconstructions in such locations are more accurate than when the normals coincide. The second pixel is from a "no bone" region near a boundary segment almost parallel to the detector. The right-hand tail of the distribution is beginning to grow due to the uncertainty.

7.3 Biomagnetic Inverse Problem: Source Localization

As explained in Chapter 6, the objective in biomagnetic inverse problems is to extract information concerning the electromagnetic activities of the human body – in particular in the heart and brain – from observations of the magnetic fields outside the body. In this chapter we consider two different aspects related to this inverse problem. The first one is localization of distributed sources.

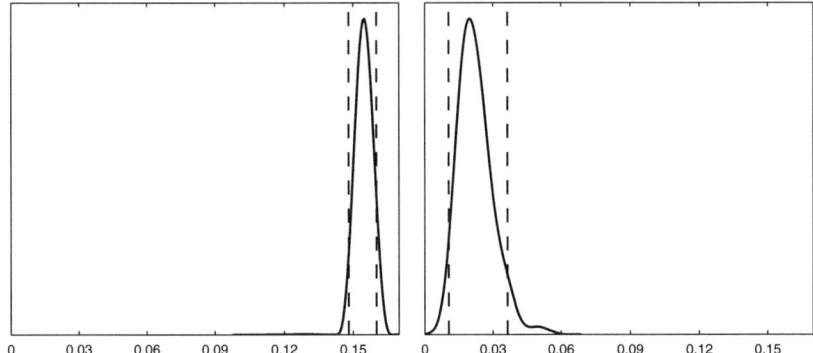

Figure 7.8. The marginal distributions of two pixels. The left distribution corresponds to a pixel in the middle of the solid bone, the right one to a pixel in a more uncertain region outside the bone tissue. The dashed lines indicate the 90% credibility intervals.

Consider the following simplified model. A planar time-independent current density flows in a known plane. Above this plane, at fixed locations, the magnetic field component perpendicular to the plane is registered. From the knowledge of the magnetic field at these locations, the task is to estimate the current density. In our model, we ignore the possible Ohmic volume currents. This choice is justified not only in vacuum but also in the case when the electric conductivity is horizontally layered. As observed in Chapter 6, the volume currents have no contribution to the data in this case. We also recall that only the horizontal component of the current density creates a magnetic field perpendicular to the plane.

Assume furthermore that our *qualitative* prior information concerning the unknown is that the current density is likely to be well localized, that is, the active sources are typically small in size. In real applications, such information could be based, for example, on physiological models of the brain activity.

A popular and classical method to deal with the localized source problem is to use dipole fitting, that is, to fit one or several dipole sources to the data in the least squares sense. Observe that in such a procedure, regularization is usually not necessary since the dimension reduction obtained by restricting the problem into a few dipole models has as a strong regularization effect. In this section, we select a different approach that demonstrates the meaning of certain prior densities in these applications.

7.3.1 Reconstruction with Gaussian White Noise Prior Model

We discretize the problem by defining a grid of size $m \times m$ over the imaging area beneath the magnetometers. The grid points are denoted by $p_j = [p_{jx}; p_{jy}]$. At each grid point, we place a current dipole $q_j = [q_{jx}; q_{jy}]$, hence assuming that the dipoles are coplanar and that the plane corresponds to the plane in which the true current density is located. Such discretization induces to a pixelized approximation of the current density. In our model, we assume that the vertical component of the magnetic field is measured at points r_ℓ, $1 \leq \ell \leq L$ above the plane of the current density. Hence, the model for the noiseless data is

$$b_\ell = \frac{\mu_0}{4\pi} \sum_{j=1}^{N} \frac{e_z \cdot (q_j \times (r_\ell - p_j))}{|r_\ell - p_j|^3}, \quad 1 \leq \ell \leq L, \tag{7.19}$$

the vector e_z being the unit vector perpendicular to the source plane. The unknown vector to be estimated in this problem consists of the dipole components,

$$x = [q_{1x}, q_{2x}, \ldots, q_{Nx}, q_{1y}, \ldots, q_{Ny}] \in \mathbb{R}^{2N}, \quad N = m^2.$$

We assume here that the measurement noise is Gaussian and additive, leading to the simple linear statistical observation model

$$Y = AX + E, \quad E \sim \mathcal{N}(0, \Gamma_{\text{noise}}).$$

The matrix $A \in \mathbb{R}^{L \times 2N}$ can be easily constructed from the formula (7.19). We assume that $\Gamma_{\text{noise}} = \sigma^2 I$, that is, the noise is white.

At first, we ignore the prior information that the sources should be localized. Instead, consider the white noise prior for X,

$$\pi_{\text{pr}}(x) \propto \exp\left(-\frac{1}{2\gamma^2} \|x\|^2\right),$$

where $\|x\|$ denotes the usual Euclidean norm. With this prior, the CM and MAP estimates coincide with the Tikhonov regularized solution,

$$x_\alpha = \begin{bmatrix} A \\ \sqrt{\alpha} I \end{bmatrix}^\dagger \begin{bmatrix} y \\ 0 \end{bmatrix}, \tag{7.20}$$

where the regularization parameter in the Tikhonov solution is $\alpha = (\sigma/\gamma)^2$. Let us see how these estimates behave. In our example, we generate numerically the data corresponding to two dipole sources. To avoid committing the most obvious inverse crime, the original dipoles are not located at any of the grid points. Also, we set one of the true dipoles slightly off the reconstruction plane.

The true current is displayed in Figure 7.9, where the locations of the magnetometers are marked by crosses. The distance of the magnetometers from the plane is 3 unit lengths, the image area having side length of 10 units. To the analytically computed data, we add zero mean Gaussian noise with standard deviation of 5% of the maximum of the noiseless signal.

Figure 7.10, shows three reconstructions corresponding to wildly different values of the Tikhonov regularization parameter α, $\alpha = 0.01$, $\alpha = 1$ and

7.3 Biomagnetic Inverse Problem: Source Localization

$\alpha = 100$. No attempt has been made to estimate the prior variance, for example, by empirical Bayes methods or discrepancy-type methods. We observe that *structurally* the reconstructions look strikingly similar, the only difference being in the absolute amplitude of the current dipoles. Hence, the result seems to suggest that the white noise prior that corresponds to the Tikhonov regularization corresponds to a structural prior information that the solution is solenoidal. Note in particular that the dipole orientations are quite insensitive to the selection of the parameter α. We shall make use of this feature below.

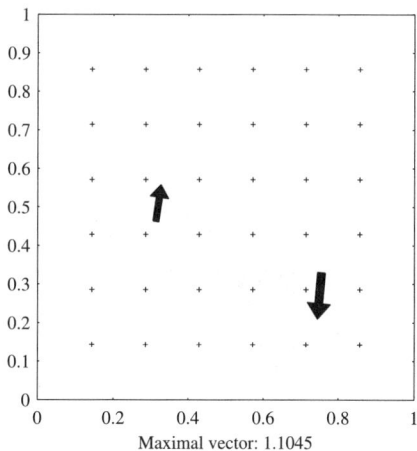

Figure 7.9. The true current density consisting of two current dipoles. The right dipole is located 0.5 unit lengths below the reconstruction plane.

7.3.2 Reconstruction of Dipole Strengths with the ℓ^1-prior Model

To implement the prior information concerning the localization of the sources, we choose the ℓ^1 prior discussed in Chapter 3. It was demonstrated by random draws that this prior favors distributions that have only a few nonvanishing components. Hence, let us write

$$[q_{jx}, q_{jy}] = t_j[\cos\theta_j, \sin\theta_j],$$

where

$$t_j = \sqrt{q_{jx}^2 + q_{jy}^2} \geq 0.$$

We choose

$$\pi_{\mathrm{pr}}(x \mid \gamma) \propto \gamma^N \pi_+(t) \exp\left(-\gamma \sum_{j=1}^N t_j\right),$$

246   7 Case Studies

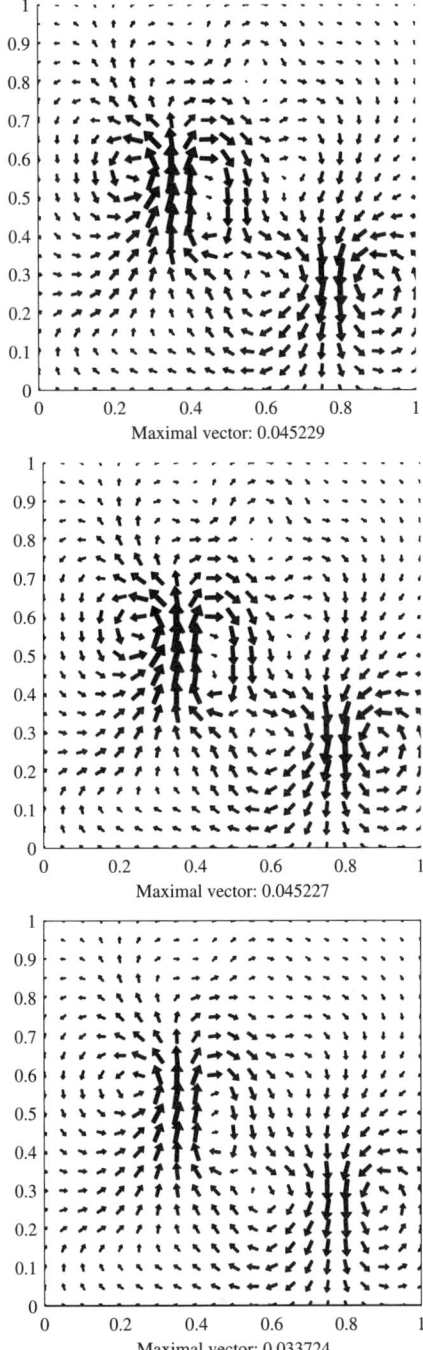

Figure 7.10. Three MAP estimates of the current density using the white noise prior. The parameter $\alpha = \sigma/\gamma$ from top to bottom is $\alpha = 0.01$, $\alpha = 1$ and $\alpha = 100$.

where π_+ is the positivity constraint, that is, $\pi_+(t) = 1$ if $t_j \geq 0$ for all j, $1 \leq j \leq N$ and $\pi_+(t) = 0$ elsewhere. Observe that the prior information here is qualitative, that is, we expect only that the source to has a small support. Therefore, we cannot assume that the prior parameter $\gamma > 0$ would be known accurately. Its determination is part of the problem.

The posterior density in this case is

$$\pi(x \mid y, \gamma) \propto \gamma^N \pi_+(t) \exp\left(-\frac{1}{2}(y - A\Theta t)^{\mathrm{T}} \Gamma_{\text{noise}}^{-1}(y - A\Theta t) - \gamma \sum_{j=1}^{N} t_j\right),$$

where Θ is the direction cosine matrix,

$$\Theta = \begin{bmatrix} \mathrm{diag}(\cos\theta_1, \ldots, \cos\theta_N) \\ \mathrm{diag}(\sin\theta_1, \ldots, \sin\theta_N) \end{bmatrix} \in \mathbb{R}^{2N \times N}.$$

To reduce the computational effort, we simplify the problem further by observing that while the white noise prior does not produce localized sources, it gives reasonable orientations for the dipoles. Hence, we fix the matrix $\Theta = \Theta_\alpha$ such that for the solution x_α given in (7.20), for some $\alpha > 0$,

$$x_\alpha = \Theta_\alpha t_\alpha,$$

and then consider the conditional density

$$\pi(t \mid y, \Theta_\alpha, \gamma) \propto \gamma^N \pi_+(t) \exp\left(-\frac{1}{2}(y - A\Theta_\alpha t)^{\mathrm{T}} \Gamma_{\text{noise}}^{-1}(y - A\Theta_\alpha t) - \gamma \sum_{j=1}^{N} t_j\right).$$

For the parameter γ, we use a Rayleigh distribution as a hyperprior,

$$\pi_{\text{hyper}}(\gamma) \propto \gamma \exp\left(-\frac{1}{2}\left(\frac{\gamma}{\gamma_0}\right)^2\right),$$

where $\gamma_0 > 0$. It turns out that the choice of γ_0 is rather immaterial as long as it is large enough. Hence, by substituting $\Gamma_{\text{noise}} = \sigma^2 I$, we obtain

$$\pi(t, \gamma \mid y, \Theta_\alpha) = \pi(t \mid y, \Theta_\alpha, \gamma) \pi_{\text{hyper}}(\gamma)$$

$$\propto \pi_+(t) \exp\left(-\frac{1}{2\sigma^2}\|y - A\Theta_\alpha t\|^2 - \gamma \sum_{j=1}^{N} t_j - \frac{1}{2}\left(\frac{\gamma}{\gamma_0}\right)^2 + (N+1)\log\gamma\right).$$

The maximum a posteriori solution $(t_{\text{MAP}}, \gamma_{\text{MAP}})$ conditioned on $\Theta = \Theta_\alpha$ is now found by minimizing the object functional

$$F(t, \gamma) = \frac{1}{2\sigma^2}\|y - A\Theta_\alpha t\|^2 + \gamma \sum_{j=1}^{N} t_j + \frac{1}{2}\left(\frac{\gamma}{\gamma_0}\right)^2 - (N+1)\log\gamma, \quad t_j \geq 0.$$

This minimization can be done, for example, by an interior point method. Similarly, the conditional mean $(x_{|y,\Theta_\alpha}, \gamma_{|y,\Theta_\alpha})$ can be estimated, for example, by using the Gibbs sampler, which is our choice in this example. The algorithm is constructed as follows:

Initialize $(t, \gamma) = (t^0, \gamma^0)$, set $k = 0$.
do until satisfactory sample size
    $(t, \gamma) = (t^k, \gamma^k)$
    for $j = 1 : N$
        draw $t_j^{k+1} \geq 0$ from $\pi(t_j \mid y, t_{-j}, \Theta_\alpha, \gamma)$
    end
    draw γ^{k+1} from $\pi(\gamma \mid y, \Theta_\alpha, t)$
    $k \leftarrow k + 1$
end

For the componentwise updating of the vector t, we write

$$\pi(t \mid y, \Theta_\alpha, \gamma) \propto \pi_+(t) \exp\left(-\frac{1}{2}(t^T B t - 2 t^T q)\right), \quad (7.21)$$

where

$$B = \frac{1}{\sigma^2} \Theta^T A^T A \Theta, \quad q = \frac{1}{\sigma^2} \Theta^T A^T y - \gamma \mathbf{1}.$$

Formula (7.21) is similar to (7.13); hence, a Gibbs sampler can be built along the same lines as in the previous section. The details are not repeated here.

To update γ, we have the conditional density

$$\pi(\gamma \mid y, \Theta_\alpha, t) \propto \exp\left(-\gamma \sum_{j=1}^N t_j - \frac{1}{2}\left(\frac{\gamma}{\gamma_0}\right)^2 + (N+1)\log\gamma\right).$$

Observe that if γ_0 is large, the quadratic term in the exponential becomes negligible. This corresponds, effectively, to the case in which no prior information about γ other than positivity is assumed. Indeed, if we choose the hyperprior to be an improper flat density, that is, $\pi_{\text{hyper}}(\gamma) \propto \pi_+(\gamma)$, we have

$$\pi(\gamma \mid y, \Theta_\alpha, t) \propto \exp\left(-\gamma \sum_{j=1}^N t_j + N \log\gamma\right). \quad (7.22)$$

In our computed example, we consider the planar geometry described above. The data correspond to the true current distribution of Figure 7.9, corrupted with additive noise of 5%. In Figure 7.11, we have plotted the CM estimate of the vector t represented as a pixel image. The estimate is computed from an MCMC sample of size 10,000. In the same figure, the estimate of the autocovarince of the vector t is also shown. In Figure 7.12, we show the histogram of the ensemble of γ values.

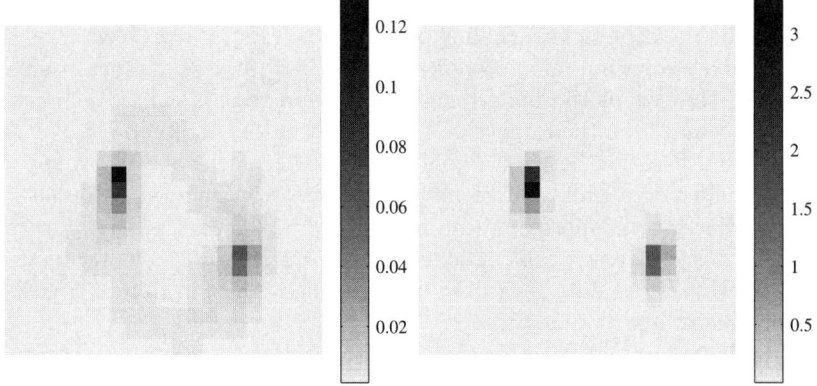

Figure 7.11. The MCMC-based CM estimate of the vector t (left) and its autocovariance (right).

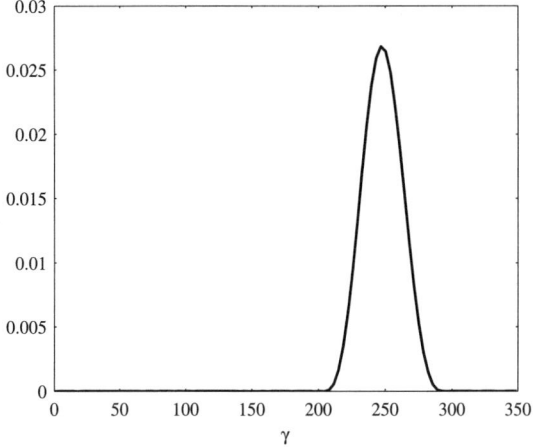

Figure 7.12. Estimated probability of the prior parameter γ.

We conclude this section by noting that the above type current estimates are called *minimum current estimates* in the biomedical literature. They are becoming increasingly popular because in contrast to the classical *minimum norm estimates* (see [53]), they can localize where the brain activity occurs. In [131], reconstructions in a realistic geometry and in vivo data are shown. The method has been shown to work well in three-dimensions also.

7.4 Dynamic MEG by Bayes Filtering

The biomagnetic inverse problems do not suffer from the complications always present in, for example, electrical impedance tomography or optical tomography, in which the data depnds nonlinearly on the parameters of interest. The problems in biomagnetics are mainly of two types: First, the problem does not have a unique solution, and second, the noise level in a single measurement is often of the size of the underlying signal. Moreover, a significant part of this noise comes from the target itself, that is, from the human body. For this reason, the distinction between signal and noise become somewhat vague.

The frequency contents of biomagnetic sources is assumed to be so low that the quasi-static approximation holds. The major part of the literature on biomagnetic inverse problems concentrates on *instantaneous* reconstruction algorithms: The data measured at a specific instant in time are used to estimate the source at that time.

To reduce the noise level in the instantaneous data, a common strategy in evoked response measurements is to repeat the measurement sequence and average over the repetitions. Such a strategy is acceptable only if the source does not alter from one measurement sequence to another. Furthermore, when measuring spontaneous events, the repetition of the measurement under identical or even similar conditions is usually not possible. For this reasons, it is desirable to develop dynamic inversion algorithms that are less sensitive to the high noise level. By dynamic algorithms, it is meant here that the time evolution of the target is included in the model.

The dynamic approach is helpful in two ways: A dynamic model can be seen as a prior model of the target, hence, it may remove the nonuniqueness of the solution and at the same time reduce the computational ill-posedness so that the sensitivity to the high measurement noise level is reduced. These ideas have been discussed in Chapter 4.

7.4.1 A Single Dipole Model

In this section we consider the most elementary case of a dynamic model for tracking a single dipole in a simple geometry. Assume that a single planar dipole moves in the plane $P = \{p = [p_1; p_2; 0]\}$, and the vertical component of the resulting magnetic field is observed above the plane. As shown in Chapter 6, the formula for the noiseless data is then

$$b(x) = [b_1(x), \ldots, b_m(x)]^T \in \mathbb{R}^m, \quad b_j(x) = \frac{\mu_0}{4\pi} \frac{e_z \cdot q \times (r_j - p)}{|r_j - p|^3}, \quad (7.23)$$

where we denoted the model parameter by

$$x = [p_1, p_2, q_1, q_2]^T \in \mathbb{R}^4.$$

Assume that we measure the magnetic fields at time instances t_j, $t_1 < t_2 < \cdots$. At $t = t_k$, the source dipole is characterized by the model parameter $x_k \in \mathbb{R}^4$. Furthermore, if we assume that the changes in the source current are relatively slow, the magnetic field can be modelled by the quasi-static model. Hence, the noiseless observation at $t = t_k$ is simply $b(x_k) \in \mathbb{R}^m$ defined by (7.23).

7.4 Dynamic MEG by Bayes Filtering

We want to apply Bayesian filtering to this problem. To this end, we define the evolution–observation model. Here, we assume that no special physiological model for the time evolution of the source is available. When no better model is available, the standard choice is to postulate a random walk model in which throughout the evolution of all components of x_k are mutually independent.

Considering the vector x at time t_k as a realization of a random variable $X_k \in \mathbb{R}^4$, we write

$$X_{k+1} = X_k + W_{k+1}, \tag{7.24}$$

where W_{k+1} is a Gaussian random vector independent of X_k with zero mean and covariance matrix

$$\Gamma_w = \mathrm{diag}(\lambda^2, \lambda^2, \delta^2, \delta^2) \in \mathbb{R}^{4 \times 4},$$

where $\lambda > 0$ controls the step sizes in the spatial evolution and $\delta > 0$ controls the steps in the amplitude evolution. In the terminology of Chapter 4, we have the Markov transition kernel

$$\pi(x_{x+1} \mid x_x) \propto \exp\left(-\frac{1}{2}(x_{x+1} - x_x)^\mathrm{T} \Gamma_v^{-1} (x_{x+1} - x_x)\right).$$

The observation model in this case is

$$Y_k = b(X_k) + E_k, \tag{7.25}$$

where we assume that the additive noise E_k is independent of X_j, $j \le k$ and Gaussian with zero mean and variance a known invertible $m \times m$ matrix Γ_n. Thus, the likelihood function is

$$\pi(y_k \mid x_k) \propto \exp\left(-\frac{1}{2}(y_k - b(x_k))^\mathrm{T} \Gamma_\mathrm{n}^{-1} (y_k - b(x_k))\right). \tag{7.26}$$

To initialize the filtering, we need to define the probability density of X_0. In our numerical examples below, we use the Gaussian probability density, that is,

$$X_0 \sim \pi_0(x_0) \propto \exp\left(\frac{1}{2}(x_0 - \overline{x}_0)^\mathrm{T} \Gamma_0^{-1} (x_0 - \overline{x}_0)\right). \tag{7.27}$$

The particle filtering algorithm for single dipole tracking can be summarized as follows.

Choose sample size N, and draw $x_0^1, \ldots, x_0^N \in \mathbb{R}^4$, from π_0 and set $k = 0$.
do
    Draw $v^1, \ldots, v^N \sim \mathcal{N}(0, \Gamma_w)$ and define $z^j = x_k^j + v^j$, $1 \le j \le N$
    Calculate the relative likelihoods, $w^j = \pi(y_k \mid z^j)/c$, $c = \sum_{j=1}^N \pi(y_k \mid z^j)$
    Draw x_{k+1}^j, $1 \le j \le N$ from $\{z^1, \ldots, z^N\}$, the probability of z^j being w^j.
    $k \leftarrow k + 1$
end

The above loop is repeated as long as new observations y_k keep arriving.

We demonstrate the algorithm with a simple example. In Figure 7.13, we have plotted a trajectory of a moving dipole. The dipole amplitudes and orientations are shown at 20 observation instances. The dipole moves counterclockwise, starting at the point nearest to the observation point (c) in the figure. The magnetic field is recorded by a 10×10 array of magnetometers above the dipole plane, locations being marked by crosses. If the distance between adjacent magnetometers is chosen to be the length unit, the distance of the magnetometers from the dipole plane is 3 length units.

To the analytically computed data, we add zero mean Gaussian random noise with standard deviation 80% of the maximum both in spatial and in temporal sense of the noiseless signal. The noisy and the noiseless signal at three measurement locations are shown in Figure 7.13.

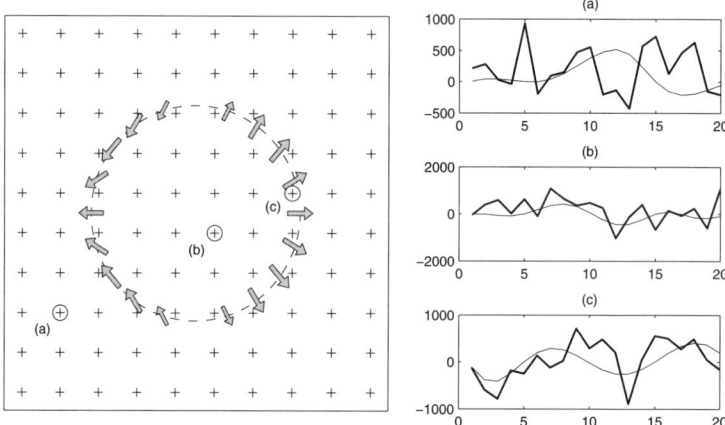

Figure 7.13. Single moving dipole model as well as the noiseless and noisy data at the locations marked with crosses.

We perform ten independent particle filter runs with the same data. The number of particles in each of these runs is 200,000. The step length in the random walk model for the dipole location is chosen to be $\lambda = 1$. Recall that the length of the image area is 10 units. The step size for the dipole amplitude is $\delta = 0.25$ units. This is about 20% of the maximum size of the true dipoles in the simulation. The selection of the step sizes can be viewd as a sort of prior information about the dynamics. Too small or too large steps fail to track the evolution. The initial density (7.27) does not have a significant effect on the performance. The only requirement is that it is wide enough so that the initial draws generate particles near the initial position.

7.4 Dynamic MEG by Bayes Filtering   253

In Figure 7.14, we have plotted the true trajectories of the dipole location and the amplitude, and in the same plot 10 independent particle filter CM estimates.

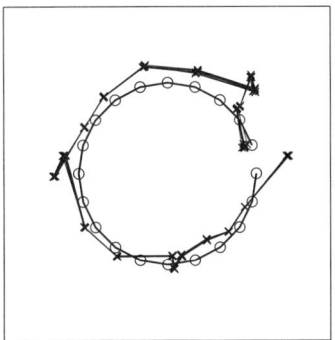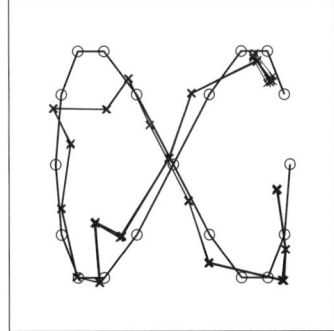

Figure 7.14. Averaged estimates of the location and amplitude. The figure shows 10 independent runs.

In Figure 7.15, we have plotted the percentages of particles that are not thrown away at the resampling stage. The figure indicates that sample impoverishment is not a crucial issue in this example.

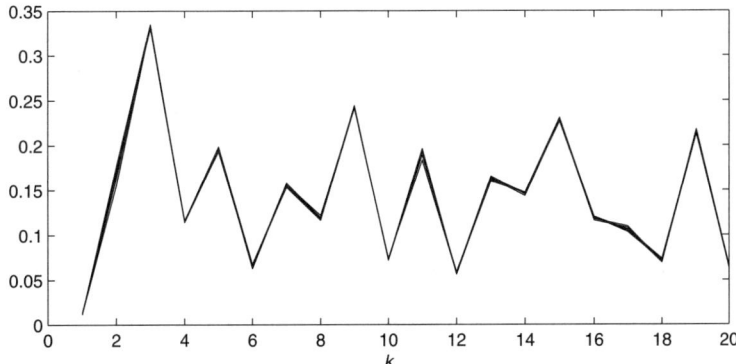

Figure 7.15. Relative number of particles that are sampled at least once at the resampling stage.

7.4.2 More Realistic Geometry

We consider next the problem of tracking a single dipole in a geometry that more realistically describes the MEG setting. Assume that the magnetometers are attached to a hemispherical surface of radius r_{meas} in the upper half-space. We assume that the electric activity is known to be confined to a volume V in the upper half-space determined by two spherical shells with radii $r_{\min}$, $r_{\max}$, with $r_{\min} < r_{\max} < r_{\text{meas}}$. The degrees of freedom are given in terms of the Cartesian components of the location and dipole moment, $[p;q] \in \mathbb{R}^6$. However, since from the radial dipole component there is no contribution to the data which consists of the radial component of the magnetic field, we restrict the transition kernels so that the radial component of the dipole moment vanishes. Let us denote $q^{\text{t}} = [q_\varphi; q_\theta]$ the tangential component of the dipole moment. We set $x = [p; q^{\text{t}}] \in \mathbb{R}^5$, and write the transition kernel as

$$\pi(x_{k+1} \mid x_k) \propto \chi_V(p_{k+1})\pi(p_{k+1} \mid p_k)\pi(q^{\text{t}}_{k+1} \mid q^{\text{t}}_k),$$

where χ_V is the characteristic function of the volume V and the partial transition kernels are Gaussian,

$$\pi(p_{k+1} \mid p_k) \propto \exp\left(-\frac{1}{2\lambda^2}\|p_{k+1} - p_k\|^2\right),$$

$$\pi(q^{\text{t}}_{k+1} \mid q^{\text{t}}_k) \propto \exp\left(-\frac{1}{2\delta^2}\|q^{\text{t}}_{k+1} - q^{\text{t}}_k\|^2\right).$$

The geometric setting is shown in Figure 7.16. In this example, we have $r_{\min} = 8\,\text{cm}$, $r_{\max} = 9\,\text{cm}$ and $r_{\text{meas}} = 11\,\text{cm}$. The number of magnetometers is 133. We add 80% Gaussian noise to the simulated noiseless signal. The steplengths in the random walk model are $\lambda = 1\,\text{cm}$ and $\delta = 0.15$, respectively. The outcome of the particle filtering with 300,000 particles is shown in Figure 7.17. The estimates are not as accurate as in the two-dimensional case. Possible reasons for this are the sparser sampling of the data and the fact that the data are not very sensitive to the distance of the source from the detector. An error in the radial direction of the dipole location can be compensated by an increased dipole moment amplitude.

7.4.3 Multiple Dipole Models

In multiple dipole models, we construct a *layered model* that allows the number of dipoles to change from one time instant to the other. Deciding how many active dipoles there are present is sometimes called a *model identification problem*.

The construction of layered multidipole models is conceptually simple. The interpretation of the results, however, requires more caution for the following reasons.

First, if the locations of two or several dipole sources coincide, they act as a single dipole source with the dipole moment equal to the resultant of the moments of the dipoles. Second, a multidipole model may contain phantom dipoles of vanishing or almost vanishing moments, in which case the model

7.4 Dynamic MEG by Bayes Filtering   255

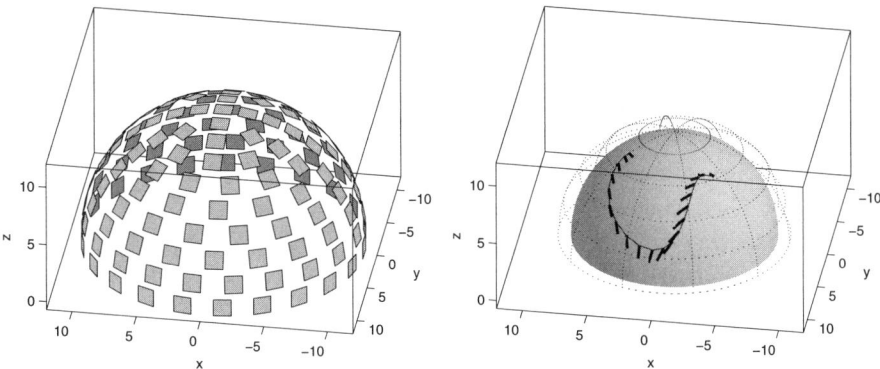

Figure 7.16. The three-dimensional measurement geometry (left) and the true dipole trajectory (right).

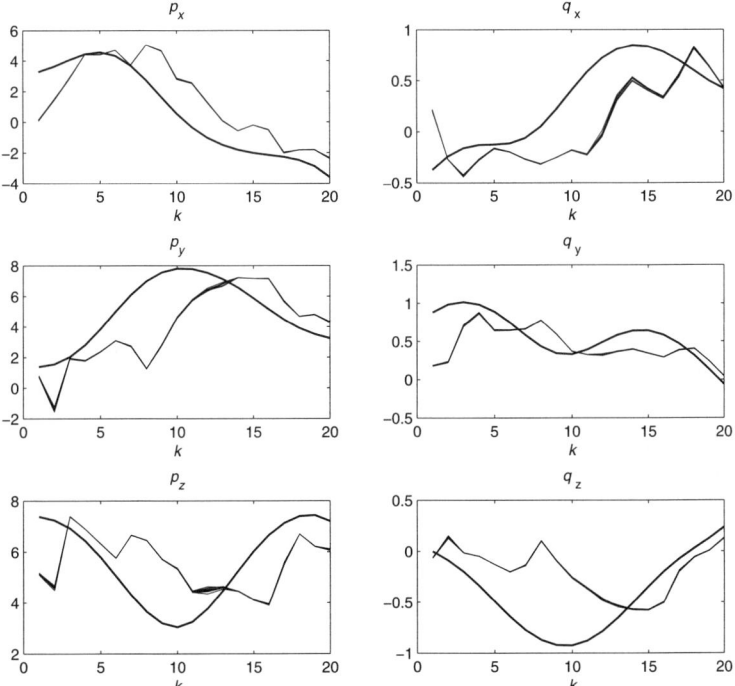

Figure 7.17. Averaged location and amplitude with 300,000 particles.

identification becomes ambiguous. The third problem is that approximate conditional means by direct averaging make in general no sense in layered models. To overcome these difficulties, we define an *activity density*.

Assume that we have a hierarchy of models with different dimensionality. We denote by $\mathcal{C} = \{C_k\}_{1 \leq k \leq K}$ a finite collection of multiple parameter subspaces, in which each C_j has a different dimensionality corresponding to a different number of active dipoles. Assume that a single dipole source may be characterized by n variables. In the previous section in our example with planar geometry, we had $n = 4$, two variables to determine the location in the plane, two to determine the dipole amplitude. Thus, the space C_k corresponding to k dipoles in the same geometry has the dimension kn. The notation $x \in \mathcal{C}$ means that x belongs to exactly one of the subspaces C_k.

We now define a state evolution model that allows switching between models of different dimensionality. Assume that the probability of the move $C_j \to C_\ell$ is given by $P_{j\ell}$ with $\sum_{\ell=1}^K P_{j\ell} = 1$. The Markov model for the time evolution is given in terms of the transition kernel

$$\pi(x_{k+1} \in \mathcal{C} \mid x_k \in C_j) = \sum_{\ell=1}^K \pi(x_{k+1} \in C_\ell \mid x_k \in C_j) P_{j\ell},$$

where the density $\pi(x_{k+1} \in C_\ell \mid x_k \in C_j)$ is the conditional transition probability density from a point in $x_k \in C_j$ to the subspace C_ℓ of ℓ dipoles. Hence, we have to specify the transition probabilities $P_{j\ell}$ between different model spaces as well as the specific jumping rules conditioned on the knowledge of the initial and final dimensionalities.

In the numerical example below, we consider again planar geometry, and we confine the discussion to the simplest possible layered model with a single- and a double-dipole state, that is, $\mathcal{C} = \{C_1, C_2\}$. Extensions to more involved multidipole models are straightforward.

We start by defining the state evolution models between subspaces with different dimensionalities. The moves within the same models, $C_1 \to C_1$ and $C_2 \to C_2$, are defined here by random walk models. Thus, if $x_k \in C_1$ is a single dipole, we define $\pi(x_{k+1} \in C_1 \mid x_k \in C_1)$ via

$$X_{k+1} = X_k + V_{k+1} \; ;$$

see (7.24) of the previous section. Similarly, if $x_k = (x_k^{(1)}, x_k^{(2)}) \in C_2$ is a double-dipole state, $x_k^{(j)} \in \mathbb{R}^4$, we define the conditional transition kernel $\pi(x_{k+1} \in C_2 \mid x_k \in C_2)$ by

$$X_{k+1} = (X_{k+1}^{(1)}, X_{k+1}^{(2)}) = (X_k^{(1)}, X_k^{(2)}) + (V_{k+1}^{(1)}, V_{k+1}^{(2)})$$

where $V_{k+1}^{(1)}$ and $V_{k+1}^{(2)}$ are identical mutually independent Gaussian random variables as in the single-dipole model.

Consider now the state transition model between C_1 and C_2. For a move $C_1 \to C_2$, a new dipole has to be created. We define dipole doubling as

$$X_{k+1} = (X_{k+1}^{(1)}, X_{k+1}^{(2)}) = (X_k + W_{k+1}^{(1)}, X_{k+1}^{(2)}),$$

where the new dipole $X_{k+1}^{(2)}$ is generated independently from a probability density $\pi_{\text{new}}(x)$. This density is chosen Gaussian density both with respect to location and amplitude.

Consider now the reverse move $C_2 \to C_1$ where one dipole is deleted. We choose randomly one of the two dipoles, delete it and propagate the remaining one according to the random walk model. Thus, the propagation step is

$$X_{k+1} = tX_k^{(1)} + (1-t)X_k^{(2)} + W_{k+1},$$

where $t \in \{0,1\}$ takes the values 0 or 1 with equal probability and W_{k+1} is Gaussian. For the transition probabilities $P_{j\ell}$, the simplest choice is to choose them equal, that is, no prior preference of any model exists. This is the choice here.

Finally, we need to define the initial distribution. This is done by postulating that

$$P\{X_0 \in C_1\} = P\{X_0 \in C_2\} = \frac{1}{2},$$

that is, the particles contain one or two dipoles with equal probability. As in the previous section with a single dipole, we draw the initial locations and amplitudes from a wide Gaussian density.

We are now ready to define the activity density. To interpret the results of a particle filter run, consider the posterior density at a given time instant $t = t_k$. Suppressing the time dependency in our notations, denote the density in $\mathcal{C} = \{C_1, C_2\}$ as $\pi = \pi_1 \oplus \pi_2$, where π_k is a density in C_k. The particle filter approximation for π_1 is

$$\pi_1(x) = \pi_1(p,q) \approx \sum_{x^n \in C_1} w^n \delta(p - p^n)\delta(q - q^n),$$

that is, the approximation is calculated using single-dipole particles. The corresponding marginal density for the single-dipole locations is then

$$\pi_1(p) = \int \pi_1(p,q)dq \approx \sum_{x^n \in C_1} w^n \delta(p - p^n).$$

In the same fashion, as C_2 consists of two copies of C_1, we may write

$$\pi_2(x) = \pi_2(p^{(1)}, q^{(1)}, p^{(2)}, q^{(2)})$$

$$\approx \sum_{x^n \in C_2} w^n \Big(\delta(p^{(1)} - (p^{(1)})^n)\delta(q^{(1)} - (q^{(1)})^n)$$

$$\oplus \delta(p^{(2)} - (p^{(2)})^n)\delta(q^{(2)} - (q^{(2)})^n) \Big),$$

and after marginalizing with respect to the dipole moments,

$$\pi_1(p^{(1)}, p^{(2)}) \approx \sum_{x^n \in C_2} w^n \bigg(\delta(p^{(1)} - (p^{(1)})^n) \oplus \delta(p^{(2)} - (p^{(2)})^n) \bigg).$$

With the help of these marginal densities, we define the *activity density* $\pi_{\mathrm{act}}(p)$ by

$$\pi_{\mathrm{act}}(p) = \pi_1(p) + \pi_2(p, p)$$

$$\approx \sum_{x^n \in C_1} w^n \delta(p - p^n) + \sum_{x^n \in C_2} w^n \bigg(\delta(p - (p^{(1)})^n) + \delta(p - (p^{(2)})^n) \bigg).$$

This density gives the posterior density of the spatial locations of all the active dipoles regardless of the models. In practice, the estimate for the activity density is calculated as the particle occurrence histogram in a prescribed pixel map of the p-plane.

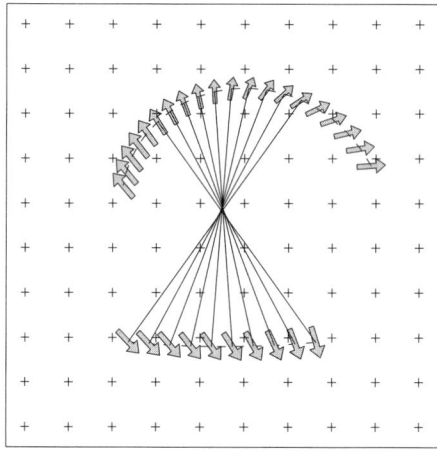

Figure 7.18. The trajectories of two moving dipoles. The adjoining lines in the figure indicate contemporaneity of the events.

As a numerical test, we consider two planar dipoles as depicted in Figure 7.18. Again, noise with standard deviation 80% of the maximum noiseless data is added to the simulated noiseless signal. The random walk step lengths in the simulation are $\lambda = 0.5$ and $\delta = 0.5$. The density π_{new} used for new dipole generation is Gaussian, $\pi_{\mathrm{new}} \sim \mathcal{N}(\overline{x}_{\mathrm{new}}, \gamma_{\mathrm{new}}^2 I)$. The algorithm is not sensitive to the choices of the parameters in this density provided that the density is wide enough so that emerging true dipoles appear in the numerical support of the density.

7.4 Dynamic MEG by Bayes Filtering

In Figure 7.19 the estimated activity density is plotted. This result indicates clearly that model identification based on the activity density is possible, while the particle distribution between single- and double-dipole states may be misleading.

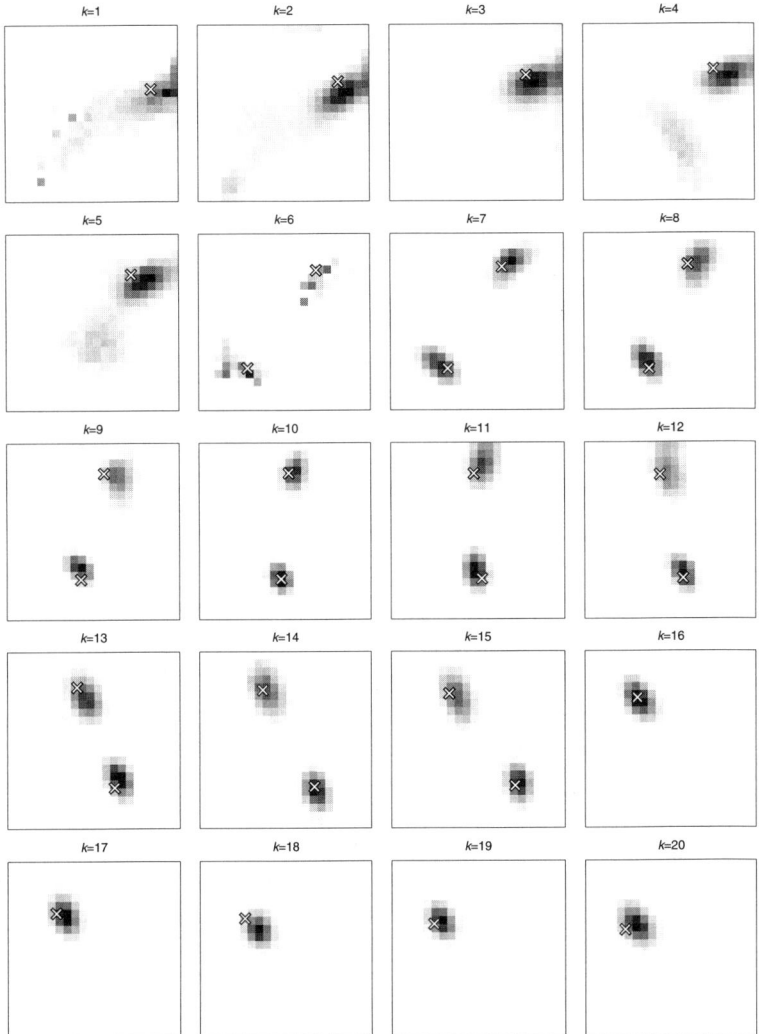

Figure 7.19. Estimated activity densities. True dipole locations are marked by crosses.

7.5 Electrical Impedance Tomography: Optimal Current Patterns

We discuss here a problem related to the optimal experiment design of electrical impedance tomography. We show how statistical theory can be applied to decide, what the optimal, or most informative measurement setting is in the light of the prior information of the target. This is also an example of *Bayesian experiment design problem* in general.

Consider the EIT problem with static current feeds, in which we have a body Ω with the conductivity distribution σ. Assume that we apply K current patterns $I^{(1)}, \ldots, I^{(K)}$ on the surface $\partial\Omega$ through L contact electrodes with known contact impedances $z_1, \ldots, z_L$ and we measure the corresponding voltages. In the absence of noise, the voltages are related to the current patterns via the impedance matrix $R(\sigma)$,

$$U^{(k)} = R(\sigma)I^{(k)}, \quad 1 \leq k \leq K,$$

where we suppressed the dependence of the impedance matrix on the contact impedances. The question we want to address in this section is:

Assuming some prior information of the conductivity distribution σ, how should one choose the current patterns $I^{(1)}, \ldots, I^{(K)}$ for maximal complementary information about the conductivity?

Above, the word *complementary* refers to information that can be described as additional information to the already known a priori information.

7.5.1 A Posteriori Synthesized Current Patterns

Before formulating the problem more precisely, let us make a few remarks. Suppose that we apply a full frame $\{I^{(k)}\}_{k=1}^{L-1}$ of current densities, and we would like to select the best possible frame. Since the space of admissible current densities is linear, we can represent any admissible current pattern as a linear combination of those actually applied. Hence, if I is the desired current pattern, in the noiseless case we can write

$$I = \sum_{k=1}^{L-1} \alpha_k I^{(k)}, \quad U = R(\rho)I = \sum_{k=1}^{L-1} \alpha_k U^{(k)}.$$

From the point of view of linear algebra, the question of optimal current patterns may seem therefore quite artificial. However, the discussion above was based on the noiseless model. Assume now that the measurements are corrupted by additive noise, that is, we have

$$V^{(k)} = U^{(k)} + E_k = R(\sigma)I^{(k)} + E_k, \quad 1 \leq k \leq L-1,$$

where E_k is the noise. Futhermore, assume for the sake of simplicity that for any measurement, the noise covariance is independent of the applied current

7.5 Electrical Impedance Tomography: Optimal Current Patterns

pattern. If we try to simulate computationally the observation corresponding to the current pattern $I = \sum \alpha_k I^{(k)}$, we would get

$$V = R(\sigma)I + E, \quad E = \sum_{j=1}^{L-1} \alpha_k E_k.$$

The computationally created noise vector E may have a higher amplitude than what the measurement noise would have, had we actually applied the pattern I. In actual applications, this effect is even worse than it may seem as the computation is done in finite precision. As an example, consider an L-electrode measurement. Assume that the frame consists of the current patterns

$$[I^{(1)}, \ldots, I^{(L-1)}] = \begin{bmatrix} 1 & 0 & \cdots & 0 & 0 \\ -1 & 1 & \cdots & 0 & 0 \\ 0 & -1 & \cdots & 0 & 0 \\ \vdots & \vdots & & \vdots & \vdots \\ 0 & 0 & \cdots & -1 & 1 \\ 0 & 0 & \cdots & 0 & -1 \end{bmatrix} \in \mathbb{R}^{L \times (L-1)},$$

that is, only two adjacent electrodes are active at a time. Suppose that we want to synthesize the current pattern

$$I = [1, 0, \ldots, 0, \overset{(\ell)}{-1}, 0, \ldots, 0],$$

that is, a two-electrode pattern with the first and ℓth electrode active. We see that

$$I = \sum_{k=1}^{\ell} I^{(k)},$$

so the computationally synthesized data, based on the measurements corresponding to the current patterns $I^{(k)}$ would be

$$V_{\text{synth}} = R(\sigma)I + \sum_{k=1}^{\ell} E_k.$$

Assuming that the noise vectors E_k are independent with covariance $\gamma^2 I$, we see that the synthesized noise has covariance

$$\mathrm{E}\{EE^\mathrm{T}\} = \sum_{k=1}^{\ell} \sum_{j=1}^{\ell} \mathrm{E}\{E_k E_j^\mathrm{T}\} = \ell \gamma^2 I,$$

that is, ℓ-fold to what the direct measurement noise would be. In practice, the situation can be even more dramatic: If the voltages corresponding to the current patterns $I^{(k)}$ have low amplitudes compared to the voltage corresponding to I, the signal-to-noise ratio for the synthesized measurement can be significantly lower than for the directly measured data, since the signal amplification is proportional to the signal amplitude, and low amplitude measurements lead to amplification of the noise level, too. In practice, this means a loss of significant digits in the data.

This observation motivates the claim that *if we believe that some current pattern gives more information than others, it is advisable to apply that current pattern rather than synthesize it computationally.*

7.5.2 Optimization Criterion

The next question that we address is what we mean by the most informative current pattern. As usual, we assume here that the conductivity is discretized and represented by a vector $\sigma \in \mathbb{R}^n$. Let the prior probability density $\pi_{\mathrm{pr}}(\sigma)$ be given. Let us assume that we are looking for m most informative current patterns. The current patterns are stacked into a single vector, denoted by $I = [I^{(1)}; \ldots; I^{(m)}] \in \mathbb{R}^{mL}$. In the following we shall refer to both I and $I^{(\ell)}$ as a current pattern.

From Bayes' formula, the posterior distribution corresponding to a used current pattern I is

$$\pi(\sigma \mid V, I) \propto \pi_{\mathrm{pr}}(\sigma) \pi(V \mid \sigma, I). \qquad (7.28)$$

We observe that in the formula (7.28), we cannot affect the shape of the posterior probability density by tampering with the prior: the prior information is determined by what we know about σ. However, we have the current density at our disposal as a control parameter, and it appears only in the likelihood part. We want to choose the current pattern from an admissible set of patterns in such a way that the likelihood becomes *complementary to the prior*, in the sense that it removes possible ambiguities of the prior density as effectively as possible. Schematically, the situation is presented in Figure 7.20.

Indeed, assume that the prior probability density is very elongated as in the figure, or even indefinite, in some subspace, or more generally, some submanifold direction $M \subset \mathbb{R}^n$. This may be the case, for example, when the prior density is a smoothness prior. For ill-posed problems, we know also that the likelihood density is typically an elongated density along some subspace or manifold $N \subset \mathbb{R}^n$. Here, N depends of course of the measurement setting, hence of the current pattern I applied. The aim in the Bayesian experiment design is now to rotate $N = N(I)$ so that the spaces M and $N(I)$ would intersect as transversally as possible, that is, the likelihood becomes a narrow density in those directions in which the prior has a large variance and is therefore uninformative.

To formulate this task more precisely, let $\widehat{\sigma} = \widehat{\sigma}(V, I)$ denote any estimate of σ based on the observations V. In Section 5.1, it was shown that the conditional mean estimate σ_{CM} coincides with the mean square estimator, that is, it minimizes the mean square error conditioned on the observation,

7.5 Electrical Impedance Tomography: Optimal Current Patterns

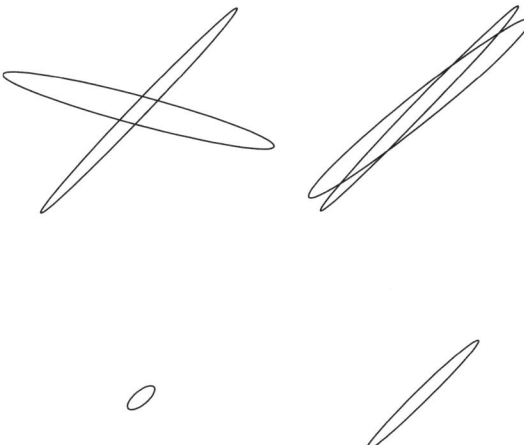

Figure 7.20. Schematic picture of different relative positions of the prior and the likelihood densities and its effect on the posterior density.

$$\sigma_{\mathrm{CM}} = \arg\min_{\widehat{\sigma}} \mathrm{E}\{\|\sigma - \widehat{\sigma}\|^2 \mid V\} = \arg\min_{\widehat{\sigma}} \left(\mathrm{Tr}\big(\mathrm{cov}(\sigma - \widehat{\sigma} \mid V)\big)\right),$$

where

$$\mathrm{cov}(\sigma - \widehat{\sigma} \mid V) = \mathrm{E}\{(\sigma - \widehat{\sigma})(\sigma - \widehat{\sigma})^{\mathrm{T}} \mid V\}$$

is the estimation error covariance of $\widehat{\sigma}$.

The above formulation of the conditional mean as a solution of a minimization problem provides a natural definition optimal current pattern. We seek to minimize the mean square error with respect to *both $\widehat{\sigma}$ and the current pattern I*. In practice, we do the minimization sequentially by looking for a current pattern that minimizes the mean square error of the CM estimator of σ. This procedure allows us to express the minimization problem in terms of the posterior covariance matrix.

It is necessary to impose two constraints on the current patterns. The first constraint is a linear one, coming from the charge conservation law

$$\sum_{j=1}^{L} I_j^{(\ell)} = 0. \tag{7.29}$$

In addition, we need to constrain the power of the current, for an obvious reason: We assume that the covariance of the observation noise is fixed, hence increasing electrode potentials would improve the signal-to-noise ratio. Since the potentials are proportional to the injected current, the optimal current pattern would be infinite. Thus, we constrain the power of the current as

$$\|I^{(\ell)}\| = 1. \tag{7.30}$$

Denote the set of m admissible current patterns by

$$\mathcal{J}_m = \{I = [I^{(1)}; \ldots; I^{(m)}] \mid I^{(\ell)} \text{ satisfies (7.29) and (7.30)}\},$$

and define

$$I_{\text{opt}} = \arg\min_{I \in \mathcal{J}_m} \left(\text{Tr}\big(\text{cov}(\sigma - \sigma_{\text{CM}}(I))\big) \right). \tag{7.31}$$

In other words, the optimal current pattern is the one that together with the optimal estimator in the MSE sense produces the least mean square estimation error. Geometrically, the trace of the error covariance measures the width of the posterior covariance. The above criterion, of course, is not the only applicable one. Another natural criterion for choosing I is to look for the current which minimizes the function

$$I \mapsto \lambda_1 \big(\text{cov}(\sigma - \sigma_{\text{CM}}(I))\big),$$

where λ_1 is the largest eigenvalue of the estimation error covariance matrix.

In general, the object functionals may depend on the unknown conductivity σ, so the optimization may be very challenging. It is possible to attack this problem statistically in all its generality, but the computational task becomes easily unfeasible. To simplify the problem we assume Gaussian models and use global linearization.

Assume that the prior density is Gaussian,

$$\pi_{\text{pr}}(\sigma) \propto \exp\left(-\frac{1}{2}(\sigma - \sigma_0)^{\text{T}} \Gamma_{\text{pr}}^{-1}(\sigma - \sigma_0)\right),$$

and that the current pattern $I^{(\ell)}$, $1 \leq \ell \leq m$ is applied, the corresponding observation $V^{(\ell)}$ is corrupted by zero mean additive Gaussian noise with covariance $\Gamma_{\text{n}}^{(\ell)}$, independent of the current pattern $I^{(\ell)}$. With these assumptions, the posterior density is of the form

$$\pi(\sigma \mid V) = \exp\left(-\frac{1}{2} Q(\sigma; I)\right), \tag{7.32}$$

where

$$Q(\sigma; I) = \sum_{\ell=1}^{m} \left(V^{(\ell)} - R(\sigma) I^{(\ell)}\right)^{\text{T}} \left(\Gamma_{\text{n}}^{(\ell)}\right)^{-1} \left(V^{(\ell)} - R(\sigma) I^{(\ell)}\right)$$

$$+ (\sigma - \sigma_0)^{\text{T}} \Gamma_{\text{pr}}^{-1} (\sigma - \sigma_0). \tag{7.33}$$

Linearizing the mapping $\sigma \mapsto R(\sigma)$ around the midpoint σ_0 of the prior we have

$$R(\sigma) I^{(\ell)} \approx R(\sigma_0) I^{(\ell)} + J_\ell (\sigma - \sigma_0),$$

where $J_\ell = J(\sigma_0, I^{(\ell)})$ is the Jacobian of the mapping $\sigma \mapsto R(\sigma) I^{(\ell)}$ at $\sigma = \sigma_0$. This approximation leads to an approximation of the posterior covariance matrix as

7.5 Electrical Impedance Tomography: Optimal Current Patterns

$$\operatorname{cov}(\sigma - \sigma_{\mathrm{CM}}(I) \mid V) \approx \Gamma_{\sigma\mid V}(I) = \left(\sum_{\ell=1}^{m} J_\ell\bigl(\Gamma_{\mathrm{n}}^{(\ell)}\bigr)^{-1} J_\ell + \Gamma_{\mathrm{pr}}^{-1}\right)^{-1}.$$

We need to solve the following nonlinear optimization problem with a nonlinear constraint,

$$I_{\mathrm{opt}} = \arg\min_{I \in \mathcal{I}_m} \left(\sum_{\ell=1}^{m} J_\ell\bigl(\Gamma_{\mathrm{n}}^{(\ell)}\bigr)^{-1} J_\ell + \Gamma_{\mathrm{pr}}^{-1}\right)^{-1}.$$

The minimization can be done by using the projected gradient method. Given a current estimate $I_{\mathrm{opt}}^{(j)}$, we calculate the gradient $\nabla_I \Gamma_{\sigma\mid V}(I_{\mathrm{opt}}^{(j)})$ by the finite-difference approximation. Let $\mathcal{D}_{\|}\Gamma_{\sigma\mid V}(I_{\mathrm{opt}}^{(j)})$ denote the orthogonal projection of the gradient onto the linear subspace of current patterns satisfying the constraint (7.29). The new iterate is the minimizer along the curve

$$t \mapsto \frac{I_{\mathrm{opt}}^{(j)} - t\mathcal{D}_{\|}\Gamma_{\sigma\mid V}(I_{\mathrm{opt}}^{(j)})}{\|I_{\mathrm{opt}}^{(j)} - t\mathcal{D}_{\|}\Gamma_{\sigma\mid V}(I_{\mathrm{opt}}^{(j)})\|}, \quad t > 0.$$

As the initial guess $I_{\mathrm{opt}}^{(0)}$ for the current patterns we use random draws from the search space defined by the two constraints.

We remark that the optimal current patterns are not unique. For example, if $I_{\mathrm{opt}}^{(\ell)}$ is an optimal current pattern, then $-I_{\mathrm{opt}}^{(\ell)}$ has the the same posterior distribution as $I_{\mathrm{opt}}^{(\ell)}$. Actually these two patterns are essentially the same, since changing the sign only changes the phase of the current, not the amplitude.

7.5.3 Numerical Examples

As a first computed example, consider the two-dimensional EIT problem in a circular body $\Omega \subset \mathbb{R}^2$ with $L = 16$ evenly distributed contact electrodes on its surface. The prior probability density is $\mathcal{N}(\sigma_0, \Gamma_{\mathrm{pr}})$, where

$$\sigma_0 = \text{constant}, \quad \Gamma_{\mathrm{pr}} = \operatorname{diag}(d), \ d \in \mathbb{R}^n.$$

The vector d represents the autocovariances of the conductivity values σ_j of pixels p_j. In this example, d is small except in a noncentered circular region $D \subset \Omega$ where the components of d are large. This means that the pixels are uncorrelated and σ is known a priori rather well outside D, while in D the uncertainty is large. More precisely, we set

$$d_j = \begin{cases} \sigma_*^2, & p_j \in D, \\ (0.05\sigma_*)^2, & p_j \notin D. \end{cases}$$

Intuitively, we expect optimal current patterns to inject as much current as possible through the domain of uncertainty D.

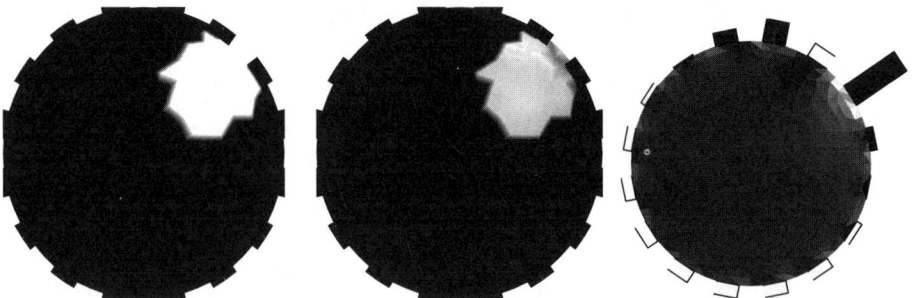

Figure 7.21. The prior autocovariance (left) and the posterior autocovariance (middle) after a measurement of a single optimized current pattern feed. The optimal pattern is represented by plotting dark columns on electrodes with positive current and light columns with negative currents (right).

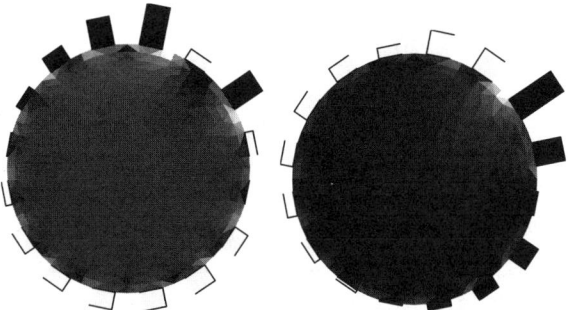

Figure 7.22. Two optimal current patterns.

Consider first the problem of finding a single optimal pattern, that is, $m = 1$. Figure 7.21 shows the prior autocovariance plotted as an image of the body Ω. The figure also shows the computed optimal current pattern as well as the posterior autocovariance corresponding to this current pattern. The result coincides well with the intuitive picture. The posterior autocovariance is plotted in the same scale as that of the prior covariance, showing that the uncertainty is reduced in the uncertainty region D. Observe that for these results, we do not need to define the true conductivity distribution.

For comparison, we consider also the case $m = 2$, that is, we determine two current patterns. The optimal patterns are shown in Figure 7.22. Notice that none of the two optimal patterns coincide with the single optimal current pattern of Figure 7.21. This reflects the nonlinearity of the minimization problem. It means that the optimal patterns cannot be calculated sequentially one at a time.

In the next example, we consider the two-dimensional model with a more complex prior. The domain Ω is partitioned into two halves, $\Omega = \Omega_{\mathrm{up}} \cup \Omega_{\mathrm{down}}$.

7.5 Electrical Impedance Tomography: Optimal Current Patterns

In Ω_{up}, we assume that the pixels are uncorrelated with all the other pixels. In Ω_{down}, we use a Markov random field model corresponding to a Gaussian smoothness prior. Enumerating the pixels so that $p_j \in \Omega_{\text{up}}$, $1 \leq j \leq n'$ and $p_j \in \Omega_{\text{down}}$ for $n'+1 \leq j \leq n$, so that the prior covariance is of the form

$$\Gamma_\sigma = \begin{bmatrix} \Gamma_{\text{up}} & 0 \\ 0 & \Gamma_{\text{down}} \end{bmatrix},$$

where $\Gamma_{\text{up}} = \alpha_{\text{up}}^2 I \in \mathbb{R}^{n' \times n'}$, and $\Gamma_{\text{down}} \in \mathbb{R}^{(n-n') \times (n-n')}$ is obtained from a Markov model

$$\pi_{\text{pr}}(\sigma_{n'+1}, \ldots, \sigma_n) \propto \exp\left(-\alpha_{\text{down}} \sum_{j=n'+1}^{n} \left(\sigma_j - \sum_{k \in \mathcal{N}_j} a_{jk} \sigma_k\right)^2\right).$$

The parameters α_{up} and α_{down} are adjusted so that the prior autocovariance is close to constant over the whole domain. Also, the mean conductivity σ_0 is assumed to be constant in Ω. Figure 7.23 shows prior correlations of a single selected pixel with other pixel values when the pixel is selected from the upper or from the lower half of the disc, respectively.

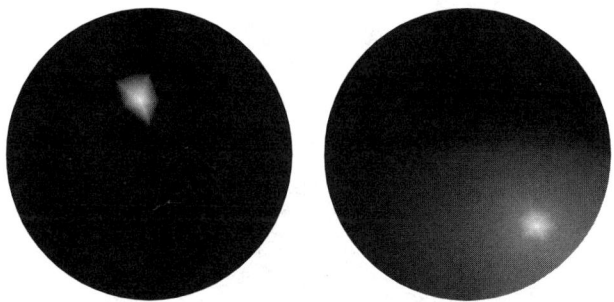

Figure 7.23. Correlations with a single pixel in the upper half having Gaussian white noise prior (left) and in the lower half where the prior is Gaussian smoothness prior (right).

We compute a single optimal current pattern corresponding to this distribution. The optimal current is shown in Figure 7.24. We observe that the current injection is concentrated on the lower half of the body. The interpretation is that such current injection reduces effectively the posterior covariance since the corresponding observation conveys information of mutually correlated variables. This is clearly visible in the resulting posterior autocovariance which is plotted in the same figure. Note that the prior autocovariance in this example is constant over the body.

268   7 Case Studies

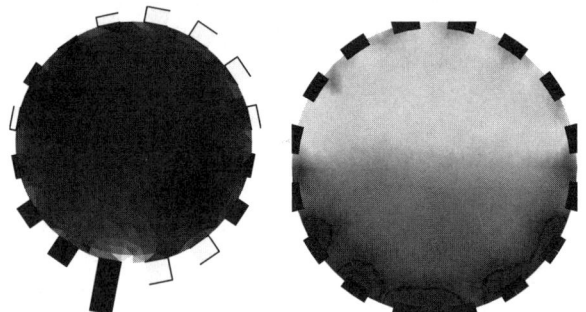

Figure 7.24. Correlations with a single pixel in the upper half having Gaussian white noise prior (left) and in the lower half where the prior is Gaussian smoothness prior (right).

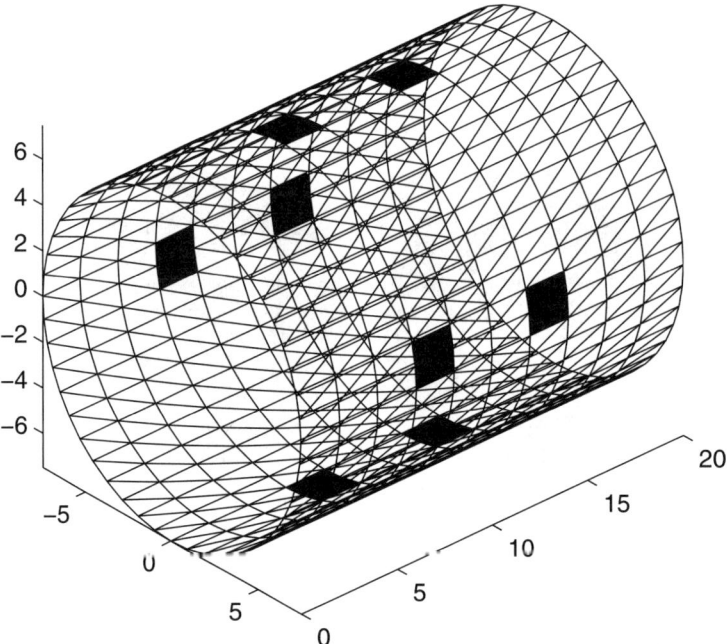

Figure 7.25. The geometry of the three-dimensional case.

As a final example, we apply the optimization to a three dimensional problem, in which the body is a cylindrical tank with a circular bottom. The number of the electrodes in this example is 8, the locations being illustrated in Figure 7.25.

The cylinder is divided into two parts along its axis. Again, in the upper half we use a white noise prior, and in the lower part a Markov field smoothness prior. The priors are adjusted so that the prior autocovariance is constant in the voxels.

As in the previous two-dimensional example, the single optimal current pattern is concentrated in the lower half of the domain where a priori smoothness is assumed.

To demonstrate the reduction of uncertainty, Figure 7.26 shows the diagonal elements of the posterior covariance matrix along three cuts across the tank. We observe again that the variances are reduced effectively only in the lower half with strong correlations.

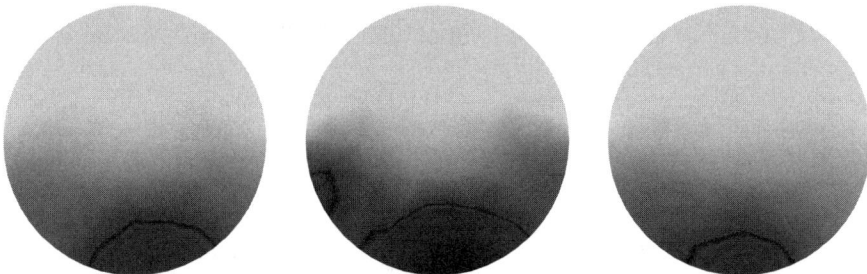

Figure 7.26. Autocovariance along three intersections of the tank. The cuts are along planes located 1/10, 1/2 and 9/10 from the left end of the tank.

7.6 Electrical Impedance Tomography: Handling Approximation Errors

An example of the effect of approximation errors on low-noise inverse problems was presented in Chapter 5. Also, a more complex case was studied earlier in this chapter. It was shown that it is possible to construct an enhanced likelihood model that accommodates approximate statistics of the approximation error in addition to the conventional additive errors. The additional error model was constructed based on two computational forward models. One model, which is the one to be employed in the inversion, exhibits reasonable computational complexity. The other one is a significantly more accurate model that is too complex to be used in real-time applications. However, the more accurate model had only to be computed once since it was used only in the construction of the second-order model for the statistics of the approximation error.

In this section we address this issue in the more difficult context of electrical impedance tomography. The added difficulty which arises from the nonlinearity of the problem is that the second-order statistics cannot be computed without simulations.

7.6.1 Meshes and Projectors

We start by defining the meshes and related forward models. In this section, let $\Omega \subset \mathbb{R}^2$ denote a unit disc, a model for the conducting body. On the boundary $\partial \Omega$ of the body, 16 equally spaced contact electrodes are attached.

We divide the body Ω into triangular elements in three different ways. Let us denote by $\mathcal{M}^k$, $k = 1, 2, 3$, the sets of elements. We refer to these sets as meshes. The individual elements of the mesh $\mathcal{M}^k$ are denoted by $\Omega_j^k \in \mathcal{M}^k$, $1 \leq j \leq n_k$. The mass center of the element Ω_j^k is denoted by r_j^k.

The meshes $\mathcal{M}^k$ are shown in Figure 7.27. They are intentionally not nested. As far as the numerical solution of the EIT forward problem is concerned, only the mesh $\mathcal{M}^1$ exhibits the desirable feature of being denser in the subregions near the electrodes where the electric potential experiences rapid changes.

Consider a conductivity distribution $\sigma : \Omega \to \mathbb{R}_+$. We represent the conductivity by its values at the mass centers of the elements of $\mathcal{M}^k$,

$$\sigma^k = [\sigma(r_1^k), \sigma(r_2^k); \ldots, \sigma(r_{n_k})]^\mathrm{T} \in \mathbb{R}^{n_k}.$$

We need mappings that permit moving between different reperesentations of the conductivity. These mappings, which we refer to as projectors,

$$P^{jk} : \mathbb{R}^{n_k} \to \mathbb{R}^{n_j}, \quad \sigma^k \mapsto \sigma^j,$$

are defined here as matrices with entries

$$P_{pq}^{jk} = \begin{cases} 1, \text{ if } |r_p^j - r_q^k| = \min_{q'} |r_p^j - r_{q'}^k|, \\ 0 \text{ otherwise.} \end{cases}$$

We remark that the minimum in the above definition is uniquely defined in the meshes under consideration. Hence, the value is copied from the nearest centerpoint of one mesh element to the other. Note that these mappings can be defined in different ways, for example, when nested meshes are used, averaging is often used in downsampling. We shall consider relatively smooth conductivity distributions, in which case the above choice seems to work consistently.

Given a conductivity σ by its representation σ^k, let the solutions of the forward problems be denoted by $U^k(\sigma^k)$. It is understood that in the forward solver, the conductivity is approximated by a piecewise constant conductivity. Two different versions for the forward model are used, which use first-

7.6 Electrical Impedance Tomography: Handling Approximation Errors

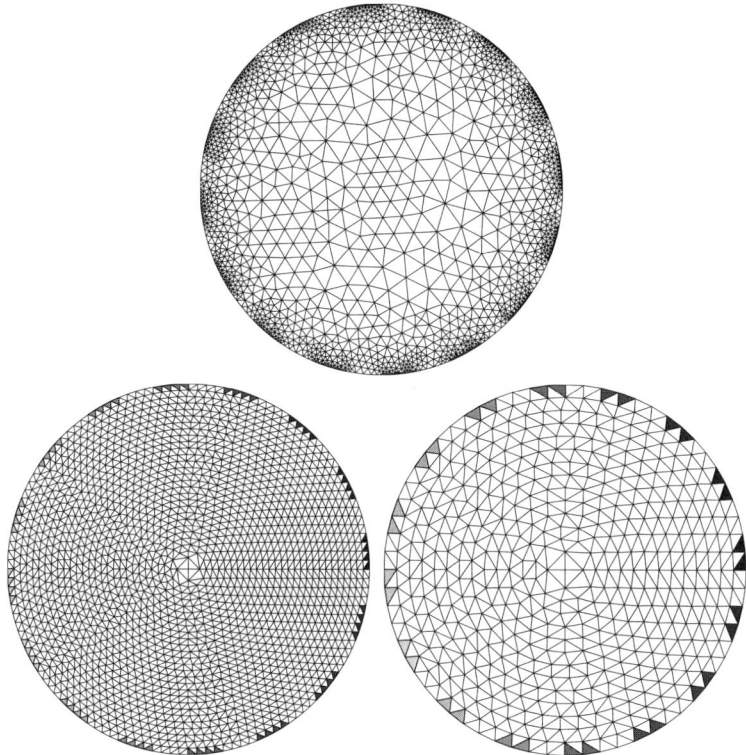

Figure 7.27. The three meshes. Top: $\mathcal{M}^1$ which is used to compute the accurate noiseless measurements, Left: $\mathcal{M}^2$ which is used to represent the true conductivity distributions. Right: $\mathcal{M}^3$ which corresponds to the model that is employed in the inversion.

and second-order finite element basis functions for the representation of the potential u in Ω. It will be indicated which version of the forward solver we use.

The error is modelled throughout as additive Gaussian zero mean noise with covariance $\Gamma_n = \lambda^2 I$, where λ varies in a range that corresponds to noise levels 0.1% and 10% of the maximum errorless measurements over a large set of simulations. While in the case of conventional EIT applications the additive noise levels can seldom be pushed below 1%, there are applications in which the measurement times are practically unlimited, yielding quite accurate data. Such cases include, for example, geophysics and material testing. In these cases the error level and structure comes almost exlusively from modelling and approximation errors.

7.6.2 The Prior Distribution and the Prior Model

In the numerical simulations, we use the mesh $\mathcal{M}_2$ to represent the true conductivity distributions. Since this mesh is independent of the measurement setting, it serves therefore as an honest basis for representing the prior information. We use a smoothness prior constructed along the lines of Section 3.4.1. For notational convenience, we write momentarily $\sigma = \sigma^2$, that is, it is understood that the conductivity in this section is represented in the mesh $\mathcal{M}^2$.

Consider first a prototype Gaussian smoothness prior,

$$\widetilde{\pi}(\sigma) \propto \exp\left(-\frac{1}{2}\alpha^2 \|L\sigma\|^2\right). \tag{7.34}$$

The matrix αL is a weighted difference matrix between adjacent elements. The above density is improper, that is, the matrix $L^\mathrm{T} L$ is not invertible. To get a proper density, select randomly a small number of elements. If $I = \{1, 2, \ldots, n_2\}$, we write a partitioning $I = I_1 \cup I_2$, where the selected indices belong to I_2. Using the density (7.34), we compute the conditional density $\widetilde{\pi}(\sigma_{I_1} \mid \sigma_{I_2})$, where σ_{I_j} contains those components whose indices are in the set I_j. This conditional density is a proper density with respect to σ_{I_1}. Define

$$\pi_\mathrm{pr}(\sigma) \propto \pi_+(\sigma - \sigma_*)\widetilde{\pi}(\sigma_{I_1} \mid \sigma_{I_2})\pi_0(\sigma_{I_2}).$$

Here, π_+ is the positivity density and σ_* is a positive lower bound for admissible conductivities. The density π_0 is a white noise prior,

$$\pi_0(\sigma_{I_2}) \propto \exp\left(-\frac{1}{2\gamma^2}\|\sigma_{I_2} - \widetilde{\sigma}\|^2\right).$$

Finally, we need to adjust the parameters. Choose $\widetilde{\sigma} = 1/400$, $\sigma_* = \widetilde{\sigma}/10$, and the variance γ^2 so that the lower bound σ_* is three standard deviations away from the center $\widetilde{\sigma}$. Furthermore, the weighting parameter α in the prototype density (7.34) is selected so that

$$\mathrm{P}\{\sigma_{I_1} > \sigma_* \mid \sigma_{I_1}\} \approx 0.99.$$

Having specified the prior $\pi_\mathrm{pr}(\sigma) = \pi_\mathrm{pr}(\sigma^2)$,

$$\pi_\mathrm{pr}(\sigma^2) \propto \pi_+(\sigma^2 - \sigma_*)\exp\left(-\frac{1}{2}(\sigma^2 - \overline{\sigma})^\mathrm{T}\Gamma_\mathrm{pr}^{-1}(\sigma^2 - \overline{\sigma})\right)$$

$$= \pi_+(\sigma^2 - \sigma_*)\mathcal{N}(\overline{\sigma}, \Gamma_\mathrm{pr}),$$

we may project it to other meshes, yielding

$$\pi_\mathrm{pr}(P^{j2}\sigma^2) = \pi_+(P^{j2}(\sigma^2 - \sigma_*))\mathcal{N}(P^{j2}\overline{\sigma}, P^{j2}\Gamma_\mathrm{pr}(P^{j2})^\mathrm{T}).$$

The simulated observations are calculated using the mesh $\mathcal{M}^1$. Evidently, this mesh is designed to produce accurate results with the forward solver. The inverse problem of estimating the conductivity distribution is based on the mesh $\mathcal{M}^3$.

7.6.3 The Enhanced Error Model

Let $\sigma = \sigma^2$ denote a model for the true conductivity distribution. As noted above, to calculate an accurate FEM approximation for the electrode voltages, we use the mesh $\mathcal{M}^1$. To this end, we need to project the conductivity to this mesh. Thus, an accurate model for the voltage in the presence of additive noise is

$$V = U^1(P^{12}\sigma^2) + E.$$

Suppose that we solve the inverse problem using the coarse mesh $\mathcal{M}^3$. The coarse meshing produces an approximation error that is not negligible, as we will demonstrate. A model using the mesh $\mathcal{M}^3$ that includes the modelling error is thus

$$\begin{aligned}V &= U^3(P^{32}\sigma^2) + \left(U^1(P^{12}\sigma^2) - U^3(P^{32}\sigma^2)\right) + E \\ &= U^3(P^{32}\sigma^2) + \epsilon + E,\end{aligned} \qquad (7.35)$$

where ϵ is the modelling error. We want to analyze the statistical properties of this error. As in Chapter 5, we call the model (7.35) the *enhanced error model*. However, we shall employ the approximate version only, in which we neglect the cross-correlation between the unknown and the modelled additive error.

Due to the nonlinearity of the mapping $\sigma \mapsto U$ the enhanced error model does not follow directly from the linear theory discussed in Chapter 5. Instead, as noted above we have to estimate the second-order statistics by simulations in the meshes $\mathcal{M}^j$.

We generate a set of $N = 50$ draws from the density $\pi_{\text{pr}}(\sigma^2)$ introduced in the previous section. Let us denote them by σ_n^2, $1 \leq n \leq N$. This is a rather small ensemble size. The limitation in size is dictated by the relatively demanding computation that ensue. Having drawn the ensemble, we compute the corresponding voltages $U^j(P^{j2}\sigma_n^2)$, $j = 1, 3$, for each σ_n^2.

We then compute estimates for the mean η_ϵ and the covariance Γ_ϵ of the approximation error ϵ,

$$\eta_\epsilon = \frac{1}{N}\sum_{n=1}^{N} \left(U^1(P^{12}\sigma_n^2) - U^3(P^{32}\sigma_n^2)\right)$$

$$\Gamma_\epsilon = \frac{1}{N}\sum_{n=1}^{N} \left(U^1(P^{12}\sigma_n^2) - U^3(P^{32}\sigma_n^2)\right)\left(U^1(P^{12}\sigma_n^2) - U^3(P^{32}\sigma_n^2)\right)^{\text{T}} - \eta_\epsilon \eta_\epsilon^{\text{T}}.$$

The covariance of the total error is

$$\Gamma_{e+\epsilon} = \Gamma_n + \Gamma_\epsilon = \lambda^2 I + \Gamma_\epsilon.$$

For later use, let us denote by $L_{e+\epsilon}$ the Cholesky factor of the inverse of the computed covariance,

$$(L_{e+\epsilon})^{\mathrm{T}} L_{e+\epsilon} = (\Gamma_{e+\epsilon})^{-1}.$$

The first thing to note is that the mean approximation error is not vanishing as in linear problems with vanishing means of the prior and noise. The estimated average corresponding to a trigonometric current pattern is plotted in Figure 7.28. It can be seen that the mean cannot be considered insignificant especially with the lower additive noise levels. Note also the significant difference between using first- and second-order bases.

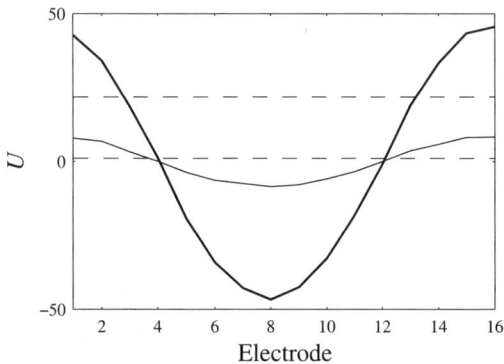

Figure 7.28. The mean of the approximation error corresponding to the first trigonometric current pattern. The bold and weak curves denote the errors when using first- and second-order basis functions, respectively. The two dashed lines denote the standard deviations of the additive noise at levels 0.1% and 2%.

The autocovariances of the approximation error, that is, the diagonal of Γ_ϵ, for the first trigonometric pattern are shown in Figure 7.29. Again, the significance of the accuracy of the forward model for the approximation error is obvious. The cross-covariances between neighboring electrodes for the first trigonometric pattern are shown in Figure 7.30.

As opposed to the linear tomography example in Section 5.8.1 in which the mean of the approximation error was small when compared to the covariance, in this case the mean of the error is substantial. Additive noise error levels around 2% are very typical. When using a first-order basis, the approximation error mean is about 4% at its maximum. Also, the standard deviations of the modelling errors are of the order of the 2% level. As in the example of the tomography problem, the errors are not homogeneous across the measurements although in this case the differences are less pronounced, partly due to the prior distribution which is smooth in this case. Again, it is clear that the nondiagonal structure of the covariance is relevant at lower additive noise levels.

7.6 Electrical Impedance Tomography: Handling Approximation Errors 275

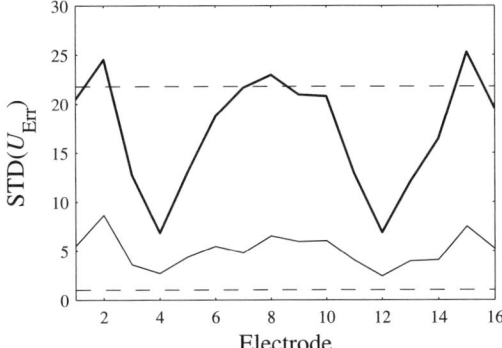

Figure 7.29. The square roots of the autocovariances (standard deviations) of the approximation error corresponding to the first trigonometric current pattern. The bold and weak curvess denote the errors when using first- and second-order basis functions, respectively. The two dashed lines denote the standard deviations of the additive noise at levels 0.1% and 2%.

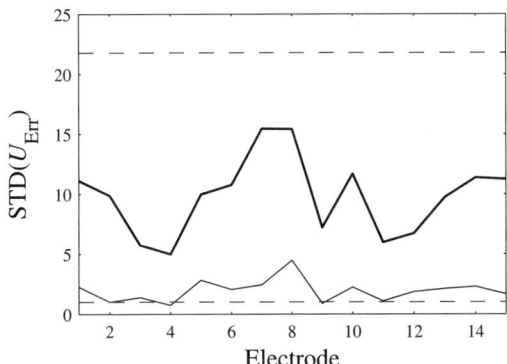

Figure 7.30. The (square roots of the absolute values of the) cross-covariances between the neighboring electrodes of the approximation error corresponding to the first trigonometric current pattern. The bold and weak curves denote the errors when using first- and second-order basis functions, respectively. The two dashed lines denote the standard deviations of the additive noise at levels 0.1% and 2%.

7.6.4 The MAP Estimates

In this section we compare the MAP estimates when the modelling error is included and when it is not. An $\mathcal{M}^3$ based model that does not include the modelling error is

$$V = U^3(\sigma) + E, \quad \sigma = \sigma^3,$$

while the corresponding enhanced error model is

$$V = U^3(\sigma) + \epsilon + E, \quad \sigma = \sigma^3,$$

The prior density for both models is

$$\sigma \sim \pi_+(\sigma - \sigma_*)\mathcal{N}(\overline{\sigma}^3, \Gamma^3),$$

where

$$\overline{\sigma}^3 = P^{32}\overline{\sigma}, \quad \Gamma^3 = P^{32}\Gamma(P^{32})^{\mathrm{T}}.$$

We denote by L_σ the Cholesky factor of the inverse of the prior covariance,

$$(L_\sigma)^{\mathrm{T}} L_\sigma = (\Gamma^3)^{-1}.$$

The MAP estimates of the above conventional and enhanced error models, respectively, are calculated as minimizers of the functionals

$$F_1(\sigma) = \|L_e(V - U(\sigma))\|^2 + \|L_\sigma(\sigma - \overline{\sigma}^3)\|^2$$

$$F_2(\sigma) = \|L_{e+\epsilon}(V - U(\sigma) - \eta_\epsilon)\|^2 + \|L_\sigma(\sigma - \overline{\sigma}^3)\|^2$$

subject to the constraint $\sigma \geq \sigma_*$.

The minimization is done using the Gauss–Newton iteration with line search. In the case of conventional error model the search direction for updating $\sigma^{(j)} \to \sigma^{(j+1)}$ is

$$d^{(j+1)} = \begin{bmatrix} L_e J(\sigma^{(j)}) \\ L_\sigma \end{bmatrix}^\dagger \begin{bmatrix} L_e(V + J(\sigma^{(j)})\sigma^{(j)}) \\ L_\sigma \overline{\sigma}^3 \end{bmatrix} - \sigma^{(j)},$$

and the formula fo the enhanced error model is analogous. Convergence is obtained typically in 4–8 and 6–10 iterations, respectively.

Let $\sigma_{n,\mathrm{MAP}}$ denote the computed MAP estimate of either one of the above models corresponding to the true conductivity σ_n^2, $1 \leq n \leq N$. To estimate the performance of the methods, we compute the expected mean square of the differences between the true conductivities σ^2 and the projections of the computed MAP estimates $\sigma_{n,\mathrm{MAP}}^3$ and average over the sample, that is,

$$\mathcal{D} = \frac{1}{N} \sum_{n=1}^{N} \|\sigma_n^2 - P^{23}\sigma_{3,\mathrm{MAP}}^3\|^2.$$

We note that taking into account the dimensions of σ^2 and the small number of draws, reliable inference is not achieved, but we have a tentative order-of-magnitude estimate.

The estimates for the mean square error as a function of the additive noise level are shown in Figures 7.31 – 7.32. The contribution of both the modelled approximation error mean and covariance are included in the lower graphs. The ruggedness of the graphs is due to the small sample size.

The benefits of employing the enhanced error model need little convincing. The estimation errors when using either the first- or second-order bases *with* the enhanced error model are essentially the same. In fact, the MAP estimates

7.6 Electrical Impedance Tomography: Handling Approximation Errors   277

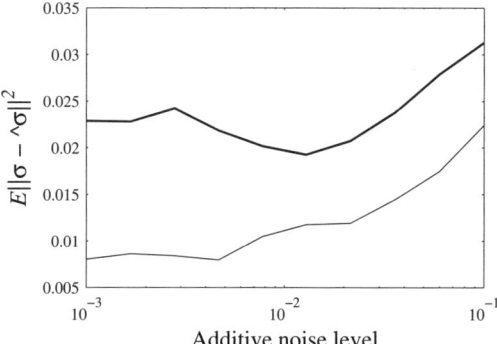

Figure 7.31. The estimates for the mean square error when using conventional (bold) and enhanced (thin) error models as a function of the additive noise level. First-order basis functions are used in the computation of the potentials.

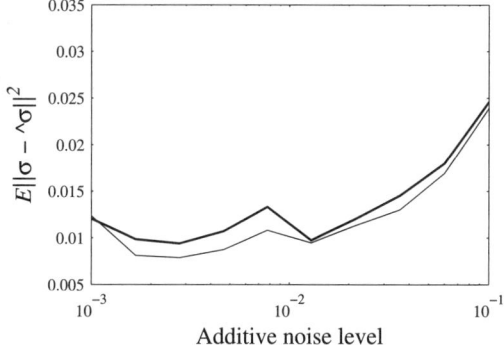

Figure 7.32. The estimates for the mean square error when using conventional (bold) and enhanced (thin) error models as a function of the additive noise level. Second-order basis functions are used in the computation of the potentials.

with the inherently very poor first-order basis, but with enhanced error model, are actually slightly better than the MAP estimates with second-order bases and the conventional error model. This is exactly what is sought in model reduction: A computationally simple model that performs as well as a more accurate and expensive method.

It is interesting to know how the error is distributed and furthermore, to divide the error into the systematic (or bias) and random parts. The means, corresponding to the systematic part and the autocovariances of the estimate errors for noise levels 0.1% and 2%, are shown in Figures 7.33 – 7.36.

With 0.1% noise level, the ringing effects of the error mean is clear when using first-order basis or conventional error model: the conductivity estimates are systematically underestimated on an annulus near the electrodes and over-

278   7 Case Studies

compensated on the following annulus and so on. The effects are also visible in the autocovariance (see Figure 7.34) although these are less pronounced. Note that the symmetry is due to the symmetry of the prior and the group symmetry of the measurements.

With a 2% noise level the errors are larger and we would expect that the advantages of using the enhanced error model would diminish since the inverse problem becomes noise dominated. Indeed, this is the case. However, the mean error when using the first-order basis or conventional error model is still negative on the outer annulus, which is broader than in the 0.1% case, and the undershoot is compensated for in an inner annulus. The autocovariance is not any worse than with the enhanced error model and when using the second-order basis.

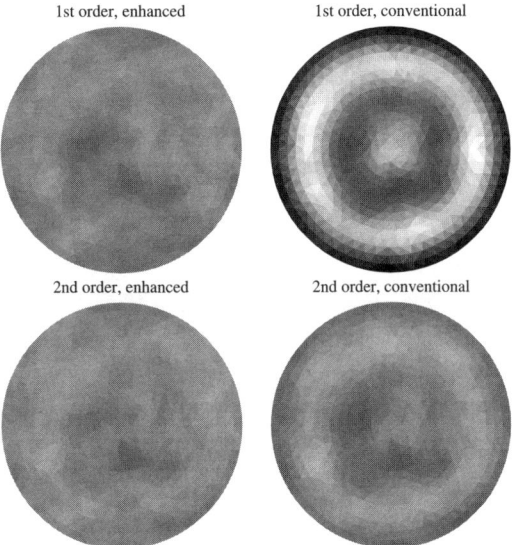

Figure 7.33. The means of the estimation error for all four cases with noise level 0.1%. Dark regions denote negative error, light regions positive errors, pale grey small errors.

7.7 Electrical Impedance Process Tomography

In this section we consider the following nonstationary electrical impedance tomography application: A set of contact electrodes are attached on the surface of a pipeline segment. Injecting currents through these electrodes and mesuring the voltages, we want to estimate the concentration of a conducting

7.7 Electrical Impedance Process Tomography 279

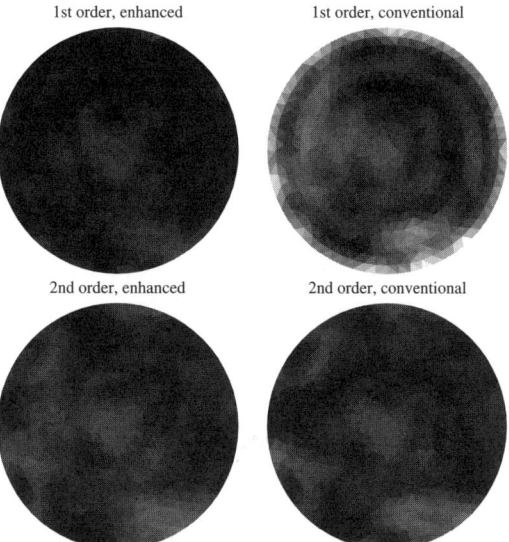

Figure 7.34. The autocovariances of the estimation error for all four cases with noise level 0.1%. Dark regions denote small error and light ones large errors.

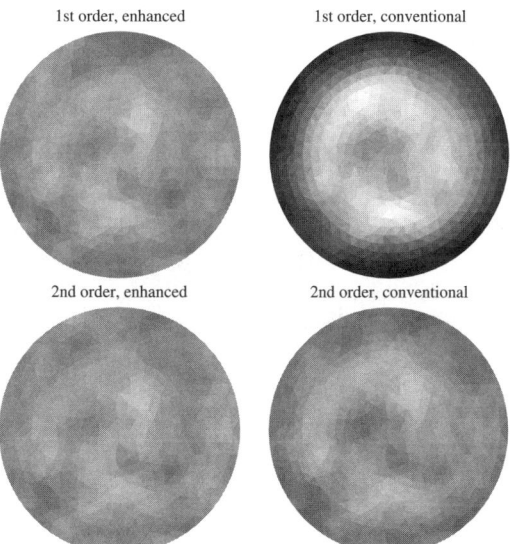

Figure 7.35. The means of the estimation error for all four cases with noise level 2%. Dark regions denote negative error, light regions positive errors pale grey small errors.

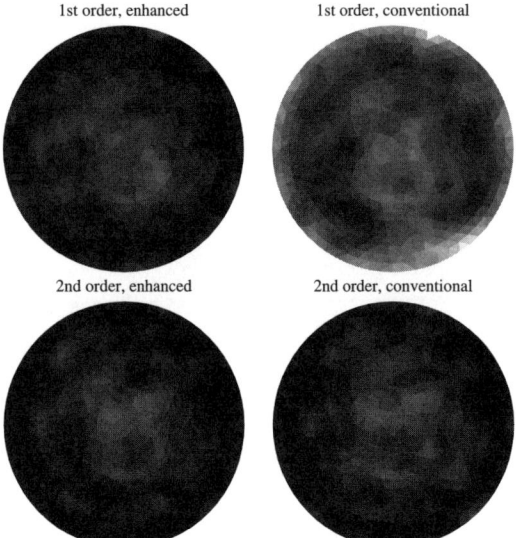

Figure 7.36. The autocovariances of the estimation error for all four cases with noise level 2%. Dark regions denote small error and light ones large errors.

substance in the pipeline. The substance flows in the pipeline, the flow velocity being so high that the target is not stationary from one current injection to other. The evolution of the concentration in the pipeline is modelled by stochastic convection–diffusion (CD) equation and the stationary incompressible Navier–Stokes (NS) flow.

The discussion contains two important aspects. First, it is demonstrated that the performance of the concentration distribution estimation depends strongly on correct modelling of the dynamics. Yet, the method is relatively robust, for example, with respect to the velocity field of the flow. This feature is essential since the information about the velocity field is always incomplete. On the other hand, despite this robustness, the data are not completely insensitive to the velocity field. We demonstrate that they contain enough information for the estimation of the velocity, too.

7.7.1 The Evolution Model

We start by descrabing the dynamic model of the flow. We use stochastic differential equations without going into technical details that are quite involved.

Let us denote by $c = c(x,t)$ the concentration distribution of a conducting substance, $x \in \Omega$, $t \in (0,T)$ where Ω is a finite segment of a pipe and $t \geq 0$. Let the concentration satisfy the convection–diffusion equation

$$\frac{\partial c}{\partial t} = \nabla \cdot \kappa \nabla c - v \cdot \nabla c + \mu, \tag{7.36}$$

where $\kappa = \kappa(x)$ is the diffusion coefficient, $v = v(x)$ is the velocity of the flow and μ is a stochastic process representing the modelling uncertainties in the CD flow model.

For the time being, we assume that the velocity field is approximately known, and thus it is not treated as a variable. In this example, the velocity field is the approximate stationary solution of the Navier–Stokes equation for incompressible fluids. In a straight pipe, the velocity profile is a parabola with vanishing velocity on the walls.

We denote the pipe walls by $\partial\Omega_\text{wall}$ and the input and the output ends of the pipe by $\partial\Omega_\text{in}$ and $\partial\Omega_\text{out}$, respectively. The natural boundary condition at the pipe walls says that there is no diffusion through the pipe walls,

$$\left.\frac{\partial c}{\partial n}\right|_{\partial\Omega_\text{wall}} = 0. \tag{7.37}$$

The boundary conditions at the ends of the pipe are clearly the primary unknowns. If these together with μ were known, we could simply solve the CD equation and would not need any measurements. Since they are not known, according to the Bayesian paradigm, they are modelled with random variables.

Consider the temporal change of the total concentration in the pipe segmant. Using the divergence theorem, the condition (7.37), the incompressibility property $\nabla \cdot v = 0$ as well as the boundary condition $\nu \cdot v = 0$ at $\partial\Omega_\text{wall}$, where ν is the exterior unit normal of $\partial\Omega_\text{wall}$, we find that

$$\frac{\partial}{\partial t}\int_\Omega c\,dx = \int_\Omega (\nabla \cdot \kappa\nabla c - v \cdot \nabla c + \mu)\,dx$$

$$= \left(\int_{\partial\Omega_\text{in}} + \int_{\partial\Omega_\text{out}}\right)\left(\kappa\frac{\partial c}{\partial u} - \nu \cdot vc\right)dS + \int_\Omega \mu\,dx$$

$$= \Phi_\text{in} - \Phi_\text{out} + \int_\Omega \mu\,dx,$$

where Φ_in and Φ_out are the fluxes at the input and output ends, respectively. We assume that μ is zero mean and that $\int_\Omega \mu\,dx$ is small when compared to the fluxes. Also, if the fluxes were modelled exactly, a nonvanishing integral would mean that substance would either be generated or would annihilate itself inside the pipe segment Ω.

We assume that the output flux is dominated by the convection term. Hence, it is reasonable to assume an approximate boundary condition

$$\left.\left(\kappa\frac{\partial c}{\partial n} - \nu \cdot vc\right)\right|_{\partial\Omega_\text{out}} = -\nu \cdot vc|_{\Omega_\text{out}} \quad \text{or} \quad \left.\frac{\partial c}{\partial n}\right|_{\partial\Omega_\text{end}} = 0. \tag{7.38}$$

For the input, we assume that the conceneration on the input boundary is an unknown Dirichlet boundary condition. Denoting by t the trace mapping to the input end (see Appendix A), we set

$$tc = c|_{\partial\Omega_\text{in}} = c_\text{in}, \tag{7.39}$$

where the input function $c_{\text{in}}(x,t)$ is a stochastic function. We assume for simplicity that $c_{\text{in}}(x,0) = 0$. Let P denote a right inverse of the trace map t, that is, $tPc_{\text{in}} = c_{\text{in}}$. We assume that $Pc_{\text{in}} = 0$ in the neighborhood of $\partial\Omega_{\text{wall}} \cup \partial\Omega_{\text{out}}$. We can now write

$$c = u + Pc_{\text{in}}, \quad tu = 0.$$

Finally, the initial value of the concentration density at $t = 0$ is denoted as $c(x,0) = c_0(x)$.

We derive the evolution model for u using analytic semigroups. Since the model contains stochastic processes, the derivation here should be taken as a formal exposition of the idea.

Denote by $\mathcal{L}$ the differential operator

$$\mathcal{L} : \mathcal{D}(\mathcal{L}) \to L^2(\Omega), \quad \mathcal{L} = \nabla \cdot \kappa \nabla - v \cdot \nabla,$$

where

$$\mathcal{D}(\mathcal{L}) = \left\{ u \in H^2(\Omega) \text{ so that } \left. \frac{\partial u}{\partial n} \right|_{\partial\Omega_{\text{wall}}} = \left. \frac{\partial u}{\partial n} \right|_{\partial\Omega_{\text{out}}} = 0, \, u|_{\partial\Omega_{\text{in}}} = 0 \right\}.$$

We seek to find $u \in \mathcal{D}(\mathcal{L})$ satisfying

$$\frac{\partial u}{\partial t} = \mathcal{L}u + q, \quad q = \left(\nabla \cdot \kappa \nabla - v \cdot \nabla - \frac{\partial}{\partial t} \right) Pc_{\text{in}} + \mu.$$

We denote by $\mathcal{U}(t)$ the operator semigroup generated by the infinitesimal generator $\mathcal{L}$. Then the solution $c = c(x,t)$ can be written as

$$c = u + Pc_{\text{in}} = \mathcal{U}(t)c_0 + \int_0^t \mathcal{U}(t-s)q(s)ds + Pc_{\text{in}}.$$

Consider now the concentration distribution sampled at discrete times t_k, $k = 0, 1, 2, \ldots$. By denoting $c_k(x) = c(x, t_k)$ and using the group property $\mathcal{U}(t_{k+1}) = \mathcal{U}(\delta t_k)\mathcal{U}(t_k)$, where $\delta t_k = t_{k+1} - t_k$, we have

$$c_{k+1} = \mathcal{U}(\delta t_k) \left(\mathcal{U}(t_k)c_0 + \int_0^{t_{k+1}} \mathcal{U}(t_k - s)q(s)ds \right) + Pc_{\text{in}}(t_{k+1})$$
$$= \mathcal{F}_k c_k + \varepsilon_{k+1}, \tag{7.40}$$

where $\mathcal{F}_k = \mathcal{U}(\delta t_k)$ and

$$\varepsilon_{k+1} = \int_{t_k}^{t_{k+1}} \mathcal{U}(t_{k+1} - s)q(s)ds - \mathcal{U}(\delta t_k)Pc_{\text{in}}(t_k) + Pc_{\text{in}}(t_{k+1})$$

which includes the contribution of the stochastic terms to the evolution of the concentration. Furthermore, we shall assume that the input c_{in} is given as a sum of a deterministic part plus a random part with zero mean,

$$c_{\text{in}} = \bar{c}_{\text{in}} + \eta.$$

In other words, we have $\bar{c}_{\text{in}} = E\, c_{\text{in}}$ which is known. By collecting all stochastic terms we obtain an evolution model

$$c_{k+1} = \mathcal{F}_k c_k + s_{k+1} + w_{k+1}, \tag{7.41}$$

where s_{k+1} is the deterministic part corresponding to $\bar{c}_{\text{in}}$ and w_{k+1} is the resulting discrete-time state noise process.

7.7.2 The Observation Model and the Computational Scheme

The observation is based on the EIT experiment. Let σ_t denote the conductivity distribution in Ω at time $t = t_k$. The dependency between the concentration distribution c_t and conductivity σ_t depend on the physical properties of the solution. For example, strong and weak electrolytes behave very differently. Here we assume a simplified model $\sigma_t = c_t$.

Let the measurement device contain L contact electrodes. Denote by $R(\sigma_t) \in \mathbb{R}^{L \times L}$ the resistance matrix at time $t = t_k$. If we inject the current pattern $I_t \in \mathbb{R}^L$, the resulting voltage pattern is $U_t = R(\sigma_t)I_t \in \mathbb{R}^L$. We linearize the mapping $\sigma_t \mapsto U_t$ at the uniform long-time mean conductivity $\bar{\sigma}$. The aveage concentration is denoted by $\bar{c}$. The noiseless linearized observation model is now

$$U_t = U_{t,0} + J_t(\bar{c})(c_t - \bar{c}),$$

or, defining $\tilde{U}_t = U_t - U_{t,0} + J_t(\bar{c})\bar{c}$,

$$\tilde{U}_t = J_t(\bar{c})c_t.$$

Here, J_t is the Jacobian of the mapping $\sigma_t \mapsto U_t$.

We discretize the model using finite elements. Let $X_t \in \mathbb{R}^n$ denote the discretized concentration density. We assume that the measurement noise is Gaussian white noise. Then, the observation model is

$$Y_t = G_t X_t + E_t, \quad E_t \sim \mathcal{N}(0, \text{diag}\,(\beta_e)),$$

where G_t is the matrix approximation of $J_t(\bar{c})$ and where the entries of the vector β_e are between 2.5%–10% of the entries of the noiseless measurement vector. This corresponds to a realistic measurement model.

We remark that the modelling errors are not taken into account in this error model. Here the modelling error consists of both the approximation error due to discretization and the linearization error.

To write a discrete evoultion model, the CD equation is solved with the conventional semidiscrete Petrov–Galerkin scheme (see [16]). The boundary conditions are modeled as Neumann boundaries except for the flow input boundary, where the boundary is a Dirichlet boundary condition. This yields

$$M\frac{\partial X_t}{\partial t} = \tilde{F}X_t + \tilde{S}_t + \tilde{W}_t,$$

where the terms correspond to those in the continuous evolution model (7.41). This equation needs to be integrated from one observation instance to the next one. It is a stiff system, and we use an implicit multistep integration. The resulting system integrated over the measurment interval yields a time evolution model

$$X_{t+1} = F_{t+1}X_t + S_{t+1} + W_{t+1}. \qquad (7.42)$$

The time indices t, $t+1$ refer to consecutive observation time instances. The matrix F_{t+1}, S_{t+1} and W_{t+1} depend on the velocity field v that is treated here as a known parameter. The state noise W_{t+1} is a zero mean state noise process, whose covariance is discussed later.

First-order finite element bases were used to solve the convection–diffusion and second-order bases for the EIT model. This holds for the forward as well as for the inverse problem. Furthermore, since the concentration is represented in piecewise linear basis, we represent the conductivity also in this basis. Note that conventionally in the EIT applications, the conductivity is expressed in the piecewise constant basis. The mesh-related issues are considered later.

The state noise W_{t+1} consists of contributions of the processes μ, η and its time derivative η'. Due to the assumptions, the state noise has zero mean, and the covariance matrix consists of three parts,

$$\Gamma_{w_{t+1}} = Y\Gamma_\mu Y^{\mathrm{T}} + D\Gamma_\eta D^{\mathrm{T}} + H\Gamma_{\eta'}H^{\mathrm{T}}$$

where Y, D and H are matrices that arise from the Petrov–Galerkin scheme. The matrices Γ_μ, Γ_η and $\Gamma_{\eta'}$ are the covariance matrices of the correpsonding discretized processes. We assume that these processes are stationary and uncorrelated. Here, we use a model

$$\Gamma_\mu = \beta_\mu I, \quad \Gamma_\eta = \beta_\eta I, \quad \Gamma_{\eta'} = 0. \qquad (7.43)$$

This model is strongly simplified. In reality, β_μ should be determined by the approximation error due to the use of a discrete model as well as by an estimate of the *overall model error* of the CD evolution model. It should account for errors in the actual velocity field and the use of a single phase model in possibly multiphase situations. As for the two other covariances in (7.43), in reality η and η' are smooth functions, and therefore the discretized values should be correlated. Furthermore, they are mutually correlated, and the correlation over time is nonvanishing. These modelling flaws are not as drastic as they may seem. The operator semigroup describing the flow is smoothing, and therefore the matrices Y, D and H in (7.43) smooth out the modelling errors. From the practical point of view, the important feature is the correct structure of the evolution noise model. The parameters β_μ and β_η are tuned with simulations. A perturbation analysis shows that the estimator performance does not depend heavily on the choice of these parameters, which is a prerequisite for the feasibility of the method.

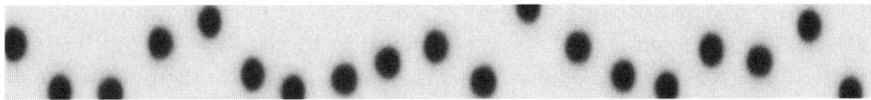

Figure 7.37. The simulated input. The abscissa is the time, and at each time instance, the Dirichlet data at the input end of the pipe segment is plotted.

The simulated time-varying Dirichlet data on Ω_{in} is shown in Figure 7.37.

The requirements for the finite-element meshes for different subproblems are somewhat different. For the CD problem and the employed velocity fields, an approximately uniform mesh is required, while for the EIT problem, the mesh should be finer near the elecrodes and especially near their boundaries, where the gradients of the potentials are large. On the other hand, far from the electrodes the potential is slowly varying and a coarser mesh is adequate. The use of different meshes requires interpolation between the meshes which further increases the modelling errors.

7.7.3 The Fixed-lag State Estimate

When deriving the evolution–observation model, we used global linearization of the voltage measurement to reduce the computational complexity, which in real-world applications is most often the deciding factor. In applications such as industrial process monitoring, the estimation is usually required on-line. Therfore, fixed interval smoothers are not feasible and we need to use either Kalman filter or fixed-lag smoothers. We consider here the fixed-lag smoothers, in which a new estimate is computed each time we obtain a new measurement, albeit for the state some time steps earlier.

Experiments show that using the lag parameter $q = 8$ yields estimates that were essentially as good as fixed-interval estimates. Simulated data are computed with a different mesh from the one used in inversion, and the meshes for the CD and EIT solvers differ from each other.

The simulation is done with a two-dimensional model. There are 16 electrodes divided into two sets and placed equidistantly along the pipeline so that about 30% of the length of the pipeline is outside the lateral electrodes on both sides. The current patterns are such that only two electrodes at the time are injecting current. The active electrodes at each time are on the opposite sides of the pipe. The current pattern order was randomized.

The whole simulation extended over 32 current pattern injections. By inspecting the target in Figure 7.38 it is evident that a conventional stationary EIT reconstruction based on the frame of 15 linearly independent current patterns does not make sense: The time evolution of the target during one frame corresponds to the first 8 rows.

In order to evaluate the sensitivity of the estimation scheme to misspecification of the mean velocity in the pipeline, we compute the estimates with the correct mean flow rate 450 cm s^{-1}, and a severely underestimated one, 200 cm s^{-1}. The fixed-lag estimates are shown in Figure 7.38. In rows 3–5 the estimation error due to the initial transient is clearly visible. By this we mean incorrect estimate of the initial state and large initial error covariance.

Also, due to the flow direction from left to right, the estimate at the left end is more affected by the uncertainties of the input flow. This is intuitively clear and can also be verified by examining the estimate error covariance. As noted earlier, the availability of spread estimates in addition to the point estimates such as the CM or MAP estimates is one of the most important assets as in the statistical inversion approach in general. The time average of the estimate error variances

$$\bar{\gamma} = \frac{1}{T} \sum_{t=k+1}^{k+T} \operatorname{diag} \Gamma_{t|t+q},$$

is shown in Figure 7.39.

We recall that the estimate error variances based on any statistical inversion scheme are themselves estimates that are correct only when all underlying models are correct. However, these estimates are useful if the models are feasible. Observe also that due to the linearization, the estimate error covariance does not depend on data. Thus, it is possible to optimize measurements, in this case the current patterns as in Section 7.5, with respect to the prior that amounts here to the evolution model.

7.7.4 Estimation of the Flow Profile

The example of the previous section shows that the EIT data carry enough information for estimating the concentration distribution, and the fixed lag estimates are not excessively sensitive to the mismodelling of the velocity field.

A further question is whether the EIT measurements are informative enough to enable the estimation of the velocity field together with the concentration. The difficulty of this estimation problem depends on the assumptions about the velocity field. In the simplest case, the velocity is stationary. This leads to what is known as a state space identification problem. These problems are usually solved by a maximum likelihood -type algorithm for the unknown parameters given all measurements.

Here we assume that the velocity field is nonstationary, $v = v(x,t)$, and it is treated as another state vector. A discrete representation of the velocity in two dimensions is given as

$$v(x,t) = \sum_{k=1}^{p} \sum_{\ell=1}^{2} b_{t,k,\ell} \phi_k(x) e_\ell, \quad t \in (0,T),$$

where the functions ϕ_k are finite-element basis functions and e_ℓ are orthogonal basis vectors. Such representation requires a prior model of the velocity field.

7.7 Electrical Impedance Process Tomography   287

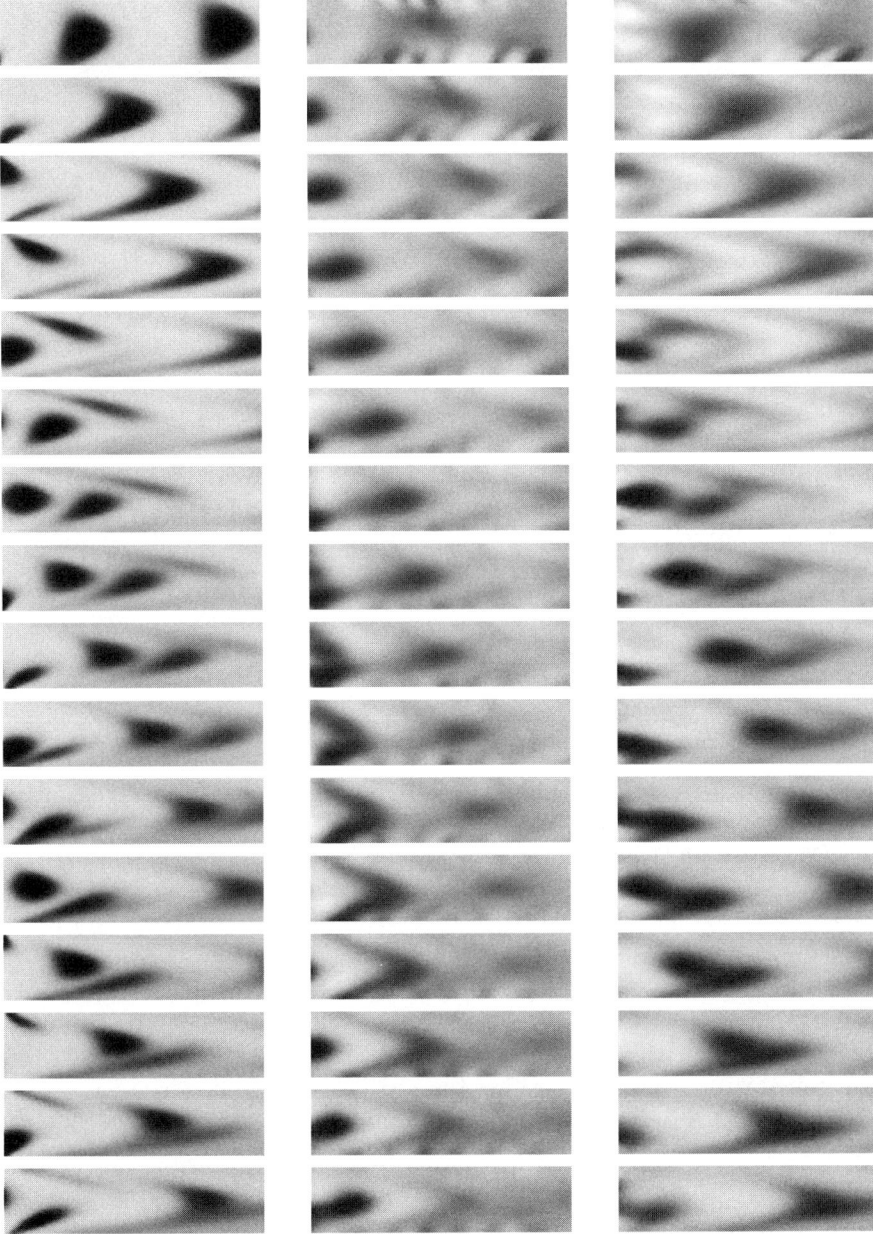

Figure 7.38. From left to right: The true conductivity distribution and the reconstructed distributions with average velocities of 200 cm s^{-1} and 450 cm s^{-1} in the evolution model. The conductivity distributions correspond to the times of the 2nd, 4th, ..., 32th current injection.

Figure 7.39. The time-averaged variances of the estimate error as given by the Kalman filter (left) and the fixed-lag smoother (right).

In the following we restrict ourselves to a simpler case, assuming that the time-varying velocity field has a vanishing component in the transversal direction. Since the pipe is straight, for an incompressible flow this means that the flow profile is the same everywhere along the pipeline, that is, $v(x,t) = v_t = b_t \phi(x_2) e_1$. Here, x_2 is the variable transversal to the pipe and the scalar function $\phi(x_2)$ is the flow profile.

The key idea in the simultaneous estimation of the concentration and the velocity is that we can define any imaginable state as long as we can write down the evolution and observation models. Here, the new state vector is a discretized version of the pair $z_t = (c_t, v_t)$. Observe that the evolution equation with respect to z_t becomes nonlinear due to the term $v \cdot \nabla c$ appearing in the CD model.

To discretize the problem, we approximate the profile using properly chosen functions,

$$v(x,t) \approx \sum_{j=1}^{q} b_{t,j} \varphi_j(x_2) \mathbf{e}_2.$$

This expression is inserted into the variational formulation of the CD equation.

The coefficient vector is denoted by $B_t = [b_{t,1}; \ldots ; b_{t,q}]$.

Corresponding to (7.42) we have now

$$X_{t+1} = \bar{F}(B_t, X_t) + S_{t+1}(B_t) + W_{t+1}(B_t) \qquad (7.44)$$
$$= F(B_t) X_t + S_{t+1}(B_t) + W_{t+1}(B_t),$$

where the covariance $\operatorname{cov}(W_{t+1})$ depends on B_t. When computing the forward problem for the simulation of the measurements, the velocity field was computed with a specified v rather than its projection onto $\operatorname{sp}\{\psi_j\}$.

For the time evolution of the coefficients B_t, we assume a random walk model,

$$B_{t+1} = B_t + W'_{t+1}, \quad \operatorname{cov}(W'_{t+1}) = \lambda I.$$

The evolution equation of Z_t is

$$Z_{t+1} = \begin{bmatrix} X_{t+1} \\ B_{t+1} \end{bmatrix} = \begin{bmatrix} \bar{F}(B_t, X_t) \\ B_t \end{bmatrix} + \begin{bmatrix} S_{t+1}(B_t) \\ 0 \end{bmatrix} + \begin{bmatrix} W_{t+1}(B_t) \\ W'_{t+1} \end{bmatrix}. \qquad (7.45)$$

Consider now the prediction step in Kalman filtering, assuming that the filtered mean $z_{t|t}$ and the covariance $\Gamma_{t|t}$ are known. To compute the mean of the predictor, $z_{t+1|t}$, in Kalman filter and the smoothers, equation (7.45) can be used and no linearization is needed. We have

$$z_{t+1|t} = \begin{bmatrix} x_{t+1|t} \\ b_{t+1|t} \end{bmatrix} = \begin{bmatrix} \bar{F}(b_{t|t}, x_{t|t}) + S_{t+1}(b_{t|t})_{t+1} \\ b_{t|t} \end{bmatrix}.$$

The approximation of the associated covariance $\Gamma_{t+1|t}$ requires linearization. The first-order Taylor approximation of the function $\bar{F}$ at a given point (x_t^*, b_t^*) gives

$$\bar{F}(X_t, B_t) \approx F(b_t^*)x_t^* + F(b_t^*)(X_t - x_t^*) + \left.\frac{\partial \bar{F}}{\partial b}\right|_{(x_t^*, b_t^*)} (B_t - b_t^*). \qquad (7.46)$$

Due to the structure imposed by the multistep implicit Euler algorithm, the computation of the derivative $F_b = \partial \bar{F}/\partial b$ is a laborious task in matrix differential calculus.

From (7.46), we only need the first-order terms for the approximation of the predictor covariance, given by

$$\Gamma_{t+1|t} = \begin{bmatrix} F(b_t^*) & F_b(x_t^*, b_t^*) \\ 0 & I \end{bmatrix} \Gamma_{t|t} \begin{bmatrix} F(b_t^*) & F_b(x_t^*, b_t^*) \\ 0 & I \end{bmatrix}^T + \begin{bmatrix} \Gamma_{w_{t+1}} & 0 \\ 0 & \lambda I \end{bmatrix}.$$

As the linearization point (x_t^*, b_t^*) in the computation of the predictor covariance, we use the best available estimates, that is,

$$x_t^* = x_{t|t} \quad \text{and} \quad b_t^* = b_{t|t}.$$

In the computation of the mean $z_{t+1|t+1}$ and covariance $\Gamma_{t+1|t+1}$ from the predictors and the observation at time $t+1$, we use the extended Kalman filter with an inner Gauss–Newton iteration. Denoting by $U_t(z_t) = U_t(c-t)$ the noiseless observation at time t and by y_t the noisy observation, we write the measurement update as a minimization problem,

$$z_{t+1|t+1} = \arg\min_z \|L_e(y_{t+1} - U_t(z))\|^2 + \|L_{t+1|t}(z - z_{t+1|t})\|^2$$

where $L_{t+1|t}$ and L_e are the Cholesky factors of the matrices $\Gamma_{t+1|t}^{-1}$ and $\text{chol}\,\Gamma_e^{-1}$, respectively.

Once the inner iteration solving the above minimization problem is complete, we compute the Jacobians and assemble the posterior covariance $\Gamma_{t+1|t+1}$ as in extended Kalman filtering. We remark that the exposition above is formal and the actual numerical scheme does not employ the inverses and factorizations.

The computation of the fixed-lag estimates are based on the Kalman predictor and filter estimates and covariances and are computed with similar modifications to the linear fixed-lag formulas.

The explained scheme is intended for an assessment of the method's feasibility. In real-time applications, it might be computationally too intensive.

290   7 Case Studies

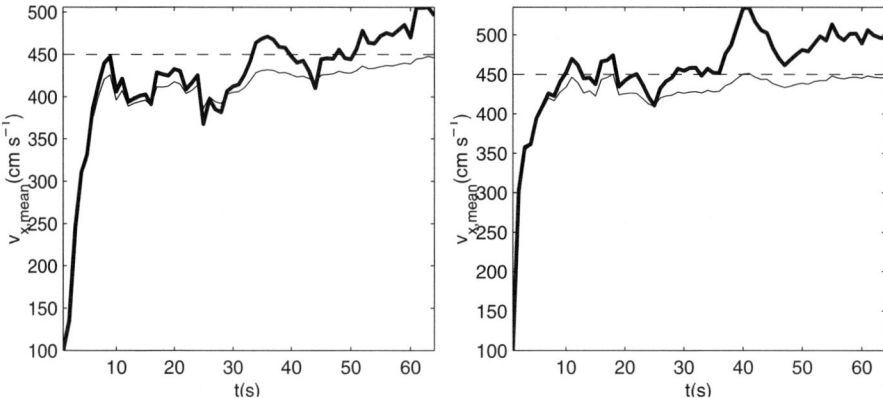

Figure 7.40. Tracking and convergence of the mean velocity. Left figure shows the performance of the Kalman filter with large λ (bold line) and with small λ (thin line). The corresponding results with the 8-lag smoother are shown in the right figure.

A lighter version based on global linearizations can be realistic, for example, when the concentration contrasts are low.

We are interested in both the tracking behavior and the stability of the estimates in a steady-state flow, that is, how fast the estimates are able to follow changes in the velocity profile and how stable the estimates are. In order to assess the tracking of the scheme we do not actually have to consider a time-varying flow. We just start from an incorrect initial guess and observe the convergence. It is clear that the parameter λ is the key parameter controlling the tracking and convergence properties.

As a numerical experiment, we consider a case in which the flow profile is asymmetric and the correct mean velocity is 450 cm s^{-1}. We start from a drastically incorrect initial value 200 cm s^{-1} of the mean velocity and with a symmetric profile. For the estimation we use three fourth-order polynomial basis functions for $v(x_2)$.

We compute the Kalman filter and the 8-lag fixed-lag smoother estimates, both with two different variances λ of the velocity profile evolution. The evolutions of the mean velocity is shown in Figure 7.40. Here the effect of λ is evident to the asymptotic variance of the estimate. What we do not yet see here is that there is a trade-off between the asymptotic variance and the convergence rate: if λ is made very small, the convergence will be slow but the variance of the asymptotic estimate will be low. However, by assuming a higher-order evolution model for v it is possible to come partially around this problem: We can have smooth evolution and fast but not abrupt changes in the estimates.

The basis functions as well as the true and estimated profiles are shown in Figure 7.41. The profile has been recovered at least qualitatively. On the

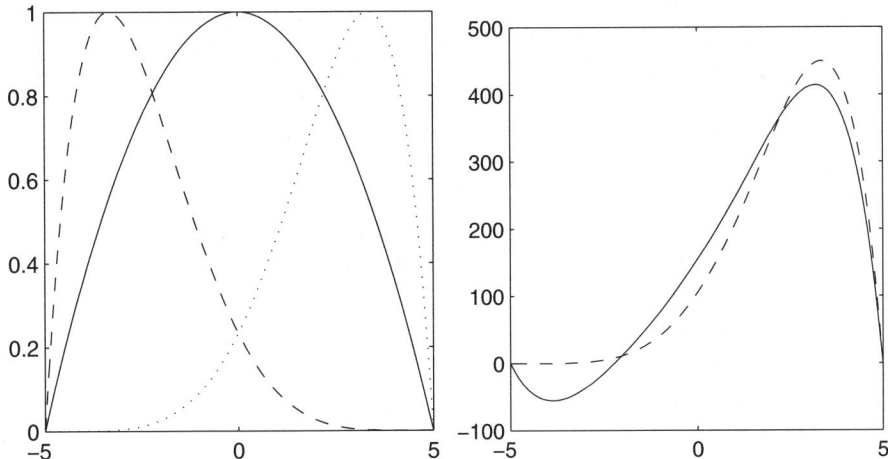

Figure 7.41. Left: The basis functions for the profile estimation. Right: The correct profile (dashed line) and the estimate at the end of the observation time interval.

other hand, it is clear that increasing the number of basis functions decreases the overall stability of the problem.

7.8 Optical Tomography in Anisotropic Media

In this section we discuss the time-harmonic optical tomography problem in strongly scattering anisotropic media. The mathematical model was derived in Section 6.4.

It is well known that material anisotropies in general are a severe obstruction for the uniqueness of inverse problems. In this sense, optical tomography is no exception. The issue of nonuniqueness is not discussed in detail here. For literature concerning these questions, see the "Notes and Comments" at the end of this section. In numerous potentially useful applications of optical tomography, the materials to be probed show significant anisotropic behavior. Examples include the gray and white brain tissue, muscles and skin. Therefore, it is vital to develop inversion methods for anisotropic media as well.

The optical anisotropy of a material is related to its scattering properties. For instance, fibrous structures favor scattering in certain directions. Hence, the parameters describing the anisotropy are related to structural information of the object. In addition to scattering, absorption affects the transmitted light distribution. The absorption is described by the scalar absorption coefficient and therefore is not directly contributing to the anisotropy.

There are several applications in which the absorption coefficient alone could give valuable information of the internal properties of the object. For instance, a cerebral hemorraghia increases the absorption of near-infrared light in the area of the brain with increased blood content, while a drop in the blood oxygenation reduces the absorption with certain wavelengths. The increased vascularization in a tumor results in an increase of absoption in that tissue. It is therefore reasonable to estimate the absorption coefficient alone from optical boundary measurement data. As it will be seen, the anisotropy cannot be ignored, but it is possible to treat it statistically as a modelling parameter that is badly known.

In this section, we restrict ourselves to the two-dimensional model, and we simplify the inverse problem by local linearization. The emphasis in this section is on the recurrent theme, a statistical treatment of modelling errors and uncertainties.

7.8.1 The Anisotropy Model

Let $\Omega \subset \mathbb{R}^2$ denote a bounded domain with a connected, smooth boundary. Assume that the body is illuminated through that boundary with a set of harmonically modulated pointlike laser sources and that the scattered and transmitted light is then measured at a fixed set of boundary points. The body is assumed to be strongly scattering so that the light propagation can be approximated by the diffusion model derived in Section 6.4. In this model, the optical properties of the medium are described by the scalar absorption coefficient μ_a and the diffusion tensor D given by (6.56). These unknowns are treated here as independent variables, while the scattering coefficient μ_s and the matrix B are the dependent variables.

To parametrize the anisotropy of the material, consider the anisotropy matrix $B = B(x) \in \mathbb{R}^{2\times 2}$ at a fixed point $x \in \Omega$. Introduce the eigenvalue decomposition of B,

$$B = USU^{\mathrm{T}},$$

where the matrices $U = U(x)$ and $S = S(x)$ are

$$U(x) = [v_1(x), v_2(x)], \quad S(x) = \begin{bmatrix} b_1(x) & 0 \\ 0 & b_2(x) \end{bmatrix},$$

and $v_j(x) \in \mathbb{R}^2$ is the normalized eigenvector associated with the eigenvalue $b_j(x) > 0$. We assume that $b_1(x) \geq b_2(x)$. The eigenvalue decomposition of the diffusion tensor $D = D(x)$ is then

$$D = U\Lambda U^{\mathrm{T}},$$

where the nonzero diagonal entries of the diagonal matrix Λ are

$$\lambda_j(x) = \frac{1}{2(\mathrm{i}k + \mu_a(x) + (1 - b_j(x))\mu_s)}.$$

The modulation frequencies used in the applications are typically low enough to allow a real approximation

$$\lambda_j(x) \approx \frac{1}{2(\mu_a(x) + (1-b_j(x))\mu_s)} > 0. \tag{7.47}$$

This will be assumed to hold also here. By fixing a Euclidian coordinate frame, we parametrize the eigenvectors $v_j(x)$ with a single direction angle $\theta = \theta(x)$,

$$v_1(x) = \begin{bmatrix} \cos\theta(x) \\ \sin\theta(x) \end{bmatrix}, \quad v_1(x) = \begin{bmatrix} -\sin\theta(x) \\ \cos\theta(x) \end{bmatrix}.$$

The orientation of the eigenvectors is chosen so that $0 \leq \theta(x) < \pi$. Observe that when the material is isotropic, that is, $\lambda_1(x) = \lambda_2(x)$, the definition of the angular parameter becomes ambiguous.

With these definitions, we parametrize the optical properties of the material by four scalar fields,

$$\Omega \to \mathbb{R}_+ \times \mathbb{R}_+ \times \mathbb{R}_+ \times [0,\pi[, \quad x \mapsto (\mu_a, \lambda_1, \lambda_2, \theta).$$

The inverse problem is to estimate these parameters from the boundary measurements. The boundary measurement consist of noisy observations of the outcoming fluxes Φ_{out} at specified points $p_j \in \partial\Omega$ corresponding to various inputs. As usual, we stack all the observations in a long vector.

Due to the complex nature of the diffusion tensor and the scalar coefficient $\mu_a' = \mu_a - ik$, the observation is also a complex vector. In practice, this means simply that one records the amplitude of the outgoing light and the phase lag with respect to the input phase. Hence, the data consists of two real observation vectors, the amplitude and the phase. Based on the diffusion model of Section 6.4, it is a straightforward matter to write the observation model for both the amplitude and the phase. Since in optical tomography the dynamical range of the amplitude of the outcoming flux is several decades, it is usually more convenient to consider the logarithm of the amplitude. We adopt this convention also here.

Consider the observation model

$$y = G(\mu_a, \lambda_1, \lambda_2, \theta) + e,$$

where $G(\mu_a, \lambda_1, \lambda_2, \theta)$ is either the logarithm of the amplitude or the phase angle and y is its noisy observation. For simplicity, we assume here additive Gaussian noise independent of the parameters to be estimated. A detailed analysis of the true noise statistics in this application is a somewhat complicated issue and is not discussed here.

For simplicity, let us consider here a purely Gaussian approximation. Hence, we assume that the variables μ_a, λ_j and θ are independent and that their prior probability densities can be approximated by a Gaussian density. This approximation ignores the constraints of the variables.

With the above assumptions the Bayes' formula gives the posterior density

$$\pi(\mu_a, \lambda_1, \lambda_2, \theta \mid y)$$

$$\propto \exp\bigg\{ -\frac{1}{2}\bigg(\big(y - G(\mu_a, \lambda_1, \lambda_2, \theta)\big)^T \Gamma_n^{-1} \big(y - G(\mu_a, \lambda_1, \lambda_2, \theta)\big)$$

$$+ \big(\mu_a - \mu_a^*\big)^T \Gamma_\mu^{-1} \big(\mu_a - \mu_a^*\big) + \big(\lambda_1 - \lambda_1^*\big)^T \Gamma_{\lambda_1}^{-1} \big(\lambda_1 - \lambda_1^*\big)$$

$$+ \big(\lambda_2 - \lambda_2^*\big)^T \Gamma_{\lambda_2}^{-1} \big(\lambda_2 - \lambda_2^*\big) + \big(\theta - \theta^*\big)^T \Gamma_\theta^{-1} \big(\theta - \theta^*\big) \bigg) \bigg\}.$$

Before proceeding, let us consider the importance of the modelling of anisotropies. More precisely, assume that the anisotropy matrix D is known, and the absorption coefficient μ_a is estimated from the data. Hence we assume that the covariace matrices $\Gamma_{\lambda_j}, \Gamma_\theta \to 0$ in the sense of quadratic forms, that is, $\lambda_j = \lambda_j^*$ and $\theta = \theta^*$ with probability one. This is the prior information concerning the anisotropy. Under this assumption, we calculate the maximum a posteriori estimate of μ_a, which is the minimizer of the functional

$$V(\mu_a) = \big(y - G(\mu_a, \lambda_1^*, \lambda_2^*, \theta^*)\big)^T \Gamma_n^{-1} \big(y - G(\mu_a, \lambda_1^*, \lambda_2^*, \theta^*)\big)$$

$$+ \big(\mu_a - \mu_a^*\big)^T \Gamma_\mu^{-1} \big(\mu_a - \mu_a^*\big).$$

The minimization of V can be done by using a Gauss–Newton minimization scheme. More precisely, assume that μ_a^j, $j = 0, 1, \ldots$ is the current estimate of the absorption coefficient. Let us denote by $G^j = G(\mu_a^j, \lambda_1^*, \lambda_2^*, \theta^*)$ the current value of the mapping G, and let J^j denote the Jacobian of the mapping $\mu_a \mapsto G(\mu_a, \lambda_1^*, \lambda_2^*, \theta^*)$ evaluated at the point μ_a^j. We write the linearized approximation of V. Using the notation $\delta\mu_a = \mu_a - \mu_a^j$,

$$V(\mu_a) \approx \widetilde{V}(\mu_a) = \big(y - G^j - J^j \delta\mu_a\big)^T \Gamma_n^{-1} \big(y - G^j - J^j \delta\mu_a\big)$$

$$+ \big(\delta\mu_a - (\mu_a^* - \mu_a^j)\big)^T \Gamma_\mu^{-1} \big(\delta\mu_a - (\mu_a^* - \mu_a^j)\big).$$

The minimum of this quadratic form occurs when

$$\delta\mu_a = \big((J^j)^T \Gamma_n^{-1} J^j + \Gamma_\mu^{-1}\big)^{-1} \big(\Gamma_n^{-1}(y - G^j) + \Gamma_\mu^{-1}(\mu_a^* - \mu_a^j)\big).$$

We update the current value of the absorption coefficient by setting

$$\mu_a^{j+1} = \mu_a^j + t\delta\mu_a,$$

where the value of the scalar t is the minimizer of the function

$$t \mapsto V(\mu_a^j + t\delta\mu_a), \quad t > 0,$$

subject to the positivity constraint

$$\mu_a^j + t\delta\mu_a \geq 0.$$

7.8 Optical Tomography in Anisotropic Media

In our numerical example, we calculate the MAP estimates corresponding to the "correct and incorrect" prior information. To generate the data, let us fix the geometry. Figure 7.42 shows the structure of the body used in data generation. The lines indicate the direction of the anisotropy. The larger eigenvalue λ_1 correspond to an eigenvector tangential to the lines of anisotropy. The shaded discs indicate regions with the absorption coefficient different from the background.

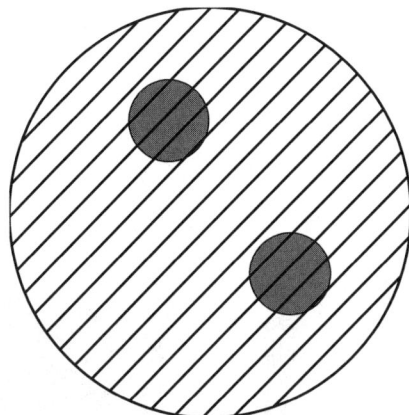

Figure 7.42. The anisotropy model. The lines are along the direction of the eigenvector corresponding to the larger eigenvalue. The eigenvalues used in the model are $\lambda_{1,\text{true}} = 0.025\,\text{cm}^{-1}$, $\lambda_{2,\text{true}} = 0.0083\,\text{cm}^{-1}$. The direction angle is $\theta_{\text{true}} = \pi/4$. The absorption coefficient in the background is $\mu_{a,\text{true}} = 0.25\,\text{cm}^{-1}$, in the shaded area $\mu_a = 1\,\text{cm}^{-1}$.

Let us denote the true values used in data generation by $\lambda_{j,\text{true}}$ and θ_{true}. The true background value of the absorption coefficient is denoted similarly by $\mu_{a,\text{true}}$. The values used in our simulation are indicated in the figure caption of Figure 7.42. We assume further that the noise is white, that is, $\Gamma_n = \sigma^2 I$. Likewise, we choose the prior density to be a white noise prior density centered at μ_a^* and covariance $\Gamma_\mu = \gamma_\mu^2 I$.

We generate the data by using the full nonlinear model. Here we consider only the data consisting of the logarithm of the amplitude. To avoid committing the most obvious inverse crime, we generate the data by using a FEM mesh different from that used in solving the inverse problem. We add normally distributed noise with standard deviation σ corresponding to 5% noise level of the noiseless amplitude.

In Figure 7.43, we have plotted the MAP estimates of the absorption coefficient in two cases. In the first case, we assume that the true values of the parameters λ_j and θ and the background absoption coefficient are in fact known, that is, $\lambda_j^* = \lambda_{j,\text{true}}$, $\theta^* = \theta_{\text{true}}$ and $\mu_a^* = \mu_{a,\text{true}}$. In this case, the estimate corresponds relatively well to the true absorption coefficient.

In the second experiment, we assume that our prior information of the anisotropy structure is flawed. We assume that $\mu_a^* = \mu_{a,\text{true}}$ but $\lambda_j^* = \lambda_{j,\text{off}} \neq \lambda_{j,\text{true}}$ and $\theta^* = \theta_{\text{off}} \neq \theta_{\text{true}}$. The values of $\lambda_{j,\text{off}}$ and θ_{off} are indicated in Figure 7.43. Using the same data as in the previous case, we repeat the MAP estimation. These results are also shown in Figure 7.43. We observe that the incorrect information of the anisotropy structure distorts the results. The MAP estimate tries to compensate the incorrectly modelled diffusion by increasing or decreasing the absoption coefficient correspondingly.

In conclusion, ignoring the possibility of anisotropic background leads to strongly erroneous estimates of the absorption coefficient. Unfortunately, the anisotropy is usually not well known a priori. In the following section we show that reasonable results can be obtained also in this case by properly modelling this lack of information.

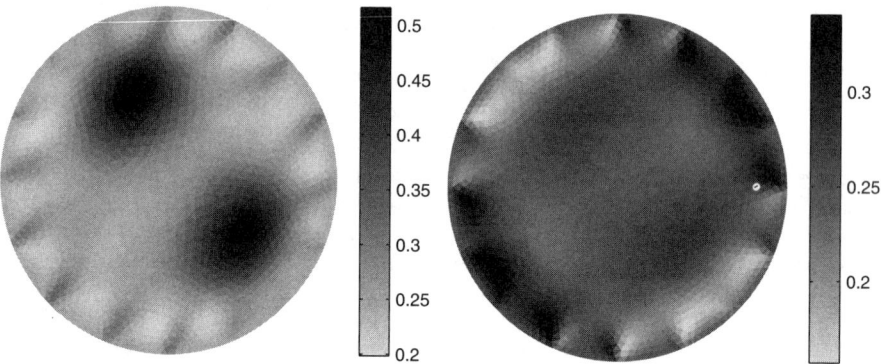

Figure 7.43. The Gauss–Newton based MAP estimate with correct (left) and incorrect (right) prior information. In the incorrect case, we used the values $\mu_a = \mu_{a,\text{true}}$, $\lambda_1^* = \lambda_{1,\text{off}} = 0.0167$, $\lambda_2^* = \lambda_{2,\text{off}} = 0.0125$ and $\theta^* = \theta_{\text{off}} = \pi/2$.

7.8.2 Linearized Model

To simplify the problem, we consider here a linearized model. The modelling errors induced by the linearization will be discussed briefly later in this section. Around some reference values

$$(\mu_a, \lambda_1, \lambda_2, \theta) = (\mu_a^*, \lambda_1^*, \lambda_2^*, \theta^*),$$

we write the approximate model

$$y \approx y^* + J_{\mu_a}\delta\mu_a + J_{\lambda_1}\delta\lambda_1 + J_{\lambda_2}\delta\lambda_2 + J_\theta\delta\theta + e,$$

where $y^* = G(\mu_a^*, \lambda_1^*, \lambda_2^*, \theta^*)$ and J_η, $\eta \in S = \{\mu_a, \lambda_1, \lambda_2, \theta\}$ denotes the Jacobian of the mapping G with respect to the variable η at the reference point.

Assume that we have specified the prior information of the variables in S, which are assumed to be mutually independent and Gaussian,

$$\eta \sim \mathcal{N}(\eta^*, \Gamma_x), \quad \eta \in S.$$

We want to estimate only the parameter μ_a from the linearized model, while the effects of the misspecified remaining parameters are treated as a modelling error, that is, their values are not estimated. Let us introduce the linearized model

$$y = y^* + J_{\mu_a} \delta\mu_a + \varepsilon,$$

where the noise term ε is

$$\varepsilon = J_{\lambda_1} \delta\lambda_1 + J_{\lambda_2} \delta\lambda_2 + J_\theta \delta\theta + e.$$

Clearly, under these approximations ε is Gaussian, zero mean and independent of the variable μ_a, and its covariance is

$$\Gamma_\varepsilon = \mathrm{E}\{\varepsilon\varepsilon^\mathrm{T}\} = J_{\lambda_1} \Gamma_{\lambda_1} J_{\lambda_1}^\mathrm{T} + J_{\lambda_2} \Gamma_{\lambda_2} J_{\lambda_2}^\mathrm{T} + J_\theta \Gamma_\theta J_\theta^\mathrm{T} + \Gamma_\mathrm{n}.$$

The CM estimate - which coincides with the MAP estimate in the linear Gaussian case - of the variable $\delta\mu_a$ is, in this case,

$$(\delta\mu_a)\big|_y = \Gamma_{\mu_a} J_{\mu_a} \left(J_{\mu_a}^\mathrm{T} \Gamma_{\mu_a} J_{\mu_a} + \Gamma_\varepsilon\right)^{-1} (y - y^*).$$

Observe that an analogous equation could have been written for any one of the variables of the set S. Also, one could estimate several or all of the variables simultaneously instead of just μ_a. This would increase the computational work.

Let us demonstrate with a numerical example how the above estimation works in practice. We use the same data corresponding to an anisotropic object as in the examples of the previous section.

In the first test, we assume that the reference values, or center points of the prior probability densities, correspond to the true values of the parameters, that is, $(\mu_a^*, \lambda_1^*, \lambda_2^*, \theta^*) = (\mu_{a,\mathrm{true}}, \lambda_{1\mathrm{true}}, \lambda_{2,\mathrm{true}}, \theta_{\mathrm{true}})$. The standard deviations for the variables are indicated in the caption of Figure 7.43. The reconstruction obtained is even better than the MAP estimate with correct information of the anisotropy. Next, we test what happens if the center point of the prior differs from the true background value. Using the same variance for the Gaussian priors as before, we move the prior center off from the true value, to $(\mu_a^*, \lambda_1^*, \lambda_2^*, \theta^*) = (\mu_{a,\mathrm{true}}, \lambda_{1\mathrm{off}}, \lambda_{2,\mathrm{off}}, \theta_{\mathrm{off}})$. The prior distributions are shown in Figure 7.44, in which the vertical lines show the true values of the parameters used in the data generation. Figure 7.45 shows also the estimated absorption coefficient. While the quality of the reconstruction is inferior

298   7 Case Studies

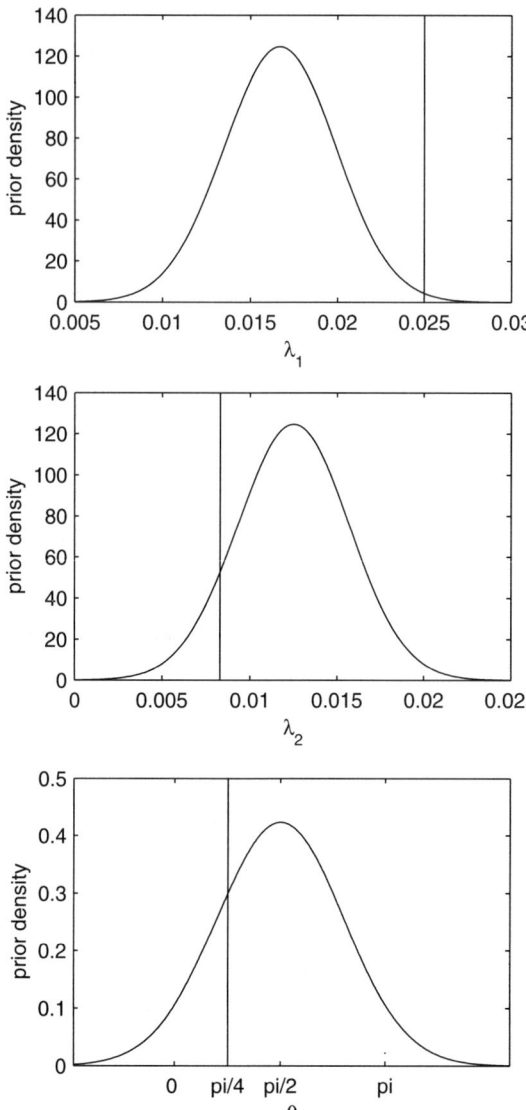

Figure 7.44. The prior distributions of the anisotropy parameters λ_j and θ. The vertical lines indicate the true coefficients.

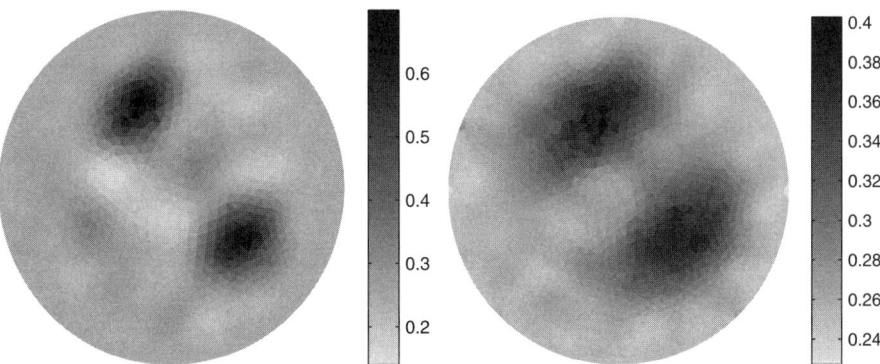

Figure 7.45. The estimates of μ_a when the unknown anisotropy is modelled as noise. In the left-hand reconstruction, the centers of the priors agree with the true parameter values. The standard deviations of λ_j is $0.0032\,\mathrm{cm}^{-1}$; that of θ is 0.735. In the right-hand reconstruction, the prior centers are different from the correct values.

to that obtained with the correctly centered priors, the correct structure is still clearly visible. Note, however, that especially $\lambda_{1\mathrm{true}}$ has a very small probability with respect to the prior for λ_1. Despite this fact the reconstruction in Figure 7.45 is relatively accurate.

Finally, let us say a few words about the linearization errors. Numerical simulations indicate that the linearization introduces errors that are not of negligible size. As demonstrated earlier, they can be taken into account by using an enhanced error model. In the present calculations, the spurious errors not properly modelled in the likelihood were compensated for by decreasing the prior variance of μ_a. The prior variance could be determined by hierarchical methods, but then either one should use the nonlinear model or an enhanced error linearized model.

7.9 Optical Tomography: Boundary Recovery

In imaging-type problems it can be argued that pixel-based priors are the most versatile ones partly because they can be constructed by discretizing from the most general representations of random fields. In many cases, however, pixel-based priors may not be feasible because the description of the prior is given as distribution of some parameters that do not fit well to pixelized representations.

In order to illustrate this problem we consider the following simple example. Let x_k, $k = 0, \ldots, N$ denote values at $t_k = k/N$ of a piecewise constant function on the interval $[0, 1]$. Define an integer-valued random variable $\ell \in [1, \ldots, N-1]$ so that $P\{\ell = k\} = 1/(N-1)$ for all $k \in [1, \ldots, N-1]$.

Let the marginal distributions of x_1 and x_N be mutually independent and Gaussian, and let $x_j \sim \mathcal{N}(\eta_j, 1)$, $j = 0, \ldots, N$ be independent of ℓ. Then set

$$P\{x_0 = \cdots = x_{\ell-1} \mid \ell\} = 1 \quad \text{and} \quad P\{x_\ell = \cdots = x_N \mid \ell\} = 1,$$

that is, the variable is constant and Gaussian on both sides of the pixel boundary $(x_{\ell-1}, x_\ell)$, the boundary location being uniformly distributed. A description of the pixelwise density of x in this case is not straightforward, and in fact the density is singular.

The implementation of an algorithm to determine the MAP estimate is difficult due to the discontinuity of the prior and the posterior, assuming, for example, a Gaussian linear observation model. On the other hand, sampling would be extremely straightforward in this case just by using the prior distribution as the candidate distribution. Dividing the interval $[0, 1]$ into two subintervals $[0, z]$ and $[z, 1]$ where $z \sim \mathcal{U}(0, 1)$ and assigning the above mutually independent Gaussian distributions for the constant parameters on both subintervals, the number of unknowns is reduced to three instead of $N + 1$ and the associated posterior is easy to handle for both MAP and sampling.

There are various alternatives to pixel-based priors. Often, however, the computation of the forward problem may be challenging. This is true in particular if the observation model is due to a partial differential equation in higher dimensions.

In this section we consider the boundary recovery problem, that is, the two-dimensional analogy to the problem discussed above. We discuss first the general elliptic case in which the forward problem is to be solved with the finite-element method. We then apply the approach to optical diffusion tomography in the frequency domain.

7.9.1 The General Elliptic Case

Let $\Omega \subset \mathbb{R}^2$ denote a bounded domain. We consider the elliptic boundary value problem

$$-\nabla \cdot a\nabla \Phi + b\Phi = f \quad \text{in } \Omega, \tag{7.48}$$

$$c\Phi + d\frac{\partial \Phi}{\partial n} = g, \quad \text{on } \partial\Omega,$$

in the special case of piecewise constant coefficients a and b. Assume that Ω is divided into $L + 1$ disjoint regions A_k

$$\Omega = \bigcup_{k=0}^{L} A_k,$$

which are bounded by smooth closed nonintersecting boundary curves. Further, let the coefficient functions a and b be constants in the subdomains, $\{a, b\}\big|_{A_k} = \{a_k, b_k\}$. Let χ_k be the characteristic function of A_k and let

7.9 Optical Tomography: Boundary Recovery

$\mathcal{C}_\ell \subset \Omega$, $\ell = 1, \ldots, L$ denote the smooth outer boundary of A_ℓ. The outer boundary of the background region A_0 is assumed to be $\partial\Omega$.

The outer boundary $\partial\Omega$ and the values $\{a_k, b_k\}$ may be either known or not known a priori, but in any case accurate geometrical information on the regions $\{A_k\}$ is missing. This missing information may be, for example, the shapes, sizes, location or in some cases even the number of regions.

We consider a typical finite-element triangulation in which the domain Ω is divided into M disjoint elements $\cup_{p=1}^{M} \Omega_p$, joined at D vertex nodes N_i. The variational formulation of (7.48)–(7.49) and subsequent application of the standard Galerkin scheme leads to the system of equations

$$(K(a) + C(b) + R(a)) = G(a) + F , \tag{7.49}$$

where the entries of the system matrices are given by

$$K_{ij} = \int_\Omega a \, \nabla\varphi_i \cdot \nabla\varphi_j \, dr , \tag{7.50}$$

$$C_{ij} = \int_\Omega b \, \varphi_i \, \varphi_j \, dr ,$$

$$R_{ij} = \int_{\partial\Omega} \frac{ac}{d} \varphi_i \, \varphi_j \, dS .$$

The terms on the right-hand side of equation (7.49) are of the form

$$F_j = \int_\Omega f \, \varphi_j \, dr, \qquad G_j = \int_{\partial\Omega} \frac{a}{d} g \, \varphi_j \, dS . \tag{7.51}$$

For details, see the references given in "Notes and Comments".

According to the above definition, we can write

$$a(r) = \sum_{k=0}^{L} a_k \chi_k(r) , \qquad b(r) = \sum_{k=0}^{L} b_k \chi_k(r). \tag{7.52}$$

Hence we get

$$K_{ij} = \sum_{k=0}^{L} \int_{\mathrm{supp}(\varphi_i\varphi_j) \cap A_k} a_k \, \nabla\varphi_i \cdot \nabla\varphi_j \, dr , \tag{7.53}$$

$$C_{ij} = \sum_{k=0}^{L} \int_{\mathrm{supp}(\varphi_i\varphi_j) \cap A_k} b_k \, \varphi_i \varphi_j \, dr, \tag{7.54}$$

where $\mathrm{supp}(\varphi_i\varphi_j)$ denotes the intersection of the supports of the basis functions φ_i and φ_j, that is, the union of the elements that contain both nodes N_i and N_j:

$$\mathrm{supp}(\varphi_i\varphi_j) = \bigcup_{\{N_i \in \bar{\Omega}_\ell,\ N_j \in \bar{\Omega}_\ell\}} \Omega_\ell.$$

302   7 Case Studies

We assume that the domains A_ℓ are simply connected and starlike, and that the boundaries $\mathcal{C}_\ell$ are sufficiently smooth so that they can be approximated by a parametrized model as

$$\mathcal{C}_\ell : \quad r_\ell(s) = \begin{bmatrix} r_{1,\ell}(s) \\ r_{2,\ell}(s) \end{bmatrix} = \sum_{n=1}^{N_\theta} \begin{bmatrix} \gamma_n^{1,\ell} \theta_n^1(s) \\ \gamma_n^{2,\ell} \theta_n^2(s) \end{bmatrix}, \quad \ell = 1, \ldots, L, \qquad (7.55)$$

where θ_n^j are periodic and differentiable basis functions with unit period.

Let γ denote the vector of all boundary shape coefficients. From now on, let $a = (a_1, \ldots, a_L)$ and $b = (b_1, \ldots, b_L)$. The next goal is to set up the computation of the system matrices, that is,

$$\gamma, a, b \mapsto K \text{ and } C.$$

This is accomplished via the following steps. First, all nodes are classified according to whether they are inside or outside of each boundary $\mathcal{C}_\ell$. Then all elements are classified according to whether they are inside, outside or are intersected by $\{\mathcal{C}_\ell\}$. For the elements that are completely inside or outside, their contribution to the integrals of the system matrices is computed as usual. For the intersected elements, the intersection boundary is approximated with a line and the element is divided into three further subelements as shown in Figure 7.46. Then the approximations of the integrals $\int \varphi_i \varphi_j \, dr$ and $\int \nabla \varphi_i \cdot \nabla \varphi_j \, dr$ are computed over these subelements with Gaussian quadrature and assembled accordingly.

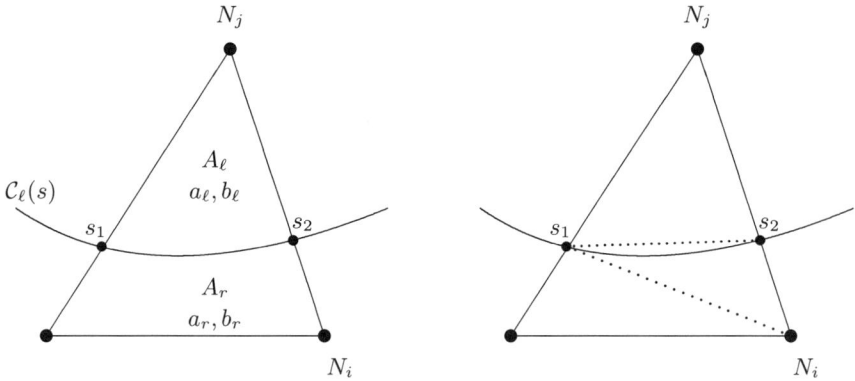

Figure 7.46. Left: A schematic representation of one FEM element Ω_m intercepted by the region boundary $\mathcal{C}_\ell(s)$. The pair (a_ℓ, b_ℓ) are the coefficients of the equation (7.48) inside the region A_ℓ, and (a_r, b_r) are the coefficients in the neighboring region A_r, respectively. $\mathcal{C}_\ell(s_1)$ and $\mathcal{C}_\ell(s_2)$ are the intersection points of the boundary $\mathcal{C}_\ell(s)$ and the element edges. Right: The division of the Ω_m into three subtriangles.

Let $F(\gamma, a, b)$ denote the sum

$$F(\gamma, a, b) = K(\gamma, a) + C(\gamma, b) . \tag{7.56}$$

Now the FEM equation (7.49) can be written in the form

$$(F(\gamma, a, b) + R(a_0))\Phi = G(a_0) + F. \tag{7.57}$$

Note that we have assumed that $\mathcal{C}_\ell, \ell > 0$ do not intersect with $\mathcal{C}_0 = \partial\Omega$ and thus $R = R(a_0)$ and $G = G(a_0)$ only. The associated Jacobians are trivial since $R(a_0) = a_0 \widetilde{R}$ and $G(a_0) = a_0 \widetilde{G}$.

Let us denote by M the measurement operator so that we have, for the errorless observations, $y = (M \circ \Phi)(\gamma, a, b) = A(\gamma, a, b)$. For the computation of the MAP estimates with the Gauss–Newton algorithm we need to compute the Jacobian of y with respect to the parameters γ, a and b. Eventually this requires the derivatives of F with respect to the individual parameters. These are

$$\left(\frac{\partial F}{\partial \gamma_n^{1,\ell}}\right)_{ij} = \sum \left\{ [b] \int_{s_1}^{s_2} \varphi_i \varphi_j \, \dot{r}_{1,\ell} \, \theta_n^1 \, ds + [a] \int_{s_1}^{s_2} \nabla\varphi_i \cdot \nabla\varphi_j \, \dot{r}_{1,\ell} \, \theta_n^1 \, ds \right\},$$

$$\left(\frac{\partial F}{\partial \gamma_n^{2,\ell}}\right)_{ij} = -\sum \left\{ [b] \int_{s_1}^{s_2} \varphi_i \varphi_j \, \dot{r}_{2,\ell} \, \theta_n^2 \, ds + [a] \int_{s_1}^{s_2} \nabla\varphi_i \cdot \nabla\varphi_j \, \dot{r}_{2,\ell} \, \theta_n^2 \, ds \right\}.$$

Above, the sums are extended over elements Ω_m that are divided by the boundary curve $\mathcal{C}_\ell$ and for which $\Omega_m \cap \mathrm{supp}(\varphi_i \varphi_j) \neq \emptyset$. The jumps of the parameters a and b across the curve are denoted by

$$[a] = a_\ell - a_r, \quad [b] = b_\ell - b_r.$$

The limits of integration s_1 and s_2 above are the curve parameters in the intersection points of the element edges and boundary curve $\mathcal{C}_\ell$ in each element Ω_m. See Figure 7.46 for further clarification. Further, $\dot{r}(s)$ denotes dr/ds.

Similarly,

$$\left(\frac{\partial F}{\partial a_k}\right)_{ij} = \int_{\mathrm{supp}(\varphi_i \varphi_j) \cap A_k} \nabla\varphi_i \cdot \nabla\varphi_j \, dr ,$$

$$\left(\frac{\partial F}{\partial b_k}\right)_{ij} = \int_{\mathrm{supp}(\varphi_i \varphi_j) \cap A_k} \varphi_i \varphi_j \, dr .$$

We now apply these results to a concrete example.

7.9.2 Application to Optical Diffusion Tomography

As discussed in Section 6.4, the diffusion approximation in the frequency domain takes the form

304   7 Case Studies

$$-\nabla \cdot a\nabla \Phi + \left(b + \frac{i\omega}{c}\right)\Phi = q, \quad \text{in } \Omega, \tag{7.58}$$

$$\Phi + 2a\vartheta\frac{\partial \Phi}{\partial n} = g, \quad \text{on } \partial\Omega. \tag{7.59}$$

In this example we set $\omega = 300\,\mathrm{MHz}$. We employ the diffuse boundary source model which is obtained by setting $q = 0$ and

$$g_j(x) = \begin{cases} -4g_0 & x \in r_{s,j} \\ 0 & x \in \partial\Omega \setminus r_{s,j} \end{cases}$$

where $r_{s,j}$ denotes the boundary patch of source j. Let Φ_j be the photon density when the source at patch j is active. The measurements are the fluxes at the boundary locations $z_i \notin r_{s,j}$,

$$y_{i,j} = -(a\nu \cdot \nabla \Phi_j)|_{x=z_i}.$$

Notice that due to the Robin boundary condition on the surface outside the source, the output flux can be expressed in terms of the Neumann data.

Due to the reparametrization, the prior is now to be written in terms of the employed parameters γ, a, b. First, it is reasonable to assume that γ and (a, b) are mutually independent. In most biomedical and industrial applications, reasonable prior models for (a, b) based, for example, on anatomical data for tissues or material samples are available. Again, if anatomical atlases are available, reasonable choices for θ and N_θ and further the prior distribution for γ could be constructed.

A problem that makes the construction of the final prior model for γ a nontrivial task is that not all $\gamma \in \mathbb{R}^{N_\theta}$ correspond to a feasible set of domain boundaries. To avoid the difficult determination of the prior of γ, we simplify the problem by assuming that N_θ so small that we can take the prior of γ to be completely uninformative. In other words, we reduce the dimensionality of the problem and rely on the measurements to be informative enough to render feasible curve parameters. Furthermore, we take $(a, b) \sim \pi_+(a, b)$.

Thus the posterior is a truncated Gaussian,

$$\pi_{\text{post}} = \pi_+(a, b)\mathcal{N}(y - A(\gamma, a, b), \Gamma_e),$$

where we take $\Gamma_e = 10^{-4}\mathrm{diag}\,(y_{i,j}^2)$. In other words, we assume that the measurement has been carried out so that the noise level of all measurements is 1%.

The test target includes two regions in addition to the background. We have a circular outer boundary with equiangularly placed and interleaved sources and detectors, 16 each. This results in $M = 256$ measurements.

For the computation of the MAP we employ Gauss–Newton iterations with a line search. The data is again computed in another mesh from the one used in the construction of the inverse problem. This applies also to the number of the basis functions. The Jacobian J of the mappings $\gamma, a, b \mapsto y$

are obtained row-wise as follows. Let f_k be any of the unknown parameters. By differentiating (7.57) we have for the jth source

$$F\frac{\partial \Phi_j}{\partial f_k} = -\frac{\partial F}{\partial f_k}\Phi_j$$

and further for the kth block of the jth column of J

$$J_{k,j} = \frac{\partial \Phi_j}{\partial f_k} = -F^{-1}\frac{\partial F}{\partial f_k}\Phi_j \ .$$

Finding reasonable initial values for the parameters for the Gauss–Newton iterations can be difficult. Here, we first compute pixelwise reconstructions using the standard Tikhonov scheme with a Gaussian smoothness prior. The initial guess is then obtained by fitting boundary curves to the curves $b \approx (b_{\max} + b_{\min})/2$.

The target as well as the pixel-based reconstruction, the initial guess and final estimates are all shown in Figure 7.47. The pixel-based scheme is able to detect the two objects with the accuracy to be expected in the case of optical tomography and the employed prior model and side constraint. In this case the targets were located well enough for the boundary recovery to converge rapidly. It is clear that the shape of targets which have only a small contrast with the background cannot be recovered accurately. This is reflected in the less accurate shape recovery of the southeast target when compared to the northwest target.

7.10 Notes and Comments

Image deblurring and anomalies. The image deblurring with anomalous prior was discussed in [65]. The material here goes beyond that article, where the effect of modelling errors and different grids were not discussed.

Limited angle tomography. This section is based on the articles [74, 116]. The discussion of exterior point optimization is based on [36]. In [116], the noise model of X-ray imaging is also discussed in more detail. We also refer to [103] concerning the sensitivity of the X-ray projection data to discontinuities in the mass absorption coefficient. The Barzilai–Borwein optimization algorithm is described in the article [11].

Note that usually an X-ray problem is considered as a limited angle problem when the available angles extend typically to 160 degrees which is only slightly less than 180 degrees, making the problem only mildly ill-posed.

Source localization in the stationary case. The idea of using ℓ^1 regularization functionals or minimum current reconstructions in MEG problem are discussed in [87] and [131]. The version in this book is based on the

306   7 Case Studies

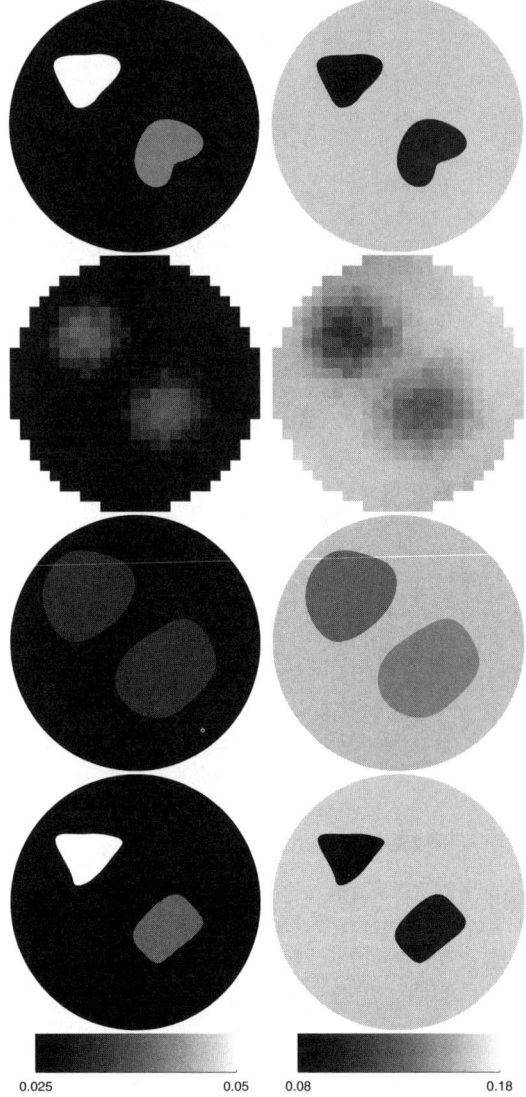

Figure 7.47. A test case with two subdomains. The left column shows images of a and the right column shows images of b. The rows from top to bottom: The actual coefficient distributions, the pixelwise reconstructions, the initial estimates and the final estimates. The number of unknowns is 40.

work behind the article [131]. In that article, the ℓ^1-criterion was taken as a Tikhonov-type penalty, and estimates were calculated by constrained linear optimization. The cited article contains also comparisons with dipole fitting algorithms from real data. The novelty of the discussion in this book is the use of the Bayesian framework with sampling and hyperprior model for the prior parameter.

Let us mention that there are full Bayesian approaches to the dipole fitting and model selection problems, too; see [51] and [111].

Source localization in the nonstationary case. The multiple-dipole dynamic problem discussed here is closely related to model determination problems that have been studied from the Bayesian point of view. The *reversible jump MCMC* methods are developed to treat moves between model spaces of different dimensions; see [48].

Optimization of current patterns. The current pattern optimization of Section 7.5 is based on [67]. The method is similar to Bayesian experiment design, for example, the D-optimality criterion [8]. Strictly speaking, to fulfill the covariance trace criterion, we should not linearize the problem at the prior mean but rather investigate the ensemble behavior with an MCMC-type approach. However, this is computationally feasible only with very small dimensional problems.

The optimal current pattern problem of Section 7.5 have been discussed in the literature from a nonstatistical point of view; see references [22, 63]. These articles are based on the concept of *distinguishability*. Assume that σ is the true conductivity of the body and $\widetilde{\sigma}$ is an a priori estimate of it. The distinguishability of a current pattern I is defined as

$$d(I) = d(I; \sigma, \widetilde{\sigma}) = \frac{\|R(\sigma)I - R(\widetilde{\sigma})I\|}{\|I\|},$$

where the norm is the usual L^2-norm. The optimal current pattern is defined in [22] as

$$I_{\text{opt}} = \arg\max_{I} \left(d(I; \sigma, \widetilde{\sigma})\right) \text{ subject to } \sum_{\ell=1}^{L} I_\ell = 0.$$

The idea is to use the current pattern that is most sensitive to the differences between the actual and the estimated conductivities. In the cited article, an adaptive algorithm for finding such a current pattern is presented. The algorithm is based on the observation that the optimal current pattern, by definition, is the eigenvector corresponding to the maximal eigenvalue of the operator $|R(\sigma) - R(\widetilde{\sigma})|$. The search of the eigenvalue can be done by power iteration [47]. Replacing $R(\sigma)I$ by the measured voltage pattern, the power iteration can be effectuated adaptively, yielding to a procedure similar to the time reversal mirror method ([37]).

The results of the Bayesian current pattern optimization presented here and the distinguishability criterion are not directly comparable since they are based on different definitions of optimality. Some comparisons can be found in [67].

The situation in nonstationary EIT is somewhat more complex than the stationary case. However, if the evolution model is linear and the observation model is linearized, the estimate covariance does not depend on the measurements. Thus, for the optimization, the Kalman recursions can be reduced to include the evolution model only for the state covariance, which fulfills a so-called Riccati equation, which is a second-order matrix difference equation. There is some evidence that there are only a few linearly independent current patterns that should be used in certain nonstationary EIT cases [76].

EIT approximation errors. The material in this section has not been previously published. The main message here is that with conventionally employed meshes, the approximation error can be even appreciably larger than with the usually assumed 1–2% level additive noise. Of course, the estimation errors have been seemingly avoided by over-regularizing the problem.

When the measurement times are not restricted, the noise level can be suppressed and the errors are almost completely due to modelling and approximation error. Examples of such experiments arise in geophysics [99] and material testing [137].

The statistical approach can also be used with error modelling when linearization or a one-step method is to be employed. For the global linearization method the implementation would simply consist of the the computation of the accurate forward problems and the linearized approximations

$$V^{(\ell)} = U(\bar{\sigma}) + J(\bar{\sigma})(\sigma^{(\ell)} - \bar{\sigma})$$

and then using these to estimate the associated mean and covariance.

Model errors in the same fashion as in the classical analysis for measurement error norm, that is, $\|\bar{A} - A\| = \delta_A$, have been considered, for example, in [141]. This analysis has the same problems as the measurement error norm consideration: it lacks the structure and variability associated with the approximation error.

Process imaging and nonstationary models with EIT The treatment of Section 7.7.1 is based on [113, 114, 115]. Apart from the relatively tedious but information-rich convection–diffusion models coupled with computational fluid models, there are situations in which much simpler evolution models may turn out to be feasible. For example, the state estimation approach has previously been studied in connection with EIT in [134] in which a low-dimensional parametrization for the conductivity was used, thus making the standard random walk feasible. The key here is the low dimensionality of the parametrization for the unknown. Similarly, the boundary reparametrization leads to much lower dimensional problems than with pixelwise parametrizations and can render the random walk model feasible; see [76] for the extension of the approach discussed in Section 7.9 and [133] for an alternative mesh perturbation approach.

The state estimation approach has also been applied to optical diffusion tomography [73] which includes a high-dimensional pixel-based parametrization and [101] that considers a more complex time series model for the unknown involving random walk and pseudoperiodic components.

Experimental evaluation of the approach in practice has been done in [112, 136] with EIT measurements.

Anisotropic optical tomography. The discussion in this section is based on the articles [58] and [59]. These articles do not discuss the modelling errors due to linearization. Numerical evidence shows that, as in EIT, the linearization generates remarkable errors that cannot be modelled with the white noise model. The discussion here should therefore be seen as a "proof of concept" of the importance of Bayesian modelling of anisotropies.

Boundary recovery in optical tomography. This section is based on the articles [71, 72, 75]. We considered only the computation of the MAP estimates. If sampling and computation of the CM estimate and spread estimates are considered, the following is to be noted. The MAP estimate of (γ, a, b) corresponds to a discontinuous estimate for the spatial coefficient distributions. When we carry out sampling to investigate the posterior distributions of a and b as function on Ω, it is clear that the parameters γ as such are not interesting. However, the parameters γ determine the subdomains A_k but the assessment of the variability of a subdomain boundary $\mathcal{C}_\ell(s)$ at point s is not straightforward. On the other hand, computing means for the samples $(\gamma, a, b)_j$ is not meaningful. Instead, the exploration of the posterior can be done as follows. Form a dense, usually rectangular, mesh to which the samples are projected so that we have the samples in the coefficient distribution form $(a(r_k), b(r_k))$, where r_k are, for example, the centers of the mesh elements. Investigating the mean and other statistics of the samples in this form makes sense. In contrast with the MAP estimate, the spatial distribution of (a, b) is not piecewise constant but exhibits transition regions across the boundaries of the subdomains which are not any more necessarily clearly defined. This could be due to the inherent uncertainty in a case where the contrast of the coefficients is small across the subdomain boundary. A similar procedure was used in Section 7.4 in which we had varying number of dipoles. In connection with EIT, a parametrization called the "colored polygon model" and the associated sampling and estimation of the posterior was carried out in [3] and [38].

As mentioned above, the MAP estimation based on extended Kalman filter in the nonstationary case is also feasible although computationally rather tedious [76]. The use of particle filters would in practice be constrained to the tracking of a single inclusion only.

A

Appendix: Linear Algebra and Functional Analysis

In this appendix, we have collected some results of functional analysis and matrix algebra. Since the computational aspects are in the foreground in this book, we give separate proofs for the linear algebraic and functional analyses, although finite-dimensional spaces are special cases of Hilbert spaces.

A.1 Linear Algebra

A general reference concerning the results in this section is [47]. The following theorem gives the singular value decomposition of an arbitrary real matrix.

Theorem A.1. *Every matrix $A \in \mathbb{R}^{m \times n}$ allows a decomposition*

$$A = U \Lambda V^{\mathrm{T}},$$

where $U \in \mathbb{R}^{m \times m}$ and $V \in \mathbb{R}^{n \times n}$ are orthogonal matrices and $\Lambda \in \mathbb{R}^{m \times n}$ is diagonal with nonnegative diagonal elements λ_j such that $\lambda_1 \geq \lambda_2 \geq \cdots \geq \lambda_{\min(m,n)} \geq 0$.

Proof: Before going to the proof, we recall that a diagonal matrix is of the form

$$\Lambda = \begin{bmatrix} \lambda_1 & 0 & \cdots & 0 & 0 \ldots 0 \\ 0 & \lambda_2 & & \vdots & \vdots \quad \vdots \\ \vdots & & \ddots & & \\ 0 & \cdots & & \lambda_m & 0 \ldots 0 \end{bmatrix} = [\mathrm{diag}(\lambda_1, \ldots, \lambda_m), 0],$$

if $m \leq n$, and 0 denotes a zero matrix of size $m \times (n-m)$. Similarly, if $m > n$, Λ is of the form

$$\Lambda = \begin{bmatrix} \lambda_1 & 0 & \cdots & 0 \\ 0 & \lambda_2 & & \vdots \\ \vdots & & \ddots & \\ & & & \lambda_n \\ 0 & \cdots & & 0 \\ \vdots & & & \vdots \\ 0 & \cdots & & 0 \end{bmatrix} = \begin{bmatrix} \mathrm{diag}(\lambda_1, \ldots, \lambda_n) \\ 0 \end{bmatrix},$$

where 0 is a zero matrix of the size $(m - n) \times n$. Briefly, we write $\Lambda = \mathrm{diag}(\lambda_1, \ldots, \lambda_{\min(m,n)})$.

Let $\|A\| = \lambda_1$, and we assume that $\lambda_1 \neq 0$. Let $x \in \mathbb{R}^n$ be a unit vector with $\|Ax\| = \|A\|$, and $y = (1/\lambda_1)Ax \in \mathbb{R}^m$, i.e., y is also a unit vector. We pick vectors $v_2, \ldots, v_n \in \mathbb{R}^n$ and $u_2, \ldots, u_m \in \mathbb{R}^m$ such that $\{x, v_2, \ldots, v_n\}$ is an orthonormal basis in $\mathbb{R}^n$ and $\{y, u_2, \ldots, u_m\}$ is an orthonormal basis in $\mathbb{R}^m$, respectively. Writing

$$V_1 = [x, v_2, \ldots, v_n] \in \mathbb{R}^{n \times n}, \quad U_2 = [y, u_2, \ldots, u_m] \in \mathbb{R}^{m \times m},$$

we see that these matrices are orthogonal and it holds that

$$A_1 = U_1^T A V_1 = \begin{bmatrix} y^T \\ u_2^T \\ \vdots \\ u_m^T \end{bmatrix} [\lambda_1 y, Av_2, \ldots, Av_n] = \begin{bmatrix} \lambda_1 & w^T \\ 0 & B \end{bmatrix},$$

where $w \in \mathbb{R}^{(n-1)}$, $B \in \mathbb{R}^{(m-1) \times (n-1)}$. Since

$$A_1 \begin{bmatrix} \lambda_1 \\ w \end{bmatrix} = \begin{bmatrix} \lambda_1^2 + \|w\|^2 \\ Bw \end{bmatrix},$$

it follows that

$$\left\| A_1 \begin{bmatrix} \lambda_1 \\ w \end{bmatrix} \right\| \geq \lambda_1^2 + \|w\|^2,$$

or

$$\|A_1\| \geq \sqrt{\lambda_1^2 + \|w\|^2}.$$

On the other hand, an orthogonal transformation preserves the matrix norm, so

$$\|A\| = \|A_1\| \geq \sqrt{\lambda_1^2 + \|w\|^2}.$$

Therefore $w = 0$.

Now we continue inductively. Let $\lambda_2 = \|B\|$. We have

$$\lambda_2 \leq \|A_1\| = \|A\| = \lambda_1.$$

If $\lambda_2 = 0$, i.e., $B = 0$, we are finished. Assume that $\lambda_2 > 0$. Proceeding as above, we find orthogonal matrices $\tilde{U}_2 \in \mathbb{R}^{(m-1)\times(m-1)}$ and $\tilde{V}_2 \in \mathbb{R}^{(n-1)\times(n-1)}$ such that

$$\tilde{U}_2^T B \tilde{V}_2 = \begin{bmatrix} \lambda_2 & 0 \\ 0 & C \end{bmatrix}$$

for some $C \in \mathbb{R}^{(m-2)\times(n-2)}$. By defining

$$U_2 = \begin{bmatrix} 1 & 0 \\ 0 & \tilde{U}_2 \end{bmatrix}, \quad V_2 = \begin{bmatrix} 1 & 0 \\ 0 & \tilde{V}_2 \end{bmatrix},$$

we get orthogonal matrices that have the property

$$U_2^T U_1^T A V_1 V_2 = \begin{bmatrix} \lambda_1 & 0 & 0 \\ 0 & \lambda_2 & 0 \\ 0 & 0 & C \end{bmatrix}.$$

Inductively, performing the same partial diagonalization by orthogonal transformations, we finally get the claim. $\square$

Theorem A.2. *Let $A = U \Lambda V^T \in \mathbb{R}^{m\times n}$ be the singular value decomposition as above, and assume that $\lambda_1 \geq \lambda_2 \geq \cdots \lambda_p > \lambda_{p+1} = \cdots = \lambda_{\min(m,n)} = 0$. By denoting*

$$U = [u_1, \ldots, u_m], \quad V = [v_1, \ldots, v_n],$$

we have

$$\mathrm{Ker}(A) = \mathrm{sp}\{v_{p+1}, \ldots, v_n\} = \mathrm{Ran}(A^T)^\perp,$$
$$\mathrm{Ker}(A^T) = \mathrm{sp}\{u_{p+1}, \ldots, u_m\} = \mathrm{Ran}(A)^\perp.$$

Proof: By the orthogonality of U, we have $Ax = 0$ if and only if

$$\Lambda V^T x = \begin{bmatrix} \lambda_1 v_1^T x \\ \vdots \\ \lambda_p v_p^T x \\ 0 \\ \vdots \\ 0 \end{bmatrix} = 0,$$

from where the claim can be seen immediately. Furthermore, we have $A^T = V D^T U^T$, so a similar result by interchanging U and V holds for the transpose.

On the other hand, we see that $x \in \mathrm{Ker}(A)$ is equivalent to

$$(Ax)^T y = x^T A^T y = 0$$

for all $y \in \mathbb{R}^m$, i.e., x is perpendicular to the range of A^T. $\square$

In particular, if we divide U and V as

$$U = [[u_1, \ldots, u_p], [u_{p+1}, \ldots, u_m]] = [U_1, U_2],$$
$$V = [[v_1, \ldots, v_p], [v_{p+1}, \ldots, v_n]] = [V_1, V_2],$$

the orthogonal projectors $P : \mathbb{R}^n \to \operatorname{Ker}(A)$ and $\tilde{P} : \mathbb{R}^n \to \operatorname{Ker}(A^{\mathrm{T}})$ assume the simple forms
$$P = V_2 V_2, \quad \tilde{P} = U_2 U_2^{\mathrm{T}}.$$

A.2 Functional Analysis

We confine the discussion to the Hilbert spaces.

Definition A.3. *Let $A : H_1 \to H_2$ be a continuous linear operator between Hilbert spaces H_1 and H_2. The operator is said to be* compact *if the sets $\overline{A(U)} \subset H_2$ are compact for every bounded set $U \subset H_1$.*

In the following simple lemmas, we denote by $X^\perp$ the orthocomplement of the subspace $X \subset H$.

Lemma A.4. *Let $X \subset H$ be a closed subspace of the Hilbert space H. Then $(X^\perp)^\perp = X$.*

Proof: Let $x \in X$. Then for every $y \in X^\perp$, $\langle x, y \rangle = 0$, i.e., $x \in (X^\perp)^\perp$.

To prove the converse inclusion, let $x \in (X^\perp)^\perp$. Denote by $P : H \to X$ the orthogonal projection. We have $x - Px \in X^\perp$. On the other hand, $Px \in X \subset (X^\perp)^\perp$, so $x - Px \in (X^\perp)^\perp$. Therefore, $x - Px$ is orthogonal with itself and thus $x = Px \in X$. The proof is complete. $\square$

Lemma A.5. *Let $X \subset H$ be a subspace. Then $\overline{X}^\perp = X^\perp$.*

The proof is rather obvious and therefore omitted.

Lemma A.6. *Let $A : H_1 \to H_2$ be a bounded linear operator. Then*
$$\operatorname{Ker}(A^*) = \operatorname{Ran}(A)^\perp, \quad \operatorname{Ker}(A) = \operatorname{Ran}(A^*)^\perp.$$

Proof: We have $x \in \operatorname{Ker}(A*)$ if and only if
$$0 = \langle A^* x, y \rangle = \langle x, Ay \rangle$$
for all $y \in H_1$, proving the claim. $\square$

Based on these three lemmas, the proof of part (i) of Proposition 2.1 follows. Indeed, $\operatorname{Ker}(A) \subset H_1$ is closed and we have
$$H_1 = \operatorname{Ker}(A) \oplus \operatorname{Ker}(A)^\perp = \operatorname{Ker}(A) \oplus (\operatorname{Ran}(A^*)^\perp)^\perp$$

by Lemma A.6, and by Lemmas A.4 and A.5,

$$(\text{Ran}(A^*)^\perp)^\perp = (\overline{\text{Ran}(A^*)}^\perp)^\perp = \overline{\text{Ran}(A^*)}.$$

Part (ii) of Proposition 2.1 is based on the fact that the spectrum of a compact operator is discrete and accumulates possibly only at the origin. If $A : H_1 \to H_2$ is compact, then $A^*A : H_1 \to H_1$ is a compact and positive operator. Denote the positive eigenvalues by λ_n^2, and let the corresponding normalized eigenvectors be $v_n \in H_1$. Since A^*A is self-adjoint, the eigenvectors can be chosen mutually orthogonal. Defining

$$u_n = \frac{1}{\lambda_n} A v_n,$$

we observe that

$$\langle u_n, u_k \rangle = \frac{1}{\lambda_n \lambda_k} \langle v_n, A^* A v_k \rangle = \delta_{nk},$$

hence the vectors u_n are also orthonormal. The triplet (v_n, u_n, λ_n) is the singular system of the operator A.

Next, we prove the fixed point theorem 2.8 used in Section 2.4.1.

Proof of Proposition 2.8: Let $T : H \to H$ be a mapping, $S \subset H$ an invariant closed set such that $T(S) \subset S$ and let T be a contraction in S,

$$\|T(x) - T(y)\| < \kappa \|x - y\| \text{ for all } x, y \in S,$$

where $\kappa < 1$. For any $j > 1$, we have the estimate

$$\|x_{j+1} - x_j\| = \|T(x_j) - T(x_{j-1})\| < \kappa \|x_j - x_{j-1}\|,$$

and inductively

$$\|x_{j+1} - x_j\| < \kappa^{j-1} \|x_2 - x_1\|.$$

It follows that for any $n, k \in \mathbb{N}$, we have

$$\|x_{n+k} - x_n\| \leq \sum_{j=1}^{k} \|x_{n+j} - x_{n+j-1}\| < \sum_{j=1}^{k} \kappa^{n+j-2} \|x_2 - x_1\|$$

$$\leq \frac{\kappa^{n-1}}{1 - \kappa} \|x_2 - x_1\|$$

by the geometric series sum formula. Therefore, (x_j) is a Cauchy sequence, and thus is convergent and the limit is in S since S is closed. □

We finish this section by giving without proof a version of the *Riesz representation theorem*.

Theorem A.7. *Let $B : H \times H \to \mathbb{R}$ be a bilinear quadratic form satisfying*

$$|B(x,y)| \leq C\|x\|\,\|y\| \quad \text{for all } x, y \in H,$$

$$B(x,x) \geq c\|x\|^2 \quad \text{for all } x \in H$$

for some constants $0 < c \leq C < \infty$. Then there is a unique linear bounded operator $T : H \to H$ such that

$$B(x,y) = \langle Tx, y \rangle \quad \text{for all } y \in H.$$

The variant of this for Hilbert spaces with complex coefficients is known as the *Lax–Milgram lemma*. See [142] for a proof.

A.3 Sobolev Spaces

In this section, we review the central definitions and results concerning Sobolev spaces.

For square integrable functions $f : \mathbb{R}^n \to \mathbb{R}$, define the L^2-norm over $\mathbb{R}^n$ as

$$\|f\| = \left(\int_{\mathbb{R}^n} |f(x)|^2 dx \right)^{1/2} < \infty. \tag{A.1}$$

The linear space of square integrable functions is denoted by $L^2 = L^2(\mathbb{R}^n)$. The norm (A.1) defines an inner product,

$$\langle f, q \rangle = \int_{\mathbb{R}^n} \overline{f(x)} g(x) dx,$$

that defines a Hilbert space structure in L^2.

The Fourier transform for integrable functions of $\mathbb{R}^n$ is defined as

$$(\mathcal{F} f)(\xi) = \widehat{f}(\xi) = \int_{\mathbb{R}^n} e^{-\mathrm{i} \langle \xi, x \rangle} f(x) dx.$$

The Fourier transform can be extended to L^2, and it defines a bijective mapping $\mathcal{F} : L^2 \to L^2$, see, e.g., [108]. Its inverse transformation is then

$$(\mathcal{F}^{-1} \widehat{f})(x) = f(x) = \left(\frac{1}{2\pi} \right)^n \int_{\mathbb{R}^n} e^{\mathrm{i} \langle \xi, x \rangle} \widehat{f}(\xi) d\xi.$$

In L^2, Fourier transform is an isometry up to a constant, and it satisfies the identities

$$\|\widehat{f}\| = (2\pi)^{n/2} \|f\| \quad \text{(Parseval)},$$
$$\langle \widehat{f}, \widehat{g} \rangle = (2\pi)^n \langle f, g \rangle \quad \text{(Plancherel)}.$$

Defining the convolution of functions as

$$f * g(x) = \int f(x - y) g(y) dy,$$

provided that the integral converges, we have

$$\mathcal{F}(f * g)(\xi) = \widehat{f}(\xi) \widehat{g}(\xi).$$

Let α be a multiindex, $\alpha = (\alpha_1, \alpha_2, \ldots, \alpha_n) \in \mathbb{N}^n$. Define the length of the multiindex, $|\alpha|$, as $|\alpha| = \alpha_1 + \cdots + \alpha_n$. We define

$$D^\alpha f(x) = (-\mathrm{i})^{|\alpha|} \frac{\partial^{\alpha_1}}{\partial x_1^{\alpha_1}} \frac{\partial^{\alpha_2}}{\partial x_2^{\alpha_2}} \cdots \frac{\partial^{\alpha_n}}{\partial x_n^{\alpha_n}} f(x) = (-\mathrm{i})^{|\alpha|} \frac{\partial^{|\alpha|}}{\partial x_1^{\alpha_1} \cdots \partial x_n^{\alpha_n}} f(x).$$

Similarly, for $x \in \mathbb{R}^n$, we define

$$x^\alpha = x_1^{\alpha_1} x_2^{\alpha_2} \cdots x_n^{\alpha_n}.$$

The Fourier transform has the property

$$\mathcal{F}(D^\alpha f)(\xi) = \xi^\alpha \widehat{f}(\xi). \tag{A.2}$$

The classical Sobolev spaces W^k over $\mathbb{R}^n$, $k \geq 0$, are defined as linear spaces of such functions $f \in L^2$ that

$$D^\alpha f \in L^2 \text{ for all multiindices } \alpha,\ |\alpha| \leq k.$$

This space is equipped with the norm

$$\|f\|_{W^k} = \left(\sum_{|\alpha| \leq k} \int_{\mathbb{R}^n} |D^\alpha f(x)|^2 dx \right)^{1/2}. \tag{A.3}$$

By using identity (A.2), the definition of Sobolev spaces can be extended to noninteger smoothness indices. For arbitrary $s \in \mathbb{R}$, we define the space $H^s = H^s(\mathbb{R}^n)$ as a linear space of those functions[1] that satisfy

$$\|f\|_{H^s} = \left(\int_{\mathbb{R}^n} |\widehat{f}(\xi)|^2 (1+|\xi|^2)^s d\xi \right)^{1/2} < \infty. \tag{A.4}$$

It turns out that $H^k = W^k$ for $k \in \mathbb{N}$, and the norms (A.3) and (A.4) are equivalent. Also, from Parseval's identity we see that $H^0 = L^2$.

Let $\Omega \subset \mathbb{R}^n$ be an open domain and assume, for simplicity, that the boundary of Ω is smooth. The Sobolev space $H^s(\Omega)$ of functions defined in Ω is

$$H^s(\Omega)\{f \mid f = g|_\Omega,\ g \in H^s(\mathbb{R}^n)\}.$$

The norm in $H^s(\Omega)$ is the infimum of the norms of all the functions g in $H^s(\mathbb{R}^n)$ whose restriction to Ω is f.

Evidently, $H^{s_1}(\Omega) \subset H^{s_2}$ for $s_1 > s_2$. The Sobolev spaces over bounded domains have the following important compact embedding property.

Theorem A.8. *Let $\Omega \subset \mathbb{R}^n$ be a smooth, bounded open set and let $s_1 > s_2$. Then the embedding*

$$I : H^{s_1}(\Omega) \to H^{s_2},\quad f \mapsto f,$$

is compact, i.e., every bounded set in $H^{s_1}(\Omega)$ is relatively compact in $H^{s_2}(\Omega)$.

---

[1] The term "function" refers here to tempered distributions. Also, the derivatives should be interpreted in the distributional sense. See [108] for exact formulation.

In particular, every bounded sequence of H^{s_1}-functions contains a subsequence that converges in H^{s_2}. For a proof of this result, see, e.g., [2].

If $\Omega \subset \mathbb{R}^n$ has a smooth boundary, we may define Sobolev spaces of functions defined on the boundary $\partial\Omega$. If $U \subset \partial\Omega$ is an open set and $\psi : U \to \mathbb{R}^{n-1}$ is a smooth diffeomorphism, a compactly supported function $f : U \to \mathbb{C}$ is in the Sobolev space $H^s(\partial\Omega)$ if

$$f \circ \psi^{-1} \in H^s(\mathbb{R}^{n-1}).$$

More generally, by using partitions of unity on $\partial\Omega$, we may define the condition $f \in H^s(\partial\Omega)$ for functions whose supports are not restricted to sets that are diffeomorphic with $\mathbb{R}^{n-1}$. The details are omitted here.

We give without proof the following *trace theorem* of Sobolev spaces. The proof can be found, e.g., in [49].

Theorem A.9. *The trace mapping $f \mapsto f\big|_{\partial\Omega}$, defined for continuous functions in $\mathbb{R}^n$, extends to a continuous mapping $H^s(\Omega) \to H^{s-1/2}(\partial\Omega)$, $s > 1/2$. Moreover, the mapping has a continuous right inverse.*

B
Appendix 2: Basics on Probability

In the statistical inversion theory, the fundamental notions are *random variables* and *probability distributions*. To fix the notations, a brief summary of the basic concepts is given in this appendix. In this summary, we avoid getting deep into the axiomatic probability theory. The reader is recommended to consult standard textbooks of probability theory for proofs, e.g., [24], [32], [43].

B.1 Basic Concepts

Let Ω be an abstract space and $\mathfrak{S}$ a collection of its subsets. We say that $\mathfrak{S}$ is a *σ-algebra* of subsets of Ω if the following conditions are satisfied:

1. $\Omega \in \mathfrak{S}$.
2. If $A \in \mathfrak{S}$, then $A^c = \Omega \setminus A \in \mathfrak{S}$.
3. If $A_i \in \mathfrak{S}, i \in \mathbb{N}$, then $\bigcup_{i=1}^{\infty} A_i \in \mathfrak{S}$.

Let $\mathfrak{S}$ be a σ-algebra over a space Ω. A mapping $\mu : \mathfrak{S} \to \mathbb{R}$ is called a *measure* if the following conditions are met:

1. $\mu(A) \geq 0$ for all $A \in \mathfrak{S}$.
2. $\mu(\emptyset) = 0$.
3. If the sets $A_i \in \mathfrak{S}$ are disjoint, $i \in \mathbb{N}$, (i.e., $A_i \cap A_j = \emptyset$, whenever $i \neq j$), then
$$\mu\left(\bigcup_{i=1}^{\infty} A_i\right) = \sum_{i=1}^{\infty} \mu(A_i).$$

The last condition is often expressed by saying that μ is *σ-additive*.

A σ-algebra is called *complete* if it contains all sets of measure zero, i.e., if $A \subset B$, and $B \in \mathfrak{S}$ with $\mu(B) = 0$, then $A \in \mathfrak{S}$. It is always possible to complete a σ-algebra by adding sets of measure zero to it.

A measure μ is called *finite* if $\mu(\Omega) < \infty$. The measure is called *σ-finite*, if there is a sequence of subsets $A_n \in \mathfrak{S}$, $n \in \mathbb{N}$, with the properties

$$\Omega = \bigcup_{n=1}^{\infty} A_n, \quad \mu(A_n) < \infty \text{ for all } n.$$

Of particular importance are the measures that satisfy

$$\mu(\Omega) = 1.$$

Such a measure is called a *probability measure*.

Let $\mathfrak{S}$ be a σ-algebra over Ω and P a probability measure. The triple $(\Omega, \mathfrak{S}, P)$ is called the *probability space*. The abstract space Ω is called the *sample space*, while $\mathfrak{S}$ is the set of *events*, and $P(A)$ is the *probability* of the event $A \in \mathfrak{S}$.

A particular example of a σ-algebra is the *Borel σ-algebra* $\mathfrak{B} = \mathfrak{B}(\mathbb{R}^n)$ over $\mathbb{R}^n$, defined as the smallest σ-algebra containing the open sets of $\mathbb{R}^n$. We say that the Borel σ-algebra is *generated* by the open sets of $\mathbb{R}^n$. In this book, $\mathbb{R}^n$ is always equipped with the Borel σ-algebra. The elements of the Borel σ-algebra are called *Borel sets*.

As another example of a σ-algebra, let I be a discrete set, i.e., a set of either finite or enumerable set of elements. A natural σ-algebra is the set $\mathfrak{S} = \{A \mid A \subset I\}$, i.e., the power set of I containing all subsets of I.

An important example of a σ-finite measure defined on the Borel σ-algebra of $\mathbb{R}^n$ is the Lebesgue measure m, assigning to a set its volume. For

$$Q = [a_1, b_1] \times \cdots \times [a_n, b_n], \quad a_j \leq b_j,$$

we have

$$m(Q) = \int_Q dx = (b_1 - a_1) \cdots (b_n - a_n).$$

In this book, we confine ourselves to consider only random variables taking values either in $\mathbb{R}^n$ or in discrete subsets. Therefore, the particular σ-algebras above are essentially the only ones needed. It is possible to develop the theory in more general settings, but that would require more measure-theoretic considerations which go beyond the scope of this book.

A central concept in probability theory is independency. Let $(\Omega, \mathfrak{S}, P)$ be a probability space. The events $A, B \in \mathfrak{S}$ are *independent* if

$$P(A \cap B) = P(A)P(B).$$

More generally, a family $\{A_t \mid i \in T\} \subset \mathfrak{S}$ with T an arbitrary index set is independent if

$$P\left(\bigcap_{i=1}^m A_{t_i}\right) = \prod_{i=1}^m P(A_{t_i})$$

B.1 Basic Concepts

for any finite sampling of the indices $t_i \in T$.

Given a probability space $(\Omega, \mathfrak{S}, P)$, a *random variable* with values in $\mathbb{R}^n$ is a *measurable mapping*
$$X : \Omega \to \mathbb{R}^n,$$
i.e., $X^{-1}(B) \in \mathfrak{S}$ for every open set $B \subset \mathbb{R}^n$. A random variable with values in a discrete set I is defined similarly.

We adopt in this section, whenever possible, the common notational convention that random variables are denoted by capital letters and their realizations by lowercase letters. Hence, we may write $X(\omega) = x$, $\omega \in \Omega$. With a slight abuse of notations, we indicate the value set of a random variable $X : \Omega \to \mathbb{R}^n$ by using the shorthand notation $X \in \mathbb{R}^n$.

A random variable $X \in \mathbb{R}^n$ generates a probability measure μ_X in $\mathbb{R}^n$ equipped with its Borel σ-algebra through the formula
$$\mu_X(B) = P(X^{-1}(B)), \quad B \in \mathfrak{B}.$$
This measure is called the *probability distribution* of X.

The *distribution function* $F : \mathbb{R}^n \to [0,1]$ of a random variable $X = (X_1, \ldots, X_n) \in \mathbb{R}^n$ is defined as
$$F(x) = P(X_1 \leq x_1, X_2 \leq x_2, \ldots, X_n \leq x_n), \quad x = (x_1, x_2, \ldots, x_n).$$

The *expectation* of a random variable is defined as
$$E\{X\} = \int_\Omega X(\omega) dP(\omega) = \int_{\mathbb{R}^n} x d\mu_X(x),$$
provided that the integrals converge. Similarly, the *correlation* and *covariance* matrices of the variables are given as
$$\mathrm{corr}(X) = E\{XX^\mathrm{T}\} = \int_\Omega X(\omega) X(\omega)^\mathrm{T} dP(\omega)$$
$$= \int_{\mathbb{R}^n} xx^\mathrm{T} d\mu_X(x) \in \mathbb{R}^{n \times n}$$
and
$$\mathrm{cov}(X) = \mathrm{corr}(X - E\{X\}) = E\{(X - E\{X\})(X - E\{X\})^\mathrm{T}\}$$
$$= \int_\Omega (X(\omega) - E\{X\})(X(\omega) - E\{X\})^\mathrm{T} dP(\omega)$$
$$= \int_{\mathbb{R}^n} (x - E\{X\})(x - E\{X\})^\mathrm{T} d\mu_X(x)$$
$$= \mathrm{corr}(X) - E\{X\}(E\{X\})^\mathrm{T},$$
assuming again that they exist.

Let $\{X_n\}$ be a sequence of random variables, and let

$$S_n = \frac{1}{n}\sum_{j=1}^{n} X_j. \tag{B.1}$$

The following theorem is a version of the *central limit theorem*.

Theorem B.1. *Assume that the random variables X_n are independent and equally distributed, and*

$$E\{X_n\} = \mu, \quad \text{cov}(X_n) = \sigma^2.$$

Then the random variables S_n in (B.1) are asymptotically normally distributed in the following sense:

$$\lim_{n\to\infty} P\left\{\frac{S_n - n\mu}{\sigma\sqrt{n}} < x\right\} = \frac{1}{\sqrt{2\pi}}\int_{-\infty}^{x} e^{-t^2/2} dt.$$

The random variable is called *absolutely continuous* if its probability distribution is absolutely continuous with respect to the Lebesgue measure over $\mathbb{R}^n$, i.e.,

$$m(B) = 0 \Rightarrow \mu_X(B) = 0,$$

where $B \in \mathfrak{B}$. A random variable and its probability distribution are *singular* if they are not absolutely continuous with respect to the Lebesgue measure.

The probability distributions of continuous random variables have an important property that follows from the Radon–Nikodym theorem, stated here without a proof.

Proposition B.2. *Let μ_1 and μ_2 be two σ-finite measures in $\mathbb{R}^n$ defined over the Borel σ-algebra such that μ_1 is absolutely continuous with respect to μ_2, denoted $\mu_1 \prec \mu_2$. Then there exists a measurable function $\pi : \mathbb{R}^n \to \mathbb{R}$ such that*

$$\mu_1(B) = \int_B \pi(x) d\mu_2(x)$$

for all $B \in \mathfrak{B}$. The function π is called the Radon–Nikodym derivative *of μ_1 with respect to μ_2, denoted by*

$$\pi = \frac{d\mu_1}{d\mu_2}.$$

It follows from the previous proposition that an absolutely continuous random variable X defines a *probability density* π_X by

$$P(X^{-1}(B)) = \mu_X(B) = \int_B \pi_X(x) dx, \quad B \in \mathfrak{B}.$$

Let $X_1 \in \mathbb{R}^n$ and $X_2 \in \mathbb{R}^m$ be two random variables defined over the same probability space Ω. The *joint probability distribution* is defined as

$$\mu_{X_1 X_2} : \mathfrak{B}(\mathbb{R}^n) \times \mathfrak{B}(\mathbb{R}^m) \to \mathbb{R}_+, \quad (B_1, B_2) \mapsto \mathrm{P}(X_1^{-1}(B_1) \cap X_2^{-1}(B_2)).$$

In other words, the joint probability distribution is simply the probability distribution of the product random variable

$$X_1 \times X_2 : \Omega \to \mathbb{R}^n \times \mathbb{R}^m, \quad \omega \mapsto (X_1(\omega), X_2(\omega)).$$

This definition can be extended to any number of random variables.

We say that the random variables X_1 and X_2 are *independent*, if

$$\mu_{X_1 X_2}(B_1, B_2) = \mu_{X_1}(B_1)\mu_{X_2}(B_2),$$

or, alternatively, if the events $X_j^{-1}(B_j) \in \mathfrak{S}$ are independent for all Borel sets B_j. If the random variables are absolutely continuous, the independence can be expressed in terms of the probability densities as

$$\pi_{X_1 X_2}(x_1, x_2) = \pi_{X_1}(x_1)\pi_{X_2}(x_2)$$

for almost all $(x_1, x_2) \in \mathbb{R}^n \times \mathbb{R}^m$.

Based on the Radon–Nikodym theorem stated in Proposition B.2, it would be possible to define the conditional probability distributions and densities of random variables. However, we postpone this general definition and consider first a more restricted class of distributions to keep the discussion more tractable.

B.2 Conditional Probabilities

A central role in statistical inversion theory is played by conditional probabilities. Therefore, we discuss this issue in slightly more detail.

The basic concept is the conditional probability of events. Hence, let $A, B \in \mathfrak{S}$ and $\mathrm{P}(B) > 0$. The *conditional probability* of A with the condition B is defined as

$$\mathrm{P}(A \mid B) = \frac{\mathrm{P}(A \cap B)}{\mathrm{P}(B)}.$$

It holds that

$$\mathrm{P}(A) = \mathrm{P}(A \mid B)\mathrm{P}(B) + \mathrm{P}(A \mid B^c)\mathrm{P}(B^c), \quad B^c = \Omega \setminus B.$$

More generally, if $\Omega = \cup_{i \in I} B_i$ with I enumerable, we have

$$\mathrm{P}(A) = \sum_{i \in I} \mathrm{P}(A \mid B_i)\mathrm{P}(B_i).$$

We want to extend the notion of conditional probability to random variables. The general definition can be done via the Radon–Nikodym theorem stated in Proposition B.2. However, to keep the ideas more transparent and better tractable, we limit the discussion first to the simple case of absolutely continuous variables with continuous probability densities. The general results are then briefly stated without detailed proofs.

Consider first two random variables $X_i : \Omega \to \mathbb{R}^{n_i}$, $i = 1, 2$. We adopt the following shorthand notation: If $B_i \subset \mathbb{R}^{n_i}$ are two Borel sets, we denote the joint probability of $B_1 \times B_2$ as

$$\mu_{X_1 X_2}(B_1, B_2) = \mu(B_1, B_2) = P(X_1 \in B_1, X_2 \in B_2),$$

i.e., we suppress the subindex from the measure and use instead subindices for the sets to remind which set refers to which random variable. Similarly, if (X_1, X_2) is absolutely continuous, we write

$$\mu(B_1, B_2) = \int_{B_1 \times B_2} \pi(x_1, x_2) dx_1 dx_2.$$

With these notations, we define first the *marginal probability distribution* of X_1 as

$$\mu(B_1) = \mu(B_1, \mathbb{R}^{n_2}) = P(X_1 \in B_1),$$

i.e., the probability of $X_1 \in B_1$ regardless of the value of X_2.

Assume now that (X_1, X_2) is absolutely continuous. Then we have

$$\mu(B_1) = \int_{B_1 \times \mathbb{R}^{n_2}} \pi(x_1, x_2) dx_1 dx_2 = \int_{B_1} \pi(x_1) dx_1,$$

where the *marginal probability density* is given as

$$\pi(B_1) = \int_{\mathbb{R}^{n_2}} \pi(x_1, x_2) dx_2.$$

Analogously to the definition of conditional probabilities of events, we give the following definition.

Definition B.3. *Assume that $B_i \subset \mathbb{R}^{n_i}$, $i = 1, 2$, are Borel sets with B_2 having positive marginal measure, i.e., $\mu(B_2) = P(X_2^{-1}(B_2)) > 0$. We define the* conditional measure *of B_1 conditioned on B_2 by*

$$\mu(B_1 \mid B_2) = \frac{\mu(B_1, B_2)}{\mu(B_2)}.$$

In particular, if the random variables are absolutely continuous, we have

$$\pi(B_1 \mid B_2) = \frac{1}{\pi(B_2)} \int_{B_1 \times B_2} \pi(x_1, x_2) dx_1 dx_2,$$

with

$$\mu(B_2) = \int_{B_2} \pi(x_2) dx_2 = \int_{\mathbb{R}^n \times B_2} \pi(x_1, x_2) dx_1 dx_2 > 0.$$

Notice that with a fixed B_2, the mapping $B_1 \mapsto \pi(B_1 \mid B_2)$ is a probability measure.

Remark: If the random variables X_1 and X_2 are independent, then

$$\mu(B_1 \mid B_2) = \mu(B_1), \quad \mu(B_2 \mid B_1) = \mu(B_2)$$

by definition, i.e., any restriction of one variable does not affect the distribution of the other.

Our goal is now to define the conditional measure conditioned on a fixed value, i.e., $X_2 = x_2$. The problem is that in general, we may have $P(X_2 = x_2) = 0$ and the conditioning as defined above fails. Thus, we have to go through a limiting process.

Lemma B.4. *Assume that the random variables X_1 and X_2 are absolutely continuous with continuous densities and $x_2 \in \mathbb{R}^{n_2}$ is a point such that*

$$\pi(x_2) = \int_{\mathbb{R}^{n_1}} \pi(x_1, x_2) dx_1 > 0.$$

Further, let $(B_2^{(j)})_{1 \leq j < \infty}$ be a decreasing nested sequence of intervals in $\mathbb{R}^k$ such that $B_2^{(j)} \downarrow \{x_2\}$, i.e., $B_2^{(j+1)} \subset B_2^{(j)}$ and $\bigcap B_2^{(j)} = \{x_2\}$. Then the limit

$$\lim_{j \to \infty} \mu(B_1 \mid B_2^{(j)}) = \mu(B_1 \mid x_2)$$

exists and it can be evaluated as the integral

$$\mu(B_1 \mid x_2) = \frac{1}{\pi(x_2)} \int_{B_1} \pi(x_1, x_2) dx_1.$$

Proof: Let us denote again by $m(B)$ the Lebesgue measure of a measurable set B. We write

$$\mu(B_1 \mid B_2^{(j)}) = \frac{1}{\mu(B_2^{(j)})} \int_{B_1 \times B_2^{(j)}} \pi(x_1, x_2') dx_1 dx_2'$$

$$= \left(\frac{1}{m(B_2^{(j)})} \int_{B_2^{(j)}} \pi(x_2') dx_2' \right)^{-1}$$

$$\frac{1}{m(B_2^{(j)})} \int_{B_2^{(j)}} \left(\int_{B_1} \pi(x_1, x_2') dx_1 \right) dx_2'.$$

By the assumption, the functions $x_2' \mapsto \pi(x_2')$ and $x_2' \mapsto \int_{B_1} \pi(x_1, x_2') dx_1$ are continuous, so

$$\lim_{j \to \infty} \frac{1}{m(B_2^{(j)})} \int_{B_2^{(j)}} \pi(x_2') dx_2' = \pi(x_2) > 0$$

and

$$\lim_{j\to\infty} \frac{1}{m(B_2^{(j)})} \int_{B_2^{(j)}} \left(\int_{B_1} \pi(x_1, x_2') dx_2 \right) dx_2' = \int_{B_1} \pi(x_1, x_2) dx_1,$$

and the claim follows. □

By the above lemma, we can extend the concept of a conditional measure to cover the case $B_2 = \{x_2\}$. Hence, we say that if $\pi(x_2) > 0$, the conditional probability of $B_1 \in \mathbb{R}^n$ conditioned on $X_2 = x_2$ is the limit

$$\mu(B_1 \mid x_2) = \int_{B_1} \frac{\pi(x_1, x_2)}{\pi(x_2)} dx_1.$$

From the above lemma, we observe that the conditional probability measure $\mu(B_1 \mid x_2)$ is defined by the probability density

$$\pi(x_1 \mid x_2) = \frac{\pi(x_1, x_2)}{\pi(x_2)}, \quad \pi(x_2) > 0.$$

As a corollary, we obtain the useful formula for the joint probability distribution.

Corollary B.5. *The joint probability density for (X_1, X_2) is*

$$\pi(x_1, x_2) = \pi(x_1 \mid x_2)\pi(x_2) = \pi(x_2 \mid x_1)\pi(x_1).$$

This formula is essentially the *Bayes formula* that constitutes the corner stone of the Bayesian interpretation of inverse problems.

Finally, we briefly indicate how the conditional probability is defined in the more general setting.

Let $B_1 \subset \mathbb{R}^{n_1}$ be a fixed measurable set with positive measure $\mu(B_1) = \mathrm{P}(X_1^{-1}(B_1))$. Consider the mapping

$$\mathfrak{B}(\mathbb{R}^{n_2}) \to \mathbb{R}_+, \quad B_2 \mapsto \mu(B_1, B_2).$$

This mapping is clearly a finite measure, and moreover

$$\mu(B_1, B_2) \leq \mu(B_2),$$

i.e., it is an absolutely continuous measure with respect to the marginal measure $B_2 \mapsto \mu(B_2)$. The Radon–Nikodym theorem asserts that there exists a measurable function $x_2 \mapsto \rho(B_1, x_2)$ such that

$$\mu(B_1, B_2) = \int_{B_2} \rho(B_1, x_2) d\mu(x_2).$$

It would be tempting to define the conditional probability $\mu(B_1 \mid x_2) = \rho(B_1, x_2)$. Basically, this is correct but not without further considerations. The problem here is that the mapping $x_2 \mapsto \rho(B_1, x_2)$ is defined only almost everywhere, i.e., there may be a set of $d\mu(x_2)$-measure zero where it is not defined. On the other hand, this set may depend on B_1. Since the number of different subsets B_1 is not countable, the union of these exceptional sets could well have a positive measure, in the worst case up to one. However, it is possible with careful measure-theoretic considerations to define a function

$$\mathfrak{B}(\mathbb{R}^{n_1}) \times \mathbb{R}^{n_2} \to \mathbb{R}_+, \quad (B_1, x_2) \mapsto \mu(B_1 \mid x_2),$$

such that

1. the mapping $B_1 \mapsto \mu(B_1 \mid x_2)$ is a probability measure for all x_2 except possibly in a set of $d\mu(x_2)$-measure zero;
2. the mapping $x_2 \mapsto \mu(B_1 \mid x_2)$ is measurable; and
3. the formula $\mu(B_1, B_2) = \int_{B_2} \mu(B_1 \mid x_2) \mu(dx_2)$ holds.

We shall not go into details here but refer to the literature.

References

1. Å. Björck. *Numerical Methods for Least Squares Problems*. SIAM, 1996.
2. R.A. Adams. *Sobolev Spaces*. Academic Press, 1975.
3. K.E. Andersen, S.P. Brooks and M.B. Hansen. Bayesian inversion of geoelectrical resistivity data. *J. R. Stat. Soc. Ser. B, Stat. Methodol.*, 65:619–642, 2003.
4. B.D.O. Anderson and J.B. Moore. *Optimal Filtering*. Prentice-Hall, 1979.
5. S. Arridge and W.R.B. Lionheart. Nonuniqueness in diffusion-based optical tomography. *Optics Lett.*, 1998.
6. S.R. Arridge. Optical tomography in medical imaging. *Inv. Probl.*, 15:R41–R93, 1999.
7. K. Astala and L. Päivärinta. Calderón's inverse conductivity problem in the plain. Preprint, 2004.
8. A.C. Atkinson and A.N. Donev. *Optimum Experimental Design*. Oxford University Press, 1992.
9. S. Baillet, J.C. Mosher, and R.M. Leahy. Electromagnetic brain mapping. *IEEE Signal Proc. Mag.*, pp. 14–30, 2001.
10. D. Baroudi, J.P. Kaipio and E. Somersalo. Dynamical electric wire tomography: Time series approach. *Inv. Probl.*, 14:799–813, 1998.
11. J. Barzilai and J.M. Borwein. Two-point step-size gradient methods. *IMA J. Numer. Anal.*, 8:141–148, 1988.
12. M. Bertero and P. Boccacci. *Introduction to Inverse Problems in Imaging*. Institute of Physics, 1998.
13. J. Besag. Spatial interaction and the statistical analysis of lattice systems. *J. Royal Statist. Soc.*, 36:192–236, 1974.
14. J. Besag. The statistical analysis of dirty pictures. *Proc. Roy. Stat. Soc.*, 48:259–302, 1986.
15. G.E.P. Box and G.C. Tiao. *Bayesian Inference in Statistical Analysis*. Wiley, 1992 (1973).
16. S.C. Brenner and L.R. Scott. *The Mathematical Theory of Finite-element Methods*. Springer-Verlag, 1994.
17. P.J. Brockwell and R.A. Davis. *Time Series: Theory and Methods*. Springer-Verlag, 1991.
18. A.P. Calderón. On an inverse boundary value problem. In W.H. Meyer and M.A. Raupp, editors, *Seminar on Numerical Analysis and Its Applications to Continuum Physics*, pp. 65–73, Rio de Janeiro, 1980. Brazilian Math. Society.

19. D. Calvetti, B. Lewis and L. Reichel. On the regularizing properties of the GMRES method. *Numer. Math*, 91:605–625, 2002.
20. D. Calvetti and L. Reichel. Tikhonov regularization of large linear problems. *BIT*, 43:263–283, 2003.
21. J. Carpenter, P. Clifford and P. Fearnhead. An improved particle filter for nonlinear problems. *IEEE Proc. Radar Son. Nav.*, 146:2–7, 1999.
22. M. Cheney and D. Isaacson. Distinguishability in impedance imaging. *IEEE Trans. Biomed. Eng.*, 39:852–860, 1992.
23. K.-S. Cheng, D. Isaacson, J.C. Newell and D.G. Gisser. Electrode models for electric current computed tomography. *IEEE Trans. Biomed. Eng.*, 3:918–924, 1989.
24. Y.S. Chow and H. Teicher. *Probability Theory: Independence, Interchangeability, Martingales*. Springer-Verlag, 1978.
25. H. Cramer. *Mathematical Methods in Statistics*. Princeton University Press, 1946.
26. R.F. Curtain and H. Zwart. *An Introduction to Infinite-dimensional Linear Systems Theory*. Springer-Verlag, 1995.
27. R. Dautray and J.-L. Lions. *Mathematical Analysis and Numerical Methods for Science and Technology*, volume 6. Springer-Verlag, 1993.
28. H. Dehgani, S.R. Arridge, M. Schweiger and D.T. Delpy. Optical tomography in the presence of void regions. *J. Opt. Soc. Am.*, 17:1659–1670, 2000.
29. R. Van der Merwe, A. Doucet, N. de Freitas and E. Wan. The unscented particle filter. http://www.cs.ubc.ca/ nando/papers/upfnips.ps, 2000.
30. D.C. Dobson and F. Santosa. Recovery of blocky images from noisy and blurred data. *SIAM J. Appl. Math.*, 56(4):1181–1198, 1996.
31. D.L. Donoho, I.M. Johnstone, J.C. Hoch and A.S. Stern. Maximum entropy and the nearly black object. *J. Roy. Statist. Ser. B*, 54:41–81, 1992.
32. J.L. Doob. *Stochastic Processes* (Reprint of the 1953 original). Wiley-Interscience, 1990.
33. A. Doucet, J.F.G. de Freitas and N.J. Gordon. *Sequential Monte Carlo Methods in Practice*. Springer-Verlag, 2000.
34. R.J. Elliot, L. Aggoun and J.B. Moore. *Hidden Markov Models. Estimation and Control*. Springer-Verlag, 1995.
35. H.W. Engl, M. Hanke and A. Neubauer. *Regularization of Inverse Problems*. Kluwer, 1996.
36. A.V. Fiacco and G.P. McCormick. *Nonlinear Programming*. SIAM, 1990.
37. M. Fink. Time-reversal of ultrasonic fields–part i: Basic principles. *IEEE Trans. Ultrason. Ferroelectr. Freq. Control*, 39:555–567, 1992.
38. C. Fox and G. Nicholls. Sampling conductivity images via MCMC. In *Proc. Leeds Annual Stat. Research Workshop 1997*, pp. 91–100, 1997.
39. M. Heath G.H. Golub and G. Wahba. Generalized cross-validation as a method for choosing a good ridge parameter. *Technometrics*, 21:215–223, 1979.
40. A.E. Gelfand and A.F.M. Smith. Sampling-based approaches to calculating marginal densities. *J. Amer. Statist. Assoc.*, 85:398–409, 1990.
41. S. Geman and D. Geman. Stochastic relaxation, Gibbs distributions and the Bayesian restoration of images. *IEEE Trans. Pattern Anal. Mach. Intell.*, 6:721–741, 1984.
42. D.B. Geselowitz. On the magnetic field generated outside an inhomogenous volume conductor by internal current sources. *IEEE Trans. Magn.*, 6:346–347, 1970.

References

43. I.I. Gikhman and A.V. Skorokhod. *Theory of Stochastic Processes I.* Springer-Verlag, 1980.
44. W.R. Gilks, S. Richardson and D.J. Spiegelhalter. *Markov Chain Monte Carlo in Practice.* Chapmann & Hall, 1996.
45. E. Giusti. *Minimal Surfaces and Functions of Bounded Variation.* Birkhauser, 1984.
46. S.J. Godsill, A. Doucet and M. West. Maximum a posteriori sequence estimation using Monte Carlo particle filters. *Ann. I. Stat. Math.*, 53:82–96, 2001.
47. G.H. Golub and C.F. van Loan. *Matrix Computations.* The Johns Hopkins University Press, 1989.
48. P.J. Green. Reversible jump Markov chain Monte Carlo computation and Bayesian model determination. *Biometrika*, 82:711–732, 1995.
49. P. Grisvard. *Elliptic Problems in Nonsmooth Domains.* Pitman Advanced Publishing, 1985.
50. C.W. Groetsch. *Inverse Problems in the Mathematical Sciences.* Vieweg, 1993.
51. M. Hämäläinen, H. Haario and M.S. Lehtinen. Inferences about sources of neuromagnetic fields using Bayesian parameter estimation. Technical Report Report TKK-F-A620, Helsinki University of Technology, 1987.
52. M. Hämäläinen, R. Hari, R.J. Ilmoniemi, J. Knuutila and O.V. Lounasmaa. Magnetoencephalography: theory, instrumentation, and applications to noninvasive studies of the working human brain. *Rev. Mod. Phys.*, 65:413–487, 93.
53. M. Hämäläinen and R.J. Ilmoniemi. Interpreting magnetic fields of the brain: Minimum norm estimates. *Med. Biol. Eng. Comput.*, 32:35–42, 1994.
54. M. Hanke. *Conjugate Gradient -Type Methods for Ill-posed Problems.* Longman, 1995.
55. P.C. Hansen and D.P. O'Learly. The use of the L-curve in the regularization of discrete ill-posed problems. *SIAM J. Comput.*, 14:1487–1503, 1993.
56. P.C. Hansen. *Rank-deficient and Discrete Ill-posed Problems: Numerical Aspects of Linear Inversion.* SIAM, 1998.
57. W. K. Hastings. Monte Carlo sampling methods using Markov chains and their applications. *Biometrika*, 57:97–109, 1970.
58. J. Heino and E. Somersalo. Modelling error approach for optical anisotropies for solving the inverse problem in optical tomography. Preprint (2004).
59. J. Heino and E. Somersalo. Estimation of optical absorption in anisotropic background. *Inverse Problems*, 18:559–573, 2002.
60. G.M. Henkin and R.G. Novikov. A multidimensional inverse problem in quantum and acoustic scattering. *Inverse Problems*, 4:103–121, 1988.
61. P.J. Huber. *Robust Statistical Procedures.* SIAM, 1977.
62. N. Hyvönen. Analysis of optical tomography with nonscattering regions. *Proc. Edinburgh Math. Soc*, 45:257–276, 2002.
63. D. Isaacson. Distinguishability of conductivities by electric current computed tomography. *IEEE Trans. Med. Imaging*, 5:91–95, 1986.
64. J.P. Kaipio, V. Kolehmainen, E. Somersalo and M. Vauhkonen. Statistical inversion and Monte Carlo sampling methods in electrical impedance tomography. *Inv. Probl.*, 16:1487–1522, 2000.
65. J.P. Kaipio and E. Somersalo. Estimating anomalies from indirect observations. *J. Comput. Phys.*, 181:398–406, 2002.
66. J.P. Kaipio, V. Kolehmainen, M. Vauhkonen and E. Somersalo. Inverse problems with structural prior information. *Inv. Probl.*, 15:713–729, 1999.

67. J.P. Kaipio, A. Seppänen, E. Somersalo and H. Haario. Posterior covariance related optimal current patterns in electrical impedance tomography. *Inv. Probl.*, 20:919–936, 2004.
68. J.P. Kaipio and E. Somersalo. Nonstationary inverse problems and state estimation. *J. Inv. Ill-Posed Probl.*, 7:273–282, 1999.
69. R.E. Kalman and R.S. Bucy. New results in linear filtering and prediction theory. *Trans. ASME J. Basic Eng.*, 83:95–108, 1961.
70. R.E. Kalman. A new approach to linear filtering and prediction problems. *Trans. ASME J. Basic Eng.*, 82D:35–45, 1960.
71. V. Kolehmainen, S.R. Arridge, W.R.B. Lionheart, M. Vauhkonen, and J.P. Kaipio. Recovery of region boundaries of piecewise constant coefficients of an elliptic PDE from boundary data. *Inv. Probl.*, 15:1375–1391, 1999.
72. V. Kolehmainen, S.R. Arridge, M. Vauhkonen and J.P. Kaipio. Simultaneous reconstruction of internal tissue region boundaries and coefficients in optical diffusion tomography. *Phys. Med. Biol.*, 15:1375–1391, 2000.
73. V. Kolehmainen, S. Prince, S.R. Arridge, and J.P. Kaipio. State-estimation approach to the nonstationary optical tomography problem. *J. Opt. Soc. Am. A*, 20(5):876–889, 2003.
74. V. Kolehmainen, S. Siltanen, S. Järvenpää, J.P. Kaipio, P. Koistinen, M. Lassas, J. Pirttilä, and E Somersalo. Statistical inversion for X-ray tomography with few radiographs II: Application to dental radiology. *Phys Med Biol*, 48:1465–1490, 2003.
75. V. Kolehmainen, M. Vauhkonen, J.P. Kaipio, and S.R. Arridge. Recovery of piecewise constant coefficients in optical diffusion tomography. *Opt. Express*, 7(13):468–480, 2000.
76. V. Kolehmainen, A. Voutilainen, and J.P. Kaipio. Estimation of non-stationary region boundaries in EIT: state estimation approach. *Inv Probl.*, 17:1937–1956, 2001.
77. K. Kurpisz and A.J. Nowak. *Inverse Thermal Problems*. Computational Mechanics Publications, Southampton, 1995.
78. J. Lampinen, A. Vehtari and K. Leinonen. Application of Bayesian neural network in electrical impedance tomography. In *Proc. IJCNN'99*, Washington DC, USA, July 1999.
79. J. Lampinen, A. Vehtari, and K. Leinonen. Using a Bayesian neural network to solve the inverse problem in electrical impedance tomography. In B.K. Ersboll and P. Johansen, editors, *Proceedings of 11th Scandinavian Conference on Image Analysis SCIA'99*, pp. 87–93, Gangerlussuaq, Greenland, June 1999.
80. S. Lasanen. Discretizations of generalized random variables with applications to inverse problems. *Ann. Acad. Sci. Fenn. Math. Dissertationes A*, 130:1–64, 2002.
81. M. Lassas and S. Siltanen. Can one use total variation prior for edge-preserving Bayesian inversion? *Inv. Probl.*, 20:1537–1563, 2004.
82. M.S. Lehtinen. On statistical inversion theory. In H. Haario, editor, *Theory and Applications of Inverse Problems*. Longman, 1988.
83. M.S. Lehtinen, L. Päivärinta and E. Somersalo. Linear inverse problems for generalized random variables. *Inv. Probl.*, 5:599–612, 1989.
84. D.V. Lindley. *Bayesian Statistics: A Review*. SIAM, 1972.
85. J.S. Liu. *Monte Carlo Strategies in Scientific Computing*. Springer-Verlag, 2003.

86. A. Mandelbaum. Linear estimators and measurable linear transformations on a Hilbert space. *Z. Wahrsch. Verw. Gebiete*, 65:385–397, 1984.
87. K. Matsuura and U. Okabe. Selective minimum-norm solution of the biomagnetic inverse problem. *IEEE Trans. Biomed. Eng.*, 42:608–615, 1995.
88. J.L. Melsa and D.L. Cohn. *Decision and Estimation Theory*. McGraw-Hill, 1978.
89. N. Metropolis, A.W. Rosenbluth, M.N. Rosenbluth, A.H. Teller and E. Teller. Equations of state calculations by fast computing machine. *J. Chem. Phys.*, 21:1087–1091, 1953.
90. K. Mosegaard. Resolution analysis of general inverse problems through inverse Monte Carlo sampling. *Inv. Probl.*, 14:405–426, 1998.
91. K. Mosegaard and C. Rygaard-Hjalsted. Probabilistic analysis of implicit inverse problems. *Inv. Probl.*, 15:573–583, 1999.
92. K. Mosegaard and A. Tarantola. Monte Carlo sampling of solutions to inverse problems. *J. Geophys. Res. B*, 14:12431–12447, 1995.
93. C. Müller. *Grundprobleme der matematischen Theorie electromagnetischer Schwingungen*. Springer-Verlag, 1957.
94. A. Nachman. Reconstructions from boundary measurements. *Annals of Math.*, 128:531–576, 1988.
95. A. Nachman. Global uniqueness for a two-dimensional inverse boundary value problem. *Annals of Math.*, 143:71–96, 1996.
96. F. Natterer. *The Mathematics of Computerised Tomography*. Wiley, 1986.
97. E. Nummelin. *General Irreducible Markov Chains and Nonnegative Operators*. Cambridge University Press, 1984.
98. B. Øksendal. *Stochastic Differential Equations*. Springer-Verlag, 1998.
99. R.L. Parker. *Geophysical Inverse Theory*. Princeton University Press, 1994.
100. J.-M. Perkkiö. Radiative transfer problem on Riemannian manifolds. Master's thesis, Helsinki University of Technology, 2003.
101. S. Prince, V. Kolehmainen, J.P. Kaipio, M.A. Franceschini, D. Boas and S.R. Arridge. Time-series estimation of biological factors in optical diffusion tomography. *Phys. Med. Biol.*, 48:1491–1504, 2003.
102. A. Pruessner and D.P. O'Leary. Blind deconvolution using a regularized structured total least norm algorithm. *SIAM J. Matrix Anal. Appl.*, 24:1018–1037, 2003.
103. T. Quinto. Singularities of the x-ray transform and limited data tomography in $\mathbf{r}^2$ and $\mathbf{r}^3$. *SIAM J. Math. Anal.*, 24:1215–1225, 1993.
104. J. Radon. Über die bestimmung von funktionen durch ihre integralwärte längs gewisser mannichfaltigkeiten. *Berichte über di Verhandlungen der Sächsischen Akademien der Wissenschaften*, 69:262–267, 1917.
105. A.G. Ramm. Multidimensional inverse problems and completeness of the products of solutions to PDE's. *J. Math.Anal.Appl*, 134:211–253, 1988.
106. A.G. Ramm and A.I Katsevich. *The Radon Transform and Local Tomography*. CRC Press, 1996.
107. Yu.A. Rozanov. *Markov Random Fields*. Springer-Verlag, 1982.
108. W. Rudin. *Functional Analysis*. McGraw-Hill, second edition, 1991.
109. Y. Saad and H. Martin. GMRES: A genralized minimal residual algorithm for solving nonsymmetric linear systems. *SIAM J. Sci. Stat. Comput.*, 7:856–869, 1986.
110. J. Sarvas. Basic mathematical and electromagnetic concepts of the biomagnetic inverse problem. *Phys. Med. Biol.*, 32:11–22, 1987.

111. D.M. Schmidt, J.S. George and C.C. Wood. Bayesian inference applied to the electromagnetic inverse problem. *Human Brain Mapping*, 7:195–212, 1999.
112. A. Seppänen, L. Heikkinen, T. Savolainen, E. Somersalo and J.P. Kaipio. An experimental evaluation of state estimation with fluid dynamical models in process tomography. In *3rd World Congress on Industrial Process Tomography, Banff, Canada*, pp. 541–546, 2003.
113. A. Seppänen, M. Vauhkonen, E. Somersalo and J.P. Kaipio. State space models in process tomography – approximation of state noise covariance. *Inv. Probl. Eng.*, 9:561–585, 2001.
114. A. Seppänen, M. Vauhkonen, P.J. Vauhkonen, E. Somersalo and J.P. Kaipio. Fluid dynamical models and state estimation in process tomography: Effect due to inaccuracies in flow fields. *J. Elect. Imaging*, 10(3):630–640, 2001.
115. A. Seppänen, M. Vauhkonen, P.J. Vauhkonen, E. Somersalo and J.P. Kaipio. State estimation with fluid dynamical evolution models in process tomography – an application to impedance tomography. *Inv. Probl.*, 17:467–484, 2001.
116. S. Siltanen, V. Kolehmainen, S.Järvenpää, J.P. Kaipio, P. Koistinen, M. Lassas, J. Pirttilä and E Somersalo. Statistical inversion for X-ray tomography with few radiographs I: General theory. *Phys. Med. Biol.*, 48: 1437-1463, 2003.
117. A.F.M. Smith and G.O. Roberts. Bayesian computation via the Gibbs sampler and related Markov chain Monte Carlo methods. *J. R. Statist. Soc. B*, 55:3–23, 1993.
118. K.T. Smith and F. Keinert. Mathematical foundations of computed tomography. *Appl. Optics*, 24:3950–3857, 1985.
119. E. Somersalo, M. Cheney and D. Isaacson. Existence and uniqueness for electrode models for electric current computed tomography. *SIAM J. Appl. Math.*, 52:1023–1040, 1992.
120. H.W. Sorenson. *Parameter Estimation: Principles and Problems*. Marcel Dekker, 1980.
121. P. Stefanov. Inverse problems in transport theory. In Gunther Uhlmann, editor, *Inside Out*, pp. 111–132. MSRI Publications, 2003.
122. S.M. Stigler. Laplace's 1774 memoir on inverse probability. *Statistical Science*, 3, 1986.
123. J. Sylvester and G. Uhlmann. A global uniqueness theorem for an inverse boundary value problem. *Ann. Math.*, 125:153–169, 1987.
124. J. Tamminen. MCMC methods for inverse problems. Geophysical Publications 49, Finnish Meteorological Institute, 1999.
125. A. Tarantola. *Inverse Problem Theory*. Elsevier, 1987.
126. A. Tarantola and B. Valette. Inverse problems = quest for information. *J. Geophys.*, pp. 159–170, 1982.
127. J.R. Thompson and R.A. Tapia. *Nonparametric Function Estimation, Modeling, and Simulation*. SIAM, 1990.
128. L. Tierney. Markov chains for exploring posterior distributions. *Ann. Statistics*, 22:1701–1762, 1994.
129. A. N. Tihonov. Solution of incorrectly formulated problems and the regularization method. *Soviet Mathematics: Doklady*, 4:1035–1038, 1963.
130. A.N. Tikhonov and V.Y. Arsenin. *Solution of Ill-posed Problems*. Wiley, 1977.
131. K. Uutela, M. Hämäläinen and E. Somersalo. Visualization og magnetoencephalographic data using minimum current estimates. *NeuroImage*, 10:173–180, 1999.

132. M. Vauhkonen, J.P. Kaipio, E. Somersalo and P.A. Karjalainen. Electrical impedance tomography with basis constraints. *Inv. Probl.*, 13:523–530, 1997.
133. M. Vauhkonen, P.A. Karjalainen and J.P. Kaipio. A Kalman filter approach applied to the tracking of fast movements of organ boundaries. In *Proc 20th Ann Int Conf IEEE Eng Med Biol Soc*, pages 1048–1051, Hong Kong, China, October 1998.
134. M. Vauhkonen, P.A. Karjalainen and J.P. Kaipio. A Kalman filter approach to track fast impedance changes in electrical impedance tomography. *IEEE Trans. Biomed. Eng.*, 45:486–493, 1998.
135. M. Vauhkonen, D. Vadász, P.A. Karjalainen, E. Somersalo and J.P. Kaipio. Tikhonov regularization and prior information in electrical impedance tomography. *IEEE Trans. Med. Imaging.*, 17:285–293, 1998.
136. P.J. Vauhkonen, M. Vauhkonen, T. Mäkinen, P.A. Karjalainen and J.P. Kaipio. Dynamic electrical impedance tomography: phantom studies. *Inv. Prob. Eng.*, 8:495–510, 2000.
137. T. Vilhunen, L.M. Heikkinen, T. Savolainen, P.J. Vauhkonen, R. Lappalainen, J.P. Kaipio and M. Vauhkonen. Detection of faults in resistive coatings with an impedance-tomography-related approach. *Measur. Sci. Technol.*, 13:865–872, 2002.
138. T. Vilhunen, M. Vauhkonen, V. Kolehmainen and J.P. Kaipio. A source model for diffuse optical tomography. Preprint, 2003.
139. C.R. Vogel. *Computational Methods for Inverse Problems*. SIAM, 2002.
140. R.L. Webber, R.A. Horton, D.A. Tyndall, and J.B. Ludlow. Tuned apertire computed tomography (TACT). Theory and application for three-dimensional dento-alveolar imaging. *Dentomaxillofacial Radiol.*, 26:53–62, 1997.
141. A. Yagola and K. Dorofeev. Sourcewise representation and a posteriori error estimates for ill-posed problems. *Fields Institute Comm.*, 25:543–550, 2000.
142. K. Yosida. *Functional Analysis*. Springer-Verlag, 1965.

Index

P_1 approximation, 212
P_k approximation, 212

absorption coefficient, 209
activity density, 256
additive noise, 56
admittivity, 196
affine estimators
    approximate enhanced error model, 183
    enhanced error model, 183
    errors with projected models, 182
    known statistics, 149
anisotropy matrix, 215, 292
approximation error model
    deconvolution problem, 225
    electrical impedance tomography, 269

Bayes cost, 147
Bayes theorem of inverse problems, 51
Bayesian credibility set, 54
Bayesian experiment design, 260
Bayesian filtering, 115
    application to MEG, 249
Biot–Savart law, 199
blind deconvolution, 59, 103
bootstrap filter, 143

central limit theorem, 107
classical regularization methods, 49
complementary information, 262
computerized tomography (CT), 190
condition number, 9
conditional covariance, 53

conditional mean (CM) estimate, 53
conjugate gradient method
    ensemble performance, 162
    inverse crime behavior, 184
constitutive relations, 195
contraction, 27

deconvolution problem, 8
diffusion matrix, 215
discrepancy principle, 14
distinguishability, 307

energy current density, 209
energy fluence, 210
evolution models
    hidden Markov model, 143
evolution–observation models, 118
    filtering problem, 120
    fixed-interval smoothing problem, 120
    fixed-lag smoothing problem, 120
    higher-order Markov models, 140
    observation, 118
    observation noise process, 119
    observation update, 122
    prediction problem, 120
    state noise process, 119
    state vector, 118
    time evolution update, 122

filtered backprojection, 192
fixed-lag and fixed-interval smoothing, *see* smoothers
fixed-point iteration, 27
fixed-point theorem, 27

Fréchet differentiability, 25
Fredholm equation, 7

Gaussian densities, 72
    conditional probabilities, 75
    posterior potential, 79
Gaussian priors
    smoothness priors, 79
Geselowitz formula, 200
Gibbs sampler, 98

Hammersley–Clifford theorem, 67, 113
hierarchical models, 108
    hyperpriors, 108

impedivity, 196
importance sampling, *see* particle filters, sampling importance resampling (SIR)
improper densities, 82
interval estimates, 52
invariant measures, 93
invariant set, 27
inverse crimes, 184
iterative methods for regularization, 27
    algebraic reconstruction technique (ART), 31, 37
    conjugate gradient least squares (CGLS), 48
    conjugate gradient method, 40
        A-conjugate, 41
    conjugate gradient normal equations (CGNE), 48
    conjugate gradient normal residual (CGNR), 48
    generalized minimal residual (GMRES), 48
    Kaczmarz iteration, 31
    Kaczmarz sequence, 32
    Krylov subspace methods, 39
    Landweber–Fridman iteration, 27

Kalman filter, 123
    extended Kalman filter (EKF), 126
    linear Gaussian case, 123
Krylov subspace, 43

Laplace transform, 8
likelihood function, 51

log-normal prior, 65

Markov chain Monte Carlo
    burn-in, 98
    convergence, 106
Markov chain Monte Carlo methods, 91
Markov chains, 92
    time-homogeneous, 92
mass absorption coefficient, 190
maximum a posteriori (MAP) estimate, 53
maximum a posteriori estimate, 149
maximum entropy method, 64
maximum likelihood, 53
maximum likelihood estimate, 146
Maxwell–Ampère law, 195
Maxwell–Faraday law, 195
mean square estimate, 147
Metropolis–Hastings algorithm, 94
minimum current estimate, 249
minimum norm solution, 14
minimum variance estimate, 148
Moore–Penrose inverse, *see* pseudoinverse
Morozov discrepancy principle, 18

nonstationary inverse problems, 115

observation models
    additive noise, 155
    Cauchy errors, 173
    electric inverse source problems, 197
    electrical impedance tomography, 202
        complete electrode model, 204
        current patterns, 203
        optimal current patterns, 260
    electrocardiography (ECG), 198
    electroencephalography (EEG), 198
    inverse source problems, 194
    limited angle tomography, 154
    magnetic inverse source problems, 198
    magnetoencephalography (MEG), 198
    noise level, 157
    optical tomography, 208
        Boltzmann equation, 208
        boundary recovery, 299
        diffusion approximation, 211
        radiation transfer equation, 208

quasi-static Maxwell equations, 194
sinogram, 153
spatial blur, 152
tomography, 153
  Radon transform, 190
X-ray tomography
  sinogram, 37
observation–evolution models
  augmented observation update, 136
  back transition density, 134
  spatial priors, 133
  weighted evolution update, 136
orthonormal white noise, 80
overregularization, 24

particle filters, 129
  layered sampling, 130
  sampling importance resampling (SIR), 130
photon flux density, 210
point estimates, 52
Poisson observation model, 60
posterior distribution, 51
prior density, 51
prior models, 62
  Cauchy prior, 63
  discontinuities, 65
  discretization invariance, 180
  entropy prior, 64
  Gaussian white noise prior, 79
  impulse or ℓ^1 prior, 62
  Markov random fields, 66
  problems related to discretization, 175
  sample-based densities, 70
  smoothness priors, 150
  structural priors, 152
  subspace priors, 71
probability transition kernel, 92
  balance equation, 95
  detailed balance equation, 95
  irreducible, 93
  periodic and aperiodic, 93
pseudoinverse, 13
  Moore–Penrose equations, 47

radiance, 209
radiation flux density, 208
resistance matrix, 204
resistivity, 196
reversible jump MCMC, 307
robust error models, 188

scattering coefficient, 209
scattering phase function, 209
Schur complements, 74
  matrix inversion lemma, 74
  Schur identity, 75
singular value decomposition, 11
  singular system, 11
  truncated singular value decomposition, 10
smoothers, 138
source localization, 242
spread estimates, 52
state space models, *see* evolution–observation models
statistical inversion, 49

Tikhonov regularization, 16
  ensemble performance, 164
  generalizations, 24
  regularization parameter, 16
  Tikhonov regularized solution, 16
time series autocovariance, 108
truncated singular value decomposition
  ensemble performance, 159

underregularization, 24
uniform probability distribution, 20

whitening matrix, 80

Applied Mathematical Sciences

(continued from page ii)

60. *Ghil/Childress:* Topics in Geophysical Dynamics: Atmospheric Dynamics, Dynamo Theory and Climate Dynamics.
61. *Sattinger/Weaver:* Lie Groups and Algebras with Applications to Physics, Geometry, and Mechanics.
62. *LaSalle:* The Stability and Control of Discrete Processes.
63. *Grasman:* Asymptotic Methods of Relaxation Oscillations and Applications.
64. *Hsu:* Cell-to-Cell Mapping: A Method of Global Analysis for Nonlinear Systems.
65. *Rand/Armbruster:* Perturbation Methods, Bifurcation Theory and Computer Algebra.
66. *Hlaváček/Haslinger/Necasl/Lovisek:* Solution of Variational Inequalities in Mechanics.
67. *Cercignani:* The Boltzmann Equation and Its Applications.
68. *Temam:* Infinite-Dimensional Dynamical Systems in Mechanics and Physics, 2nd ed.
69. *Golubitsky/Stewart/Schaeffer:* Singularities and Groups in Bifurcation Theory, Vol. II.
70. *Constantin/Foias/Nicolaenko/Temam:* Integral Manifolds and Inertial Manifolds for Dissipative Partial Differential Equations.
71. *Catlin:* Estimation, Control, and the Discrete Kalman Filter.
72. *Lochak/Meunier:* Multiphase Averaging for Classical Systems.
73. *Wiggins:* Global Bifurcations and Chaos.
74. *Mawhin/Willem:* Critical Point Theory and Hamiltonian Systems.
75. *Abraham/Marsden/Ratiu:* Manifolds, Tensor Analysis, and Applications, 2nd ed.
76. *Lagerstrom:* Matched Asymptotic Expansions: Ideas and Techniques.
77. *Aldous:* Probability Approximations via the Poisson Clumping Heuristic.
78. *Dacorogna:* Direct Methods in the Calculus of Variations.
79. *Hernández-Lerma:* Adaptive Markov Processes.
80. *Lawden:* Elliptic Functions and Applications.
81. *Bluman/Kumei:* Symmetries and Differential Equations.
82. *Kress:* Linear Integral Equations, 2nd ed.
83. *Bebernes/Eberly:* Mathematical Problems from Combustion Theory.
84. *Joseph:* Fluid Dynamics of Viscoelastic Fluids.
85. *Yang:* Wave Packets and Their Bifurcations in Geophysical Fluid Dynamics.
86. *Dendrinos/Sonis:* Chaos and Socio-Spatial Dynamics.
87. *Weder:* Spectral and Scattering Theory for Wave Propagation in Perturbed Stratified Media.
88. *Bogaevski/Povzner:* Algebraic Methods in Nonlinear Perturbation Theory.
89. *O'Malley:* Singular Perturbation Methods for Ordinary Differential Equations.
90. *Meyer/Hall:* Introduction to Hamiltonian Dynamical Systems and the N-body Problem.
91. *Straughan:* The Energy Method, Stability, and Nonlinear Convection, 2nd ed.
92. *Naber:* The Geometry of Minkowski Spacetime.
93. *Colton/Kress:* Inverse Acoustic and Electromagnetic Scattering Theory, 2nd ed.
94. *Hoppensteadt:* Analysis and Simulation of Chaotic Systems, 2nd ed.
95. *Hackbusch:* Iterative Solution of Large Sparse Systems of Equations.
96. *Marchioro/Pulvirenti:* Mathematical Theory of Incompressible Nonviscous Fluids.
97. *Lasota/Mackey:* Chaos, Fractals, and Noise: Stochastic Aspects of Dynamics, 2nd ed.
98. *de Boor/Höllig/Riemenschneider:* Box Splines.
99. *Hale/Lunel:* Introduction to Functional Differential Equations.
100. *Sirovich (ed):* Trends and Perspectives in Applied Mathematics.
101. *Nusse/Yorke:* Dynamics: Numerical Explorations, 2nd ed.
102. *Chossat/Iooss:* The Couette-Taylor Problem.
103. *Chorin:* Vorticity and Turbulence.
104. *Farkas:* Periodic Motions.
105. *Wiggins:* Normally Hyperbolic Invariant Manifolds in Dynamical Systems.
106. *Cercignani/Illner/Pulvirenti:* The Mathematical Theory of Dilute Gases.
107. *Antman:* Nonlinear Problems of Elasticity, 2nd ed.
108. *Zeidler:* Applied Functional Analysis: Applications to Mathematical Physics.
109. *Zeidler:* Applied Functional Analysis: Main Principles and Their Applications.
110. *Diekmann/van Gils/Verduyn Lunel/Walther:* Delay Equations: Functional-, Complex-, and Nonlinear Analysis.
111. *Visintin:* Differential Models of Hysteresis.
112. *Kuznetsov:* Elements of Applied Bifurcation Theory, 3d ed.
113. *Hislop/Sigal:* Introduction to Spectral Theory: With Applications to Schrödinger Operators.
114. *Kevorkian/Cole:* Multiple Scale and Singular Perturbation Methods.
115. *Taylor:* Partial Differential Equations I, Basic Theory.
116. *Taylor:* Partial Differential Equations II, Qualitative Studies of Linear Equations.

(continued on next page)

Applied Mathematical Sciences

(continued from previous page)

117. *Taylor:* Partial Differential Equations III, Nonlinear Equations.
118. *Godlewski/Raviart:* Numerical Approximation of Hyperbolic Systems of Conservation Laws.
119. *Wu:* Theory and Applications of Partial Functional Differential Equations.
120. *Kirsch:* An Introduction to the Mathematical Theory of Inverse Problems.
121. *Brokate/Sprekels:* Hysteresis and Phase Transitions.
122. *Gliklikh:* Global Analysis in Mathematical Physics: Geometric and Stochastic Methods.
123. *Le/Schmitt:* Global Bifurcation in Variational Inequalities: Applications to Obstacle and Unilateral Problems.
124. *Polak:* Optimization: Algorithms and Consistent Approximations.
125. *Arnold/Khesin:* Topological Methods in Hydrodynamics.
126. *Hoppensteadt/Izhikevich:* Weakly Connected Neural Networks.
127. *Isakov:* Inverse Problems for Partial Differential Equations.
128. *Li/Wiggins:* Invariant Manifolds and Fibrations for Perturbed Nonlinear Schrödinger Equations.
129. *Müller:* Analysis of Spherical Symmetries in Euclidean Spaces.
130. *Feintuch:* Robust Control Theory in Hilbert Space.
131. *Ericksen:* Introduction to the Thermodynamics of Solids, Revised ed.
132. *Ihlenburg:* Finite Element Analysis of Acoustic Scattering.
133. *Vorovich:* Nonlinear Theory of Shallow Shells.
134. *Vein/Dale:* Determinants and Their Applications in Mathematical Physics.
135. *Drew/Passman:* Theory of Multicomponent Fluids.
136. *Cioranescu/Saint Jean Paulin:* Homogenization of Reticulated Structures.
137. *Gurtin:* Configurational Forces as Basic Concepts of Continuum Physics.
138. *Haller:* Chaos Near Resonance.
139. *Sulem/Sulem:* The Nonlinear Schrödinger Equation: Self-Focusing and Wave Collapse.
140. *Cherkaev:* Variational Methods for Structural Optimization.
141. *Naber:* Topology, Geometry, and Gauge Fields: Interactions.
142. *Schmid/Henningson:* Stability and Transition in Shear Flows.
143. *Sell/You:* Dynamics of Evolutionary Equations.
144. *Nédélec:* Acoustic and Electromagnetic Equations: Integral Representations for Harmonic Problems.
145. *Newton:* The N-Vortex Problem: Analytical Techniques.
146. *Allaire*: Shape Optimization by the Homogenization Method.
147. *Aubert/Kornprobst:* Mathematical Problems in Image Processing: Partial Differential Equations and the Calculus of Variations.
148. *Peyret:* Spectral Methods for Incompressible Viscous Flow.
149. *Ikeda/Murota:* Imperfect Bifurcation in Structures and Materials: Engineering Use of Group-Theoretic Bifucation Theory.
150. *Skorokhod/Hoppensteadt/Salehi:* Random Perturbation Methods with Applications in Science and Engineering.
151. *Bensoussan/Frehse:* Regularity Results for Nonlinear Elliptic Systems and Applications.
152. *Holden/Risebro:* Front Tracking for Hyperbolic Conservation Laws.
153. *Osher/Fedkiw:* Level Set Methods and Dynamic Implicit Surfaces.
154. *Bluman/Anco:* Symmetry and Integration Methods for Differential Equations.
155. *Chalmond:* Modeling and Inverse Problems in Image Analysis.
156. *Kielhöfer:* Bifurcation Theory: An Introduction with Applications to PDEs.
157. *Kaczynski/Mischaikow/Mrozek:* Computational Homology.
158. *Oertel:* Prandtl's Essentials of Fluid Mechanics, 2nd ed.
159. *Ern/Guermond:* Theory and Practice of Finite Elements.
160. *Kaipio/Somersalo:* Statistical and Computational Inverse Problems.